Reasonable Books – Reasonable Prices

Irina Kleyman

Exceptional for Home Schooling

Math Help for University

(Basic, Algebra, Geometry)

AF875416

First Edition

Reasonable Books = Reasonable Prices

Irina Kleyman

Exceptional for Home Schooling

Math Help for University (Basic, Algebra, Geometry)

I dedicate this book to my children and grandchildren.

Acknowledgments
Special thanks to Michael G LaFratta, Vladimir Kleyman, who has assisted in bringing all 10 books to market.

Author's note: In the many years that I've taught mathematics, I have been informed by many of my students that my methods for teaching have helped them tremendously in learning math. I have written these books in order to continue to help others learn math effectively.

Math Help Grade 1
Math Help Grade 2
Math Help Grade 3
Math Help Grade 4
Math Help Grade 5
Math Help Grade 6
Math Help Grade 7
Math Help Grade 8
Math Help for College (Summary Grades 1 – 5)
Math Help for University (Basic, Algebra, Geometry)

ISBN: 978-1-7349699-9-3 Paperback

Math Help for University
(Basic, Algebra, Geometry)

Syllabus includes topics, what students need to know.

Math Help for University (Basic, Algebra, Geometry)

Table 1 Example

		Grades	Average
	Diagnostic Test	68%	D+
1)	Test 1	70%	
2)	Test 2	72%	Av1=(T1+T2)/2 = (70+72)/2 = 71% C
3)	Test 3	74%	
4)	Test 4	76%	Av2=(T1+T2+T3+T4)/4= (70+72+74+76)/4 = 73% C
5)	Test 5	78%	
6)	Test 6	80%	Av3=(T1+T2+T3+T4+T5+T6)/6 = 75% C+
7)	Test 7	82%	
8)	Test 8	84%	Av4=(T1+T2+T3+T4+T5+T6+T7+T8)/8=77% C+
9)	Test 9	86%	
10)	Test 10	88%	Av5=(T1+T2+T3+T4+T5+T6+T7+T8+T9+T10) /10= 79% C+
11)	Test 11	90%	
12)	Final Exam	92%	Av6=(T1+ T2+ …+T9+T10+T11 +FE)/12= 81% B

Math Help for University
(Basic, Algebra, Geometry)
Table 2 Your Grades

		Grades	Average
	Diagnostic Test		
1)	Test 1		
2)	Test 2		Av1=
3)	Test 3		
4)	Test 4		Av2=
5)	Test 5		
6)	Test 6		Av3=
7)	Test 7		
8)	Test 8		Av4=
9)	Test 9		
10)	Test 10		Av5=
11)	Test 11		
12)	Final Exam		Av6=

Parents, Tutors, Students Middle School, High School, Colleges, University can help primary school students. Please read slowly math books to primary school students

Smart Exercises!!!!

Relax and read very slowly. I have prepared good exercises for you. Mental math exercises help improve your attention. Try to work in scratch paper if you need, but I recommend you later to do this exercise mentally, without paper and pen. Teacher can Mental Math exercises put on plastic for work in class.
If you memorize rules, properties and formulas, you have exercises for memory which improves your memory. Review exercises before the test, and test exercises after each chapter gives you good knowledge.
My practice exercises will teach you to solve any test.
You have to show all steps of your work, and not just answers. Do not miss any steps. You will improve your logic thinking. You can take make up test B to improve your grades. Tables help you in the future.
If you forgot something you can refresh your knowledge.
You will need in future for any subject and any job, good attention, memory, knowledge, logic thinking. If students try their best, ask questions, practice, and wish to understand, they will be good.
Do not be afraid, work with book.

Basic and Advanced Problems

This book for students High school, Colleges, University to refresh knowledge. Also, this book good for teachers, tutors, parents, who want to help students. This book for college students, who never understand math in schools. You have two kinds of problems. There are basic problems: whole numbers, properties of whole numbers, decimals, fractions and mixed numbers, converting decimals and fractions, rational numbers, summary, and extra reviews. Also, you have advanced problems: equations, word problems, geometry. Basic and advanced problems help students improve attention, memory and logical thinking and get good knowledge that helps students in any job.

This book help people to learn math. If you never understand math and have not any knowledge in math you got chance to know this subject. If you were a small boy or girl or in school some problems you got very early for you, some problems were difficult for you, now you got chance to get knowledge. I have not enough time like teacher in school to explain all information. I am sure you want to know math and make your life better and you want to work with book and solve the problems. If you do not understand today, read again tomorrow and some day you feel better. I have some books in philosophy and I do not always understand material, but later I will be read again and again for understanding. Not every day we are smart in any reason. Some day you feel good, some day not. Some people got master degree in math for 5 years, but I got master degree about 6 years. Nothing wrong, I know my math very good. You need to read book then makes notes, rewrite in your notebook information what you do not know or cannot remember.

Our brain has the same measurement in 18 centuries, in 19 centuries, in 20 centuries, and now in 21 centuries. We have much more information now than before; this is way we need to work very slow. I am very slow person. I repeat rules in my book for you many times and if you work slow you will be known material. I played in piano 5 years, from 4 grade to 8 grade, if I repeat many times my lesson, I played every day better and better. First book is Math for College for students who has grade C in math. Second book for University for students who has grade A or B, which is summary more important topics, rules, examples. If you want to be in university, work with this book, read book, rules, work with examples, solve the problems, reviews, tests, final exams.

Psychology helps students.

Many schools want to use the calculator, but many colleges do not. This book helps you to be ready for college and teaches you how to work without a calculator, just use your brain. Do your best, but try hard to solve problems. After that, check the answers, which you have after end of each chapter. To improve the student's attention, you need to rewrite problems from the book to their notebook, because 30% of our brain has a connection to the hands. Show all steps of solution gives you full credits. Students, be very, very, very, very patient with yourselves. You need to love yourself, care about your health, behavior, and education. Respect yourself, and respect others.

It would help to know some information from psychology.
Every 23 days, our physical condition changes.
Every 28 days, our emotional condition changes.
Every 33 days, our intellectual conditional changes.
We are smart, but not every day. You can make some mistakes.
Students understand material better if they are in good physical, emotional, and intellectual condition. When the students are ready to learn, the book can give students knowledge.

Students, we remember on average …
20 % of what we read, 30% of what we hear
40% of what we see with attention, 50% of what we say, read loudly
60% of what do, what we write, 70% of what we repeat rules
80% of what we practice, 90% of what we try to memorize
100% of what we read, hear, see, say, write, and teach others.
Students read the textbook, notes, tables, listen to the teacher, and watch with all your attention. Read loudly, write notes, and repeat many times to improve your memory. Practice, memorize and make it your goal to improve your logical thinking. You will soon be proud of yourself and happy with the result. Many students did it and you can do it too.

Just 10 problems from each lesson!

This book is the second part of mathematics where students learn how to do operations with numbers, equations, geometry, word problems. This book contains lessons, rules, examples, exercises for practice, tests, and answers for exercises. This a book designed to help students improve their attention, memory, and logical thinking.

Students, you do not need to solve all of the problems in first step, first check your knowledge from numbers in contents. Students need to solve a minimum of 10 problems from each lesson of list of contents to get good knowledge in math, a maximum of 15 problems. Evaluate your work in percent. For example, a test has 10 problems, and a student solves correctly 8 of the 10 problems, this is will be $8 \div 10 \times 100\ \% = 80\ \%$.

If students taking the test got a grade, more than 80% (80% - 100%), they pass the chapter and can move to next chapter or to the final exam. If students taking the test got a grade less than 80%, they need to repeat rules and solve another 5 examples from each lesson. Then check your mistakes in the Test A or in the Final Exam A. After that, try to solve Test B or Final Exam B to improve your grades.

If students took the test and got a grade less than 80%, never be upset, and just tell yourself positive things like "I tried my best!", "I just need to practice more" or "I need to ask more questions." Look at all of your grades on the test and find the average. For example, a student has 4 tests and Final Exam with the grades 60%, 70%, 80%, 90%, and 100%. The average is (60 + 70 + 80 + 90 + 100) and divided by 5, their average would be 80%.

Math classes in elementary, middle, high schools, colleges, university.

1) Primary School:Basic four operations with whole numbers.
2) Middle School: College Mathematics, Prealgebra
3) 9th grade: Introductory Algebra (1), Intermediate Algebra (2)
4) 10th grade: Geometry and Trigonometry
5) 11th grade: College Algebra (3) and Precalculus
6) 12th grade: Calculus 1 and Calculus 2

How to be a successful college student in Math!

1) Be respectful to students and instructor. Be polite, be patient.

2) Cell phones should have the ringer turned off during class. Using cell phones (iPods and other electronic devices) is not allowed.

3) Try not to miss class, but if you missed a class, work hard to cover missed material.

4) Come to class on time, don't be late.

5) Review notes from previous classes before each class.

6) It's very important to have an organized folder.

7) Prepare notebook with graph paper, and bring a couple of erasable pens.

8) Listen to your teacher and pay attention. Take notes every class.

9) Try your best when you practice and work hard.

10)Be open to new ideas if you want to understand the subject.

11) Practicing on a lot of problems. This increases your speed and confidence.

12) Ask your questions after the teacher finishes explaining every problem.

How to be a successful college student in Math!

13) Always analyze the whole test, and start from problems that are easy for you.

14) Show all your knowledge on tests, quizzes, or homework – this means; show all steps for full credit (not just answers, so you do not miss any steps.) Otherwise, your work may not be accepted.

15) Try to find a quiet place for studying, and nothing but studying.

16) Look over class-work and read textbook, notes, example problems from tables before attempting the homework problems. Turn all homework on time.

17) Plan your schedule so, that you have at least 2 hours of homework per 1 hour of class-work. After 50 min of work, take a 10 min break.

18) If you get stuck on a problem or cannot find your mistake - leave it. Go back later and redo it on a separate sheet of paper. Prepare questions for your instructor for the next class.

19) Try to find time for attending tutoring center or your instructor's office for help if you need it.

20) Don't bring food to class; only bottled water.

Students!

**You need to improve,
your attention, memory,
logic thinking and
have knowledge.**

**Say to yourself:
I have a good attention.
I have a good memory.
I have a good logic thinking.**

**I have a good knowledge.
You need this in life
and in any job.
Try not to repeat mistakes.**

Students!

Think only positive.
Say only positive.
Do only positive.
You will be happy.

Respect for Yourselve.
Respect for Others
Responsibility for
All you're Actions!

Choose a job you love,
and you will never
have to work
a day in your life!

BEAUTY OF MATHEMATICS!
IF THE LETTERS IS REPRESENTED AS THE NUMBERS!

A = 1	B = 2	C = 3
D = 4	E = 5	F = 6
G = 7	H = 8	I = 9
J = 10	K = 11	L = 12
M =13	N = 14	O = 15
P = 16	Q = 17	R = 18
S = 19	T = 20	U = 21
V = 22	W = 23	X =24
Y = 25	Z = 26	

WHAT EQUALS 100% IN LIFE?

KNOWLEDGE=11+14+15+23+12+5+4+7+5=96%
HARDWORK=8+1+18+4+23+15+18+11=98%
ATTITUDE=1+20+20+9+20+21+4+5=100%

IT'S THE ATTITUDE, HARDWORK, KNOWLEDGE, THAT WILL PUT YOU ON THE TOP.

Parents and Kids!

Parents, you need to help your kids. If kids feel your love and care they will be kind and smart. Encourage and hug kids. Tell them; I love you. I understand you. I forgive you. I proud of you. Forgive me, please. You can do anything. Try again. You did a great job. Thank you, dear. Be patient with your kids. Parents like God for kids until the age of 12. Kids of ages 13–19 need more help, because in this time they not adults. Parents, please open the internet and ask the question: how I can help a girl or boy of ages 13-19? It is very important when parents love and care about children. Children respect parents and be grateful to the them.

An average, adults works 8 hours a day. For kids, it is enough to work in school 6 hours a day from 8:00 am to 2:00 pm. After that kids have to play.Students have to learn diiferent sciences and have different activities. Knowledge and health is very important for kids. Many doctors agree.

Schools and Kids!

Example for elementary School

1-4 grades, 8:00 am–1:30 pm, 30 mins break-play-time after each class

1st lesson 8:00-8:30 English	4th lesson 11:00–11:30 Science
2nd lesson 9:00-9:30 Spanish	5th lesson 12:00–12:30 Dance
3rd lesson 10:00–10:30 Math	6th lesson 1:00–1:30 Volleyball

Example for Middle School

5–8 grades, 8:00 am–1:45 pm,	15 mins break after each class
1st lesson 8:00–8:45 English	4th lesson 11:00–11:45 Geography
2nd lesson 9:00–9:45 Spanish	5th lesson 12:00–12:45 History
3rd lesson 10:00–10:45 Math	6th lesson 1:00-1:45 Dance

Example for High School

9–12 grades, 8:00 am–1:50 pm,	10 mins break after each class
1st lesson 8:00–8:50 English	4th lesson 11:00–11:50 Physics
2nd lesson 9:00–9:50 Spanish	5th lesson 12:00–12:50 Chemistry
3rd lesson 10:00 – 10:50 Math	6th lesson 1:00 – 1:50 Biology

Evaluation – Student feedback

Students need to read contents and work with book. When you finish chapter 10, you can evaluate the lessons all chapters from 1 to 10.
I try to give students not just knowledge, I try to improve your attention, memory, logic thinking, which you need in any job in future.
Send me please your evaluation on my site
www.mathbooksbyirina.com
Evaluate activities what helping you get the competence of lessons?
Yes or no.
1)Do you like repeat lesson slow or twice?
2)Do you like the tables for the summary of material?
3)Do you like to practice exercises that give you good knowledge?
4)Do you like to practice more to improve your knowledge?
5)How you like Mental Math exercises for improving your attention?
6)Do you like memorize rules and formulas for improving your memory?
7)Do you like answers for all numbers to help you check your solution?
8)Do you like to do review before each test to help you be ready for the test?
9)Do you like make up test B to improve your grades?
10)Do you like my grading scale without minuses?
11)Do you show all steps for full credits, not just answer, for improving logic thinking?
12)Do you like mini tests A of 10 problems, after each chapter, which give you knowledge?
How can lessons be improved?
Would you recommend lessons to a friend?

My Grading Scale without minuses

A=91–100%	B+=86–90%	B=81–85%	C+=76–80%
C=71–75%	D+=66–70%	D=61–65 %	F=0 - 60%

Tentative Schedule
Basic, Algebra, Geometry
MW 115 min
Chapter – Ch, Lesson - L

1) 02/03 Ch 1, L1 – L 5
2) 02/05 Ch 1, L 6 – L 9

3) 02/10 Ch 1, Test 1, L 10
4) 02/12 Ch 2, L 1 – L 5

5) 02/17 Ch 2, L 6 – L 9
6) 02/19 Ch 2, Test 2, L 10

7) 02/24 Ch 3, L 1 – L 5
8) 02/26 Ch 3, L 6 – L 9

9) 03/02 Ch 3, Test 3, L 10
10) 03/04 Ch 4, L 1 – L 5

11) 03/09 Ch 4, L 6 – L 9
12) 03/11 Ch 4, Test 4, L 10

13) 03/16 Ch 5, L 1 – L 5
14) 03/18 Ch 5, L 6 – L 9

15) 03/23 Ch 5, Test 5, L 10
16) 03/25 Ch 6, L 1 – L 5

17) 04/01 Ch 6, L 6 – L 9
18) 04/06 Ch 6, Test 6, L 10

19) 04/08 Ch 7, L 1 – L 5
20) 04/13 Ch 7, L 1 – L 5

21) 04/15 Ch 7, Test 7, L 10
22) 04/20 Ch 8, L 1 – L 5

23) 04/22 Ch 8, L 6 – L 9
24) 04/27 Ch 8, Test 8, L 10

25) 04/29 Ch 9, L 1 – L 5
26) 05/04 Ch 9, L 6 – L 9

27) 05/06 Ch 9, Test 9, L 10
28) 05/11 Final Exam A

29)05/13 College Placement Test A
30)05/18 College Placement Test B

31) 05/20 Extra 1 (Optional)
32) 05/25 Extra 2 (Optional)

Syllabus Basic, Algebra, Geomertry

Syllabus includes topics, what students need to know.

Basic: Chapter 1 Whole numbers, Basic Problems, Lessons 1 –10

Lesson 1 Pretest, College Placement Test
Lesson 2 Review 1 A, Review 1 B
Lesson 3 Comparison and Rounding of whole numbers
Lesson 4 Addition and Subtraction of whole numbers
Lesson 5 Multiplication of whole numbers,
Memorize table 1 of multiplication
Lesson 6 Multiplication Review 1 A, Review 1 B
Lesson 7 Division of whole numbers
Lesson 8 Division Review 2 A, Review 2 B
Lesson 9 Review 1 A, 1 B for Test 1
Lesson 10 Test 1 A, Test 1 B. Answers for Chapter 1

Basic: Chapter 2 Properties of whole numbers,
Basic Problems, Lessons 1 - 10

Lesson 1 Table Multiplication, Pretest, Mental Math 1 – 4
Lesson 2 Power of Whole numbers, Table 2 Perfect Square Numbers,
Prime Factorization
Lesson 3 Tests for division
Lesson 4 Divisors and Multiples
Lesson 5 Order of five operations
Lesson 6 Laws for whole numbers 1
Lesson 7 Laws for operations 2
Lesson 8 Common factor
Lesson 9 Review 2 A, Review 2 B
Lesson 10 Test 2 A, Test 2 B. Answers Chapter 2

Basic: Chapter 3 Decimals, Basic Problems, Lessons 1 - 10

Lesson 1 Table Multiplication, Pretest
Lesson 2 Addition and Subtraction of decimals
Lesson 3 Multiplication, Division and Power of Decimals
Lesson 4 Table 3, Mental Math # 5, #6
Lesson 5 Order of five operations with decimals
Lesson 6 Worksheet 1 A, Worksheet 1 B for Final Exam
Lesson 7 Laws for decimals
Lesson 8 Review 1 A, Review 1 B
Lesson 9 Review 3 A, Review 3 B
Lesson 10 Test 3 A, Test 3 B. Answers Chapter 3

Algebra: Chapter 7 Equations. Advanced Problems. Lesson 1 – 10
Lesson 1 Pretest
Lesson 2 Algebraic Expressions, Identities
Lesson 3 Equations with Like Terms
Lesson 4 Finding Unknowns
Lesson 5 Equations with Decimals and Fractions
Lesson 6 Conditional Equations
Lesson 7 Proportions
Lesson 8 Worksheet 5 A, Worksheet 5 B for Final Exam
Lesson 9 Review 7 A, Review 7 B
Lesson 10 Test 7 A, Test 7 B
Answers Chapter 7
Algebra: Chapter 8 Word Problems Advanced Problems, Lesson 1 – 12
Lesson 1 Pretest
Lesson 2 Operations
Lesson 3 Worth
Lesson 4 Decimals
Lesson 5 Fractions
Lesson 6 Direct Proportionality
Lesson 7 Uniform Motion
Lesson 8 Percent
Lesson 9 Tables 11, 12, 13, 14, Solve Mental Math 17, 18 odd
Lesson 10 Worksheet 6 A, Worksheet 6 B
Lesson 11 Review 8 A, Review 8 B
Lesson 12 Tests 8 A, Test 8 B
Answers for Chapter 8
Chapter 9 Geometry Advanced Problems, Lesson 1 – 12
Lesson 1 Pretest
Lesson 2 Segments
Lesson 3 Angles
Lesson 4 Triangles
Lesson 5 Right triangles
Lesson 6 Perimeter of polygons
Lesson 7 Area of Polygons
Lesson 8 Circles
Lesson 9 Solid Geometry
Lesson 10 Table 9, 10, Mental Math 15, 16
Lesson 11 Reviews 9 A, Review 9 B
Lesson 12 Tests 9 A, Test 9 B
Answers for Chapter 9

<u>Chapter 10 Summary. Advanced Problems. Lesson 1 – 14</u>

College Placement Test
Lesson 1 Work with Tables 1, 2, 3, 4, 5, 6
Lesson 2 Work with Tables 7, 8, 9, 10, 11, 12
Lesson 3 Work with Tables 13, 14, 15, 16, 17, 18
Lesson 4 Work with Tables 19, 20, 21, 22, 23, 24
Lesson 5 Work with Tables 25, 26, 27, 28, 29, 30
Lesson 6 Work with Mental Math 1, 2, 3, 4, 5, 6
Lesson 7 Work with Mental Math 7, 8, 9, 10, 11, 12
Lesson 8 Work with Mental Math 13, 14, 15, 16
Lesson 9 Work with Mental Math 17, 18, 19, 20
Lesson 10 Review for Final Exam 10 -A, B, C, D
Lesson 11 Worksheets 7A, 7 B
Lesson 12 Worksheets 8 A, 8 B
Lesson 13 College Placement Test A # 1 - # 10 all
College Placement Test B # 1 - # 10 all
Lesson 14 Solve Final Exam 1 A # 1 - # 10 all
Solve Final Exam 1 B # 1 - # 10 all
Answers for Chapter 10
Memorize Exponents

<u>Chapter 11 Extra Review. Basic Problems Lesson 1 - 10</u>

College Placement Test
Lesson 1 Review 1 A – Review 4 A
Lesson 2 Review 5 A – Review 8 A
Lesson 3 Review 1 B – Review 4 B
Lesson 4 Review 5 B – Review 8 B
Lesson 5 Review 1 C – Review 4 C
Lesson 6 Review 5 C – Review 8 C
Lesson 7 Review 1 D – Review 4 D
Lesson 8 Review 5 D – Review 8 D
Lesson 9 Worksheet 9 # 1 - # 10 all
Lesson 10 Worksheet 10 # 1 – 10 all
Answers for Chapter 11

Basic

Chapter 1

Whole

Numbers

Students!

Students, you are smarter than a calculator!
Students, you are smarter than a computer!
Students, you need to believe in yourself.
Students, be patient and do your best.

Do not be afraid to make mistakes.
Tell yourself positive words.
I am young! I am smart!
I am strong! I can do anything!

Smart students use their brain!
Your brain is always with you!
A calculator is not always with you.
Mathematics is work for the brain.

Be open to new ideas in math!
Work hard and try your best!
Check answers and ask questions!
Practice! Practice! Practice!

Basic Chapter 1 Lessons
Whole numbers
Lessons 1 –10
Basic Problems

Lesson 1 Pretest. College Placement Test

Lesson 2 Review 1 A, Review 1 B

Lesson 3 Comparison and Rounding of whole numbers

Lesson 4 Addition and Subtraction of whole numbers

Lesson 5 Multiplication of whole numbers.
Memorize table of multiplication

Lesson 6 Multiplication Review 1 A, Review 1 B

Lesson 7 Division of whole numbers

Lesson 8 Division Review 2 A, Review 2 B

Lesson 9 Review 1 A, Review 1 B for Test 1

Lesson 10 Test 1 A, Test 1 B

Answers for Chapter 1

Basic Chapter 1 Contents
Whole numbers

Contents have topics and exercises, which students need to solve before tests.1)Work with lesson 1, read rules and examples, 2) Solve the problems from contents, 3) Check answers in the end of chapter 1, 4) Work with the next lessons 2, 3, 4, 5, …, 5) Solve Test 1 A, 6) Grading Scale without minuses

A=90–100% B+=85–89% B=80–84% C+=75–79%
C=70–74% D+=65–69% D=60–64 % F=0 - 59%

7) If you solve 10 problems correctly, you get an A, 8) If have an A, you can move on to next chapter, 9) If your grade is lower than an A, work with the mistakes, then solve Test 1B, 10) When you finish all of the chapters, solve Final Exam in last chapter.

 # 1 - # 20 odd or # 1, # 3, # 5, # 7, # 9, # 11, # 13, # 15, # 17, # 19
Lesson 1 Pretest. College Placement Test.
Solve the Pretest 1 # 1, # 2, # 3, # 4, # 5
College Placement Test # 1, # 2, # 3, # 4, # 5, # 6, # 7, # 8, # 9, # 10
Lesson 2 Review 1 A, Review 1 B
Solve the problems Review 1 A # 1, # 2, # 3, # 4, # 5
Solve the problems Review 1 B # 1, # 2, # 3, # 4, # 5

Lesson 3 Comparison and Rounding of whole numbers
Solve the problems Review 1 A # 1 - # 20 odd
Solve the problems Review 1 B # 1 - # 20 odd
Lesson 4 Addition and Subtraction of whole numbers.

Lesson 5 Multiplication of whole numbers.
Memorize table 1 of multiplication
Lesson 6 Multiplication Review 1 A # 1 – # 10 all.
Review 1 B # 1 – # 10 all

Lesson 7 Division of whole numbers
Review 1 A # 1 - # 10 all. Review 1 B # 1 - # 10 all
Lesson 8 Review 2 A # 1 - # 20 odd. Review 2 B # 1 - # 20 odd

Lesson 9 Solve the problems: Review 1 A, Review 1 B # 1 – # 10 all
Lesson 10 Solve the problems Test 1 A, Test 1 B # 1 – # 10 all
Answers Chapter 1

Basic Chapter 1 Lesson 1
Pretest 1 (25 minutes)

Date **Name**

1) Multiply

$$\begin{array}{r} 807 \\ \underline{\times 504} \end{array}$$

1)_______

2) Divide $245427 \div 403$

2)______

3) Evaluate $420420 \div 70 \times 2$

3)______

4) Evaluate $640640 - 320320 \div 4 \times 5$

4)______

5) Evaluate $180 \div 6 \times 2 - 66 \div 6 \times 3$

5)_______

How do you feel about Math?
a)Love b) Like c) Nothing d) Do not Like e) Hate

Basic Chapter 1 Lesson 1
College Placement Test (50 min)

Date **Name**

Evaluate

1) $18000 \div (75^2 - 45^2)$

2) $5.6 \times 4 + 2.7 \times 4 - 8.3 \times 4$

3) $3\frac{3}{5} \times 15 - 2\frac{1}{5} \times 15$

4) $-3\frac{2}{3} \div \frac{5}{6} + 1\frac{1}{3} \div \frac{5}{6}$

5) Simplify $\frac{3}{7}(14x-21y)-15(\frac{2}{3}x-\frac{1}{3}y)$

6) Solve the equation $-2(3.1x - 1) + 3(1.2x + 1) = -14.5$

7) Find the perimeter of right triangle if it hypotenuse is $8\frac{1}{2}$ ft and another leg is 1.3 ft.

8) Find the area of trapezoid, if it height is $6\frac{2}{3}$ in, bases are $8\frac{1}{2}$ in and $9\frac{1}{2}$ in.

9) A bus drove 48 mi at a speed of 20 mi/h.
How much time bus spend?

10) Of 16 children, 12.5% were boys. How many were girls?

Basic Chapter 1 Lesson 2
Review 1 A (25 minutes)

Date **Name**

Read directions carefully. **Answers must be in the indicated place.**
Show all steps for full credits. **Each problem is worth 5 points.**
Do only 1 operation per 1 step. **Solve 1 problem 5 minutes**

1) **Multiply** $\begin{array}{r} 706 \\ \underline{\times 605} \end{array}$

1)_______

2) **Divide** $243714 \div 302$

2)_______

3) **Evaluate** $240240 \div 80 \times 3$

3)_______

4) **Evaluate** $560560 - 480480 \div 8 \times 5$

4)_______

5) **Evaluate** $120 \div 3 \times 2 - 60 \div 4 \times 3$

5)_______

Basic Chapter 1 Lesson 2
Review 1 B (25 minutes)

Date **Name**

Read directions carefully. **Answers must be in the indicated place.**
Show all steps for full credits. **Each problem is worth 5 points.**
Do only 1 operation per 1 step. **Solve 1 problem 5 minutes**

1) Multiply $\begin{array}{r} 605 \\ \underline{\times 403} \end{array}$

1)_______

2) Divide $182508 \div 201$

2)____

3) Evaluate $180180 \div 90 \times 2$

3)_______

4) Evaluate $480480 - 360360 \div 6 \times 5$

4)_______

5) Evaluate $150 \div 5 \times 2 - 60 \div 5 \times 3$

5)_______

Basic Chapter 1 Lesson 3
Chapter 1 Operations with Whole Numbers
Rounding and Comparison of Whole Numbers

This book based on mathematics, which studies properties of operations with numbers. All numbers are made up of the digits: 0, 1, 2, 3, 4, 5, 6, 7, 8, and 9. Basic operations are: addition, subtraction, multiplication, and division. Natural (counting) numbers are numbers that use for calculations 1, 2, 3 … Whole numbers are natural numbers and zero: 0, 1, 2, 3, 4, 5 … Odd numbers are 1, 3, 5, 7, 9, 11, 13, 15…
Even numbers are 0, 2, 4, 6, 8, 10, 12, 14, 16, 18 …
Whole numbers can have represented with points on a number line
0 1 2 3 4 5 The first operation with numbers is addition, which combines two or more numbers together.

The first number in addition is called the addend	**2**
The second number in addition is called the addend	**3**
The result of addition is called the sum	**5**
Addend + Addend = Sum	$2 + 3 = 5$

The second operation with numbers is subtraction, which means to take something away.

The first number in subtraction is called the minuend	**5**
The second number in subtraction is called the subtrahend	**2**
The result of subtraction is called the difference	**3**
Minuend – Subtrahend = Difference	$5 - 2 = 3$

Rounding

Using an approximate value for number called rounding.
A rounded number is always rounded to a given place value
Numbers can round to any digit: one, ten, and hundred …
If the digit to the right of the given place value is less than 5 (0, 1, 2, 3, 4), then that digit and all digits to the right are replaced by zeros. Rounding a number to the ten: $283 \approx 280$
If the digit to the right of the given place value is greater than or equal to 5 (5, 6, 7, 8, 9), increase the digit in the given place value by 1, and replace all other digits to the right by zeros. Rounding numbers to the ten: $238 \approx 240$

Comparison numbers

Determine which of number greater than (>) or smaller than (<) another number, or numbers are equal (=). $2 < 7$ $8 > 3$ $5 = 5$
Any whole number greater than zero $6 > 0$**. Number zero smaller than any whole number** $0 < 5$**. To compare numbers, compare digits from left to right** $1234 = 1234$ $1234 < 1235 (4 < 5)$ $1238 > 1236$ $(8 > 6)$

Basic Chapter 1 Lesson 3

Review 1 A

Round the numbers to the nearest ten (to the 10)

1) 123
2) 234
3) 345
4) 456
5) 567
6) 678
7) 789
8) 898
9) 987
10) 876

Round the numbers to the nearest hundred (to the 100)

11) 1234
12) 2345
13) 3456
14) 4567
15) 5678
16) 6789
17) 7898
18) 8987
19) 9876
20) 8765

Review 1 B

Compare the numbers

(Put Symbols <, >, or = between numbers)

1) 1 ? 14
2) 345 ? 23
3) 456 ? 1234
4) 12345 ? 2345
5) 2345 ? 12345
6) 23456 ? 9876
7) 3456 ? 23456
8) 34567 ? 8765
9) 9876 ? 9876
10) 8765 ? 8765
11) 9876 ? 9080
12) 1234 ? 1325
13) 1001 ? 1001
14) 1234 ? 1543
15) 9876 ? 9767
16) 1234 ? 1243
17) 3456 ? 3465
18) 3465 ? 3456
19) 1243 ? 1234
20) 2345 ? 2354

Basic Chapter 1 Lesson 4
Addition and Subtraction of Whole Numbers

The standard form of numbers is 7, 76, and 765
The expanded form of number is $76 = 70 + 6 = 7\times10 + 6$
$765 = 700 + 60 + 5 = 7\times 100 + 6\times10 + 5$
For large numbers, commas used to separate digits into groups of three called periods. Each period has three digits.The one-digit number is 7, the two-digit number is 76, and the three-digit number is 765. The four-digit number is 7,654; the five-digit number is 76,543; and six-digit number is 765,432. The first period that has three digits is called ones, for example, in the number 765. The place value of the first period, read from left to right, are: hundreds is 7, tens is 6, ones is 5. The second period has six digits and called thousands, for example the number 765, 432. The place value of the second period thousands, read from left to right are: hundred-thousand is 7, ten-thousand is 6, thousands is 5, hundreds is 4, tens is 3, and ones is 2.
Six-digit number contains six digits: 100,000; 100,001; …999,999. The least six-digit number is 100,000; greatest six-digit number is 999,999. The third new period has nine digits and is called millions: 987, 654, 321 The place value of the third period, read from left to right are: hundred million is 9, ten-millions is 8, millions is 7, hundred thousand is 6, ten-thousands is 5, thousands is 4, hundred is 3, tens is 2, and ones is 1.
Seven- digit numbers contain seven digits: 1,000,000 … 9,999,999. The name of seven- digit numbers from left to right are millions, Hundred thousand, ten-thousand, thousand, hundred, ten, one.
A one-digit number is always less than a two, three, four, five, six or seven- digit number. A seven-digit number is always greater than a six, five, four, three, two, or one-digit number. To compare numbers, compare digits from left to right: hundred thousand, ten-thousand, thousand, hundred, ten, and one:
$123456 = 123456$ $123456 < 123459$ $(6 < 9)$ $123458 > 123450$ $(8 > 0)$

Basic Chapter 1 Lesson 4

Properties of Addition and Subtraction

Letter can represent any numbers. A and B can be any numbers.

Changing the order of two addends does not change the sum.

A + B = B + A   2000 + 3000 = 5000   3000 + 2000 = 5000

The sum of any number and zero is the same number.

A + 0 = 0 + A = A   5000 + 0 = 5000   0 + 5000 = 5000

The difference between any number and zero is the same number.

A – 0 = A   5000 – 0 = 5000

The difference between two of the same numbers is zero.

A – A = 0   5000 – 5000 = 0

To check addition use subtraction

200 + 300 = 500   500 – 200 = 300   500 – 300 = 200

To check subtraction use addition or subtraction.

500 – 200 = 300   500 – 300 = 200   200 + 300 = 500

Remember

10 = 9 + 1 = 8 + 2 = 7 + 3 = 6 + 4 = 5 + 5	**15 = 9 + 6 = 8 + 7**
11 = 9 + 2 = 8 + 3 = 7 + 4 = 6 + 5	**16 = 9 + 7 = 8 + 8**
12 = 9 + 3 = 8 + 4 = 7 + 5 = 6 + 6	**17 = 9 + 8**
13 = 9 + 4 = 8 + 5 = 7 + 6	**18 = 9 +9**
14 = 9 + 5 = 8 + 6 = 7 + 7	**19 = 10 + 9**

Addition of six or seven- digit numbers: (work from right to left) add the ones, add the tens, add the hundreds, add the thousands, add the ten thousand, add the hundred thousand, add the millions.

Subtraction of seven- digit numbers: (work from right to left) subtract the ones, subtract the tens, subtract the hundreds, subtract the thousands; subtract the ten thousand,

subtract the hundred, subtract thousand, subtract the millions

Add or subtract numbers

```
   1111        8 17 16 15 14 13
  456789         9 8 7 6 5 3
+ 345678     -   7 8 9 7 8 9
  802467         1 9 7 8 6 4

    1 1            10 12
 1020304         6081012
+5060708        -1020304
 6081012         5060708
```

Basic Chapter 1 Lesson 5
Multiplication of Whole Numbers

The third operation with numbers is multiplication $\times$ or $\cdot$
Multiplication is the operation of repeated addition of the same numbers $2\times 3 = 2+2+2 = 6$ **or** $2\cdot 3 = 2+2+2 = 6$

The first number in multiplication is called the factor 2
The second number in multiplication is called the factor 3
The result of multiplication is called the product 6
Factor $\times$ Factor = Product $2\times 3 = 6$

Properties of Multiplication

Changing the order of two factors does not change the product
$AB = BA$ $2\times 3 = 2+2+2 = 6$ $3\times 2 = 3+3 = 6$

The product of any number and zero is zero
$A\times 0 = 0\times A = 0$ $5\times 0 = 0$ $0\times 5 = 0$

The product of any number and one is the same number
$A\times 1 = 1\times A = A$ $5\times 1 = 5$ $1\times 5 = 5$

Multiplying Numbers by 10, 100 …

To multiply numbers by 10, add zero to the end of a number:
$5\times 10 = 50$

To multiply numbers by 100, add two zeros to end of a number:
$5\times 100 = 500$

Table 1
Multiplication and Division

$A + 0 = A$ $0 + A = A$ $A - 0 = A$ $A - A = 0$

$A \times 1 = A$ $1 \times A = A$ $A \div 1 = A$ $A \div A = 1$

$A \times 0 = 0$ $0 \times A = 0$ $0 \div A = 0$ $A \div 0 = \emptyset$

$A + B = B + A$ $A \times B = B \times A$ $\emptyset$ - no solutions

Number 5 from the left column multiply by number 4 from first row to get answer 20. $5 \times 4 = 20$ $20 \div 5 = 4$ $20 \div 4 = 5$

Memorize table of multiplication

×	1	2	3	<u>4</u>	5	6	7	8	9	10
1	1	2	3	4	5	6	7	8	9	10
2	2	4	6	8	10	12	14	16	18	20
3	3	6	9	12	15	18	21	24	27	30
4	4	8	12	16	20	24	28	32	36	40
<u>5</u>	5	10	15	<u>20</u>	25	30	35	40	45	50
6	6	12	18	24	30	36	42	48	54	60
7	7	14	21	28	35	42	49	56	63	70
8	8	16	24	32	40	48	56	64	72	80
9	9	18	27	36	45	54	63	72	81	90
10	10	20	30	40	50	60	70	80	90	100

Basic Chapter 1 Lesson 6

To multiply two numbers put in first place number with more digits for easy multiplication.

First Way of Multiplication

Example 1

Top 3

$$\begin{array}{r} 405 \\ \times 106 \\ \hline 2430 \text{ (P1)} \\ +\ 40500 \text{ (P2)} \\ \hline 42930 \text{ (P)} \end{array}$$

1) $6 \times 5 = 30$
2) Put 0 below 6 (P1)
3) Put 3 above 0 (top)
4) $6 \times 0 = 0$
5) $0 + 3 = 3$
6) Put 3 ahead 0 (P1)
7) $6 \times 4 = 24$
8) Put 24 ahead 30 (P1)
9) $405 \times 100 = 40500$ (P2)
10) $2430 + 40500 = 42930$ (P)

Example 2

Top 1 2

$$\begin{array}{r} 6304 \\ \times 15 \\ \hline 31520 \text{ (P1)} \\ +63040 \text{ (P2)} \\ \hline 94560 \text{ (P)} \end{array}$$

1) $5 \times 4 = 20$
2) Put 0 below 5 (P1)
3) Put 2 above 0 (top)
4) $5 \times 0 = 0$
5) $0 + 2 = 2$
6) Put 2 ahead 0 (P1)
7) $5 \times 3 = 15$
8) Put 5 ahead 2 (P1)
9) Put 1 above 6 (top)
10) $5 \times 6 = 30$
11) $30 + 1 = 31$
12) Put 31 ahead 5 (P1)
13) $6304 \times 10 = 63040$ (P2)
14) 94560 (P)

Second Way of Multiplication

Example 3

$$\begin{array}{r} 405 \\ \times 106 \\ \hline 2430 \text{ (P1)} \\ +40500 \text{ (P2)} \\ \hline 42930 \text{ (P)} \end{array}$$

1) $106 = 100 + 6$

2) $\begin{array}{r} 405 \\ \times 6 \\ \hline 2430 \text{ (P1)} \end{array}$

3) $\begin{array}{r} 405 \\ \times 100 \\ \hline 40500 \text{ (P2)} \end{array}$

Example 4

$$\begin{array}{r} 6304 \\ \times 15 \\ \hline 31520 \text{ (P1)} \\ +63040 \text{ (P2)} \\ \hline 94560 \text{ (P)} \end{array}$$

1) $15 = 10 + 5$

2) $\begin{array}{r} 6304 \\ \times 5 \\ \hline 31520 \text{ (P1)} \end{array}$

3) $\begin{array}{r} 6304 \\ \times 10 \\ \hline 63040 \text{ (P2)} \end{array}$

4) $31520 + 63040 = 94{,}560$ (Product=P)

Basic Chapter 1 Lesson 6

Review 1 A

To multiply two numbers put in first place number with more digit for easy multiplication. Multiply numbers

1) 15×4

2) 169×6

3) 1046×9

4) 12008×6

5) 9×14006

6) 86×39

7) 203×16

8) 3006×15

9) 304×607

10) 19×4006

Review 1 B

To multiply two numbers put in first place number with more digit for easy multiplication. Multiply numbers

1) 16×6

2) 178×7

3) 1057×8

4) 13009×7

5) 8×15007

6) 76×48

7) 304×17

8) 4002×16

9) 406×706

10) 18×5007

Basic Chapter 1 Lesson 7
Division of Whole Numbers

The fourth operation with numbers is division ÷ or :
Division means to separated number by some equal groups.

The first number in division is called the dividend	**6**
The second number in division is called the divisor	**2**
The result of division is called the quotient	**3**
Dividend ÷ Divisor = Quotient	$6 \div 2 = 3$

Properties of Division

Any number divided by one is the same number:	$5 \div 1 = 5$
Any number divided by itself is one:	$5 \div 5 = 1$
Zero divided by any number is zero:	$0 \div 5 = 0$
Division by zero is not allowed	$5 \div 0 = \emptyset$

Formulas: 1) $A \div 1 = A$ 2) $A \div A = 1$
3) $0 \div A = 0$ 4) $A \div 0 = \emptyset$ $\emptyset$ **- no solutions**

To check multiplication use division
$2 \times 3 = 6$ Check $6 \div 2 = 3$ or $6 \div 3 = 2$

To check division use multiplication or division
$6 \div 2 = 3$ Check $2 \times 3 = 6$ or $6 \div 3 = 2$

Division by 10
To divide numbers by 10, delete one zero: $40 \div 10 = 4$

Division by 100
To divide numbers by 100, delete two zeros: $400 \div 100 = 4$

Even numbers 16, 48
Numbers where last digit are even 0, 2, 4, 6, 8 called even numbers

Odd numbers 31, 27
Number where last digit are odd 1, 3, 5, 7, 9, called odd numbers

Basic Chapter 1 Lesson 7

Average

The quotient of division sum of numbers, by number of addend, called the average of numbers.

Find average of numbers 50, 60, 70, 80, 90
$(50 + 60 + 70 + 80 + 90) \div 5 = 350 \div 5 = 70$

Find average of numbers 10, 11, 12, 13, 14
$(10 + 11 + 12 + 13 + 14) \div 5 = 60 \div 5 = 12$

Exact Division without remainder R = 0

$$\begin{array}{r} 5 \\ 7\overline{)\ 35} \\ \underline{-35} \\ 0 \end{array}$$

35 Dividend 7 Divisor 5 Quotient 0 Remainder

Division with remainder R > 0

$$\begin{array}{r} 5 \\ 7\overline{)\ 36} \\ \underline{-35} \\ 1 \end{array}$$

Remainder must be less than divisor. If 36 divides by 7, we get: Dividend - 36, Divisor - 7, Partial Quotient - 5, Remainder – 1

To find the dividend in division with remainder, multiply divisor by the partial quotient and add remainder.

Dividend = Divisor× Partial Quotient + Remainder $36 = 7 \times 5 + 1$

Basic Chapter 1 Lesson 7

Divide the Numbers   Example 1

$6 \div 2 = 3$ dividend 6   divisor 2   quotient 3

Every time bring down just one digit

$$\begin{array}{r} 5070 \\ 607\overline{)\,3071420} \\ \underline{-3035} \\ 3642 \\ \underline{-3642} \end{array}$$

1) $607 \times 5 = 3035$
2) $3071 - 3035 = 36$
3) Bring down the 4
4) $607 \times \underline{0} = 0$
5) $364 - 0 = 364$
6) Bring down the 2
7) $607 \times \underline{6} = 3642$
8) $3642 - 3642 = 0$
9) Put last 0 from dividend in quotient

Example 2

Every time bring down just one digit

$$\begin{array}{r} 607 \\ 5060\overline{)\,3071420} \\ \underline{-30360} \\ 35420 \\ \underline{-35420} \end{array}$$

1) $5060 \times \underline{6} = 30360$
2) $30714 - 30360 = 354$
3) Bring down the 2
4) $5060 \times \underline{0} = 0$
5) $3542 - 0 = 3542$
6) Bring down the 0
7) $5060 \times \underline{7} = 35420$
8) $35420 - 35420 = 0$

Basic Chapter 1 Lesson 7

Review1 A

Find the average of numbers

1) 60, 70

2) 80, 90

3) 60, 70, 80

4) 70, 80, 90

5) 50, 60, 70, 80

6) 60, 70, 80, 90

7) 55, 60, 65, 70, 75

8) 75, 80, 85, 90, 95

9) 40, 50, 60, 70, 80, 90

10) 50, 55, 60, 60, 65, 70

Review 1 B

Complete the Table of Division 1 by using Table of Multiplication

$5 \times 4 = 20$ $20 \div 4 = 5$ $20 \div 5 = 4$

		a	b	c	d	e	f	g	h
	×	2	?	4	?	6	?	8	?
1)	1								9
2)	?								18
3)	3							24	
4)	?							32	
5)	5			20			35		
6)	?					36			
7)	7				35				
8)	?			32					
9)	9		27						
10)	?	20							

Basic Chapter 1 Lesson 8

Review 2 A

Divide the numbers

1) 250 ÷ 25
2) 500 ÷ 5
3) 750 ÷ 25
4) 1000 ÷ 5
5) 1250 ÷ 25
6) 7500 ÷ 125
7) 8750 ÷ 125
8) 10000 ÷ 125
9) 11250 ÷ 125
10) 12500 ÷ 125
11) 816 ÷ 8
12) 1624 ÷ 8
13) 2432 ÷ 8
14) 3240 ÷ 8
15) 4048 ÷ 8
16) 4856 ÷ 8
17) 5664 ÷ 8
18) 6472 ÷ 8
19) 7264 ÷ 8
20) 6456 ÷ 8

Review 2 B

Divide the numbers

1) 16032 ÷ 16
2) 32048 ÷ 16
3) 48064 ÷ 16
4) 64080 ÷ 16
5) 80096 ÷16
6) 96112 ÷ 16
7) 112128 ÷ 16
8) 128144 ÷ 16
9) 144128 ÷ 16
10) 128112 ÷ 16
11) 32064 ÷ 32
12) 64096 ÷ 32
13) 96128 ÷ 32
14) 128160 ÷ 32
15) 160192 ÷ 32
16) 192224 ÷ 32
17) 224256 ÷ 32
18) 256192 ÷ 32
19) 288256 ÷ 32
20) 256224 ÷ 32

Basic Chapter 1 Lesson 8
Four Operations with numbers

1) Multiplication numbers: for memory (x, x, any)
To multiply numbers, work in any order
$9 \times 8 \times 7 = 72 \times 7 = 504$ $\quad$ $9 \times 8 \times 7 = 9 \times 56 = 504$

2) Work from left to right: for memory (L → R)
a)Multiplication and Division numbers: (× or ÷)
To multiply and divide numbers work from left to right (L → R)
$9 \times 6 \div 3 = 54 \div 3 = 18$ $\quad$ $4 \times 81 \div 9 = 324 \div 9 = 36$
b)Division and Multiplication numbers: (÷ or ×)
To divide and multiply numbers work from left to right (L → R)
$24 \div 6 \times 2 = 4 \times 2 = 8$ $\quad$ $81 \div 9 \times 4 = 9 \times 4 = 36$
c)Division numbers: (÷, ÷)
To divide numbers, work from left to right (L → R)
$24 \div 4 \div 2 = 6 \div 2 = 3$ $\quad$ $810 \div 9 \div 3 = 90 \div 9 = 10$

3) Do Multiplication or Division before Addition or Subtraction
for memory (× or ÷) before (+ or -)
a)Multiplication and Addition numbers: 1) × 2) +
To multiply and add numbers always first multiply than add
$9 \times 6 + 9 = 54 + 9 = 63$ $\quad$ $9 + 9 \times 9 = 9 + 81 = 90$
b)Multiplication and Subtraction numbers: 1) × 2) −
To multiply and subtract numbers always first multiply than subtract
$9 \times 9 - 9 = 81 - 9 = 72$ $\quad$ $90 - 9 \times 9 = 90 - 81 = 9$
c)Division and Addition numbers: 1) ÷ 2) +
To divide and add numbers always first divide than add
$50 \div 5 + 39 = 10 + 39 = 49$ $\quad$ $88 + 171 \div 9 = 88 + 19 = 107$
d)Division and Subtraction numbers: 1) ÷ 2) −
To divide and subtract numbers always first divide than subtract
$108 \div 6 - 9 = 18 - 9 = 9$ $\quad$ $500 - 126 \div 7 = 500 - 18 = 482$

Basic Chapter 1 Lesson 9
Reviews for Test 1
Review 1 A

Date **Name**

Read directions carefully. **Answers must be in the indicated place.**
Show all steps for full credits. **Each problem is worth 5 points.**
Do only 1 operation per 1 step. **Solve 1 problem 5 minutes**

Evaluate

1) $\begin{array}{r} 900000 \\ \underline{-\ 456789} \end{array}$

1)______

2) $\begin{array}{r} 987 \\ \underline{\times 806} \end{array}$

2)______

3) $\begin{array}{r} 908 \\ \underline{\times 605} \end{array}$

3)______

4) $437658 \div 543$

4)______

5) $304920 \div 504$

5)______

Basic Chapter 1 Lesson 9
Review 1 A

Date **Name**

Read directions carefully. **Answers must be in the indicated place.**
Show all steps for full credits. **Each problem is worth 5 points.**
Do only 1 operation per 1 step. **Solve 1 problem in 5 minutes**

Evaluate

6) $300300 \div 60 \times 5$

6)______

7) $9250 - 650 \times 5$

7)______

8) $90081 - 81072 \div 9$

8)______

9) $(50 \times 300) \div 6$

9)______

10) $361 \div (117 - 98)$

10)______

Basic Chapter 1 Lesson 9

Review 1 B

Date **Name**

Read directions carefully. **Answers must be in the indicated place.**

Show all steps for full credits. **Each problem is worth 5 points.**

Do only 1 operation per 1 step. **Solve 1 problem 5 minutes**

Evaluate

1)
$$\begin{array}{r} 800000 \\ \underline{-\ 345678} \end{array}$$

1)______

2)
$$\begin{array}{r} 876 \\ \underline{\times 705} \end{array}$$

2)______

3)
$$\begin{array}{r} 807 \\ \underline{\times 706} \end{array}$$

3)______

4) $304560 \div 432$

4)______

5) $284518 \div 403$

5)______

Basic Chapter 1 Lesson 9
Review 1 B

Date **Name**

Read directions carefully. **Answers must be in the indicated place.**
Show all steps for full credits. **Each problem is worth 5 points.**
Do only 1 operation per 1 step. **Solve 1 problem 5 minutes**

Evaluate

6) $280280 \div 70 \times 4$

6)______

7) $8960 - 740 \times 4$

7)______

8) $72064 - 64056 \div 8$

8)______

9) $(40 \times 420) \div 7$

9)______

10) $324 \div (105 - 87)$

10)______

Basic Chapter 1 Lesson 10
Tests for Chapter 1
Test 1 A

Date **Name**

Read directions carefully. **Answers must be in the indicated place.**
Show all steps for full credits. **Each problem is worth 5 points.**
Do only 1 operation per 1 step. **Solve 1 problem 5 minutes**

Evaluate

1) $\begin{array}{r} 700000 \\ -\ 234567 \\ \hline \end{array}$

1)______

2) $\begin{array}{r} 765 \\ \times 604 \\ \hline \end{array}$

2)______

3) $\begin{array}{r} 706 \\ \times 605 \\ \hline \end{array}$

3)______

4) $193884 \div 321$

4)______

5) $243714 \div 302$

5)______

Basic Chapter 1 Lesson 10

Test 1 A

Date **Name**

Read directions carefully. **Answers must be in the indicated place.**

Show all steps for full credits. **Each problem is worth 5 points.**

Do only 1 operation per 1 step. **Solve 1 problem 5 minutes**

Evaluate

6) $240240 \div 80 \times 3$

6)______

7) $7490 - 830 \times 3$

7)______

8) $56049 - 49042 \div 7$

8)______

9) $(30 \times 560) \div 8$

9)______

10) $289 \div (93 - 76)$

10)______

Basic Chapter 1 Lesson 10
Test 1 B

Date **Name**

Read directions carefully. **Answers must be in the indicated place.**

Show all steps for full credits. **Each problem is worth 5 points.**

Do only 1 operation per 1 step. **Solve 1 problem 5 minutes**

Evaluate

1) $\begin{array}{r} 600000 \\ \underline{-\ 123456} \end{array}$

1)______

2) $\begin{array}{r} 654 \\ \underline{\times 503} \end{array}$

2)______

3) $\begin{array}{r} 605 \\ \underline{\times 403} \end{array}$

3)______

4) $105630 \div 210$

4)______

5) $182508 \div 201$

5)______

Basic Chapter 1 Lesson 10
Test 1 B

Date **Name**

Read directions carefully. **Answers must be in the indicated place.**
Show all steps for full credits. **Each problem is worth 5 points.**
Do only 1 operation per 1 step. **Solve 1 problem 5 minutes**

Evaluate

6) $180180 \div 90 \times 2$

6)______

7) $6840 - 920 \times 2$

7)______

8) $42036 - 36030 \div 6$

8)______

9) $(20 \times 720) \div 9$

9)______

10) $256 \div (81 - 65)$

10)______

Basic Answers Chapter 1

Lesson 1 Pretest 1

1) 406728 2) 609 3) 12012 4) 240240 5) 27

Lesson 1 College Placement Test

1) 5 2) 0 3) 21 4) - 2.8 5) - 4x – 4y
6) 7.5 7) 18.2 ft 8) 60 in^2 9) 2.4 hr. 10) 14 girls

Lesson 2 Review 1 A

1) 427,130 2) 807 3) 9,009 4) 260,260 5) 35

Lesson 2 Review 1 B

1) 243,815 2) 908 3) 4,004 4) 180,180 5) 24

Lesson 3 Review 1 A

1) 120	2) 230	3) 350	4) 460	5) 570
6) 680	7) 790	8) 900	9) 990	10) 880
11) 1,200	12) 2,300	13) 3,500	14) 4,600	15) 5,700
16) 6,800	17) 7,900	18) 9,000	19) 9,900	20) 8,800

Lesson 3 Review 1 B

1) 1< 14	2) 345 > 23	3) 456 < 1,234
4) 12,345 > 2,345	5) 2,345 < 12,345	6) 23,456 > 9,876
7) 3,456 < 23,456	8) 34,567 > 8,765	9) 9,876 = 9,876
10) 8,765 = 8,765	11) 9,876 > 9,080	12) 1,234 < 1,325
13) 1,001 = 1,001	14) 1,234 < 1,543	15) 9,876 > 9,767
16) 1,234 < 1,243	17) 3,456 < 3,465	18) 3,465 > 3,456
19) 1,243 > 1,234	20) 2,345 < 2,354	

Lesson 6 Review 1 A

1) 60 2) 1,014 3) 9,414 4) 72,048 5) 126,054
6) 3,354 7) 3,248 8) 45,090 9) 184,528 10) 76,114

Lesson 6 Review 1 B

1) 96 2) 1,246 3) 8,456 4) 91,063 5) 120,056
6) 3,648 7) 5,168 8) 64,032 9) 286,636 10) 90,126

Basic Answers Chapter 1

Lesson 7 Review 1 A

1) 65	2) 85	3) 70	4) 80	5) 65
6) 75	7) 65	8) 85	9) 65	10) 60

Lesson 8 Review 2 A

1) 10	2) 100	3) 30	4) 200	5) 50
6) 60	7) 70	8) 80	9) 90	10) 100
11) 102	12) 203	13) 304	14) 405	15) 506
16) 607	17) 708	18) 809	19) 908	20) 807

Lesson 8 Review 2 B

1) 1,002	2) 2,003	3) 3,004	4) 4,005	5) 5,006
6) 6,007	7) 7,008	8) 8,009	9) 9,008	10) 8,007
11) 1,002	12) 2,003	13) 3,004	14) 4,005	15) 5,006
16) 6,007	17) 7,008	18) 8,006	19) 9,008	20) 8,007

Lesson 9 Review 1 A

1) 443211	2) 795522	3) 549340	4) 806	5) 605
6) 25025	7) 6000	8) 81073	9) 2500	10) 19

Lesson 9 Review 1 B

1) 454322	2) 617580	3) 569742	4) 705	5) 706
6) 16016	7) 6000	8) 64057	9) 2400	10) 18

Lesson 10 Test 1 A

1) 465433	2) 462060	3) 427130	4) 604	5) 807
6) 9009	7) 5000	8) 49043	9) 2100	10) 17

Lesson 10 Test 1 B

1) 476544	2) 328962	3) 243815	4) 503	5) 908
6) 4004	7) 5000	8) 36031	9) 1600	10) 16

Basic

Chapter 2

Properties of Whole Numbers

Students!

Students, you are smarter than a calculator!
Students, you are smarter than a computer!
Students, you need to believe in yourself.
Students, be patient and do your best.

Do not be afraid to make mistakes.
Tell yourself positive words.
I am young! I am smart!
I am strong! I can do anything!

Smart students use their brain!
Your brain is always with you!
A calculator is not always with you.
Mathematics is work for the brain.

Be open to new ideas in math!
Work hard and try your best!
Check answers and ask questions!
Practice! Practice! Practice!

Basic Chapter 2 Lessons
Properties of whole numbers
Lessons 1 - 10
Basic Problems

Lesson 1 Table of Multiplication. Pretest. Mental Math 1 – 4

Lesson 2 Power of Whole numbers. Perfect Square Numbers. Table 2. Prime Factorization

Lesson 3 Tests for division

Lesson 4 Divisors and Multiples

Lesson 5 Order of five operations

Lesson 6 Laws for whole numbers 1

Lesson 7 Laws for operations 2

Lesson 8 Common factor

Lesson 9 Review 2 A, Review 2 B

Lesson 10 Test 2 A, Test 2 B

Answers Chapter 2

Basic Chapter 2 Contents
Properties of whole numbers

Contents have topics and exercises, which students need to solve before tests.1)Work with lesson 1, read rules and examples, 2) Solve the problems from contents, 3) Check answers in the end of chapter 1, 4) Work with the next lessons 2, 3, 4, 5, …, 5) Solve Test 1 A, 6) Grading Scale without minuses
A=90–100%   B+=85–89%   B=80–84%   C+=75–79%
C=70–74%   D+=65–69%   D=60–64 %   F=0 - 59%
7) If you solve 10 problems correctly, you get an A, 8) If have an A, you can move on to next chapter, 9) If your grade is lower than an A, work with the mistakes, then solve Test 1B, 10) When you finish all of the chapters, solve Final Exam in last chapter.

Lesson 1 Memorize table of multiplication. Solve the Pretest 1 # 1 – # 5
Mental Math 1 – 4 odd

Lesson 2 Power of whole numbers. Perfect Square Numbers
Memorize Table 2. Prime Factorization.

Lesson 3 Tests for division
Lesson 4 Divisors and Multiples

Lesson 5 Order of five operations
Lesson 6 Laws for whole numbers 1
Review 1 A # 16 – # 25. Review 1 B # 16 – # 25

Lesson 7 Laws for operations 2
Review 1 A # 1 – # 20 odd. Review 1 B # 1 – # 20 odd
Lesson 8 Common factor
Review 1 A # 6 – # 10 all, Review 1 B # 11 – # 20 all

Lesson 9 Solve the problems
Review 2 A # 1 – # 10 all. Review 2 B # 1 – # 10 all
Lesson 10 Solve the problems
Test 2 A # 1 – # 10 all. Test 2 B # 1 – # 10 all

Answers Chapter 2

Basic Chapter 2 Lesson 1
Table 1
Multiplication and Division

$A + 0 = A$	$0 + A = A$	$A - 0 = A$	$A - A = 0$
$A \times 1 = A$	$1 \times A = A$	$A \div 1 = A$	$A \div A = 1$
$A \times 0 = 0$	$0 \times A = 0$	$0 \div A = 0$	$A \div 0 = \emptyset$
$A + B = B + A$		$A \times B = B \times A$	$\emptyset$ - no solutions

Number 5 from the left column multiply by number 4 from first row to get answer 20. $5 \times 4 = 20$ $20 \div 5 = 4$ $20 \div 4 = 5$

Memorize Table 1

×	1	2	3	<u>4</u>	5	6	7	8	9	10
1	1	2	3	4	5	6	7	8	9	10
2	2	4	6	8	10	12	14	16	18	20
3	3	6	9	12	15	18	21	24	27	30
4	4	8	12	16	20	24	28	32	36	40
<u>5</u>	5	10	15	<u>20</u>	25	30	35	40	45	50
6	6	12	18	24	30	36	42	48	54	60
7	7	14	21	28	35	42	49	56	63	70
8	8	16	24	32	40	48	56	64	72	80
9	9	18	27	36	45	54	63	72	81	90
10	10	20	30	40	50	60	70	80	90	100

Basic Chapter 2 Lesson 1
Pretest

Date **Name**

Evaluate

1) $8 \cdot 225 - 8 \cdot 75$

1)___

2) $12912 \div 16 - 8064 \div 16$

2) ______

3) $3^4 - 4^3 \div 2^3$

3) ______

4) $70200 \div (95^2 - 35^2)$

4) _______

5) $\{3 \cdot 8^2 - 2^2[2^3 + (1 + 3)^2]\} + 2^3$

5) ______

Basic Chapter 2 Lesson 1

Mental Math 1 Add

1) 9 + 1
2) 8 + 2
3) 7 + 3
4) 6 + 4
5) 5 + 5
6) 9 + 2
7) 8 + 3
8) 7 + 4
9) 6 + 5
10) 9 + 3
11) 8 + 4
12) 7 + 5
13) 6 + 6
14) 9 + 4
15) 8 + 5
16) 7 + 6
17) 9 + 5
18) 8 + 6
19) 7 + 7
20) 9 + 6
21) 8 + 7
22) 9 + 7
23) 8 + 8
24) 9 + 8
25) 9 + 9

Mental Math 2 Subtract

1) 10 – 1
2) 10 – 2
3) 10 – 3
4) 10 – 4
5) 10 – 5
6) 11 - 2
7) 11 – 3
8) 11 – 4
9) 11 – 5
10) 12 – 3
11) 12 – 4
12) 12 – 5
13) 12 – 6
14) 13 – 4
15) 13 – 5
16) 13 – 6
17) 14 – 5
18) 14 – 6
19) 14 – 7
20) 15 – 6
21) 15 – 7
22) 16 – 7
23) 16 – 8
24) 17 – 8
25) 18 - 9

Basic Chapter 2 Lesson 1

Mental Math 3 Multiply

1) 5×6
2) 6×6
3) 7×6
4) 8×6
5) 9×6
6) 5×7
7) 6×7
8) 7×7
9) 8×7
10) 9×7
11) 5×8
12) 6×8
13) 7×8
14) 8×8
15) 9×8
16) 5×9
17) 6×9
18) 7×9
19) 8×9
20) 9×9
21) 5×10
22) 6×10
23) 7×10
24) 8×10
25) 9×10

Mental Math 4 Divide

1) $30 \div 6$
2) $36 \div 6$
3) $42 \div 6$
4) $48 \div 6$
5) $54 \div 6$
6) $35 \div 7$
7) $42 \div 7$
8) $49 \div 7$
9) $56 \div 7$
10) $63 \div 7$
11) $40 \div 8$
12) $48 \div 8$
13) $56 \div 8$
14) $64 \div 8$
15) $72 \div 8$
16) $45 \div 9$
17) $54 \div 9$
18) $63 \div 9$
19) $72 \div 9$
20) $81 \div 9$
21) $50 \div 10$
22) $60 \div 10$
23) $70 \div 10$
24) $80 \div 10$
25) $90 \div 10$

Basic Chapter 2 Lesson 2

Power of Whole Numbers.

The four arithmetic operations with numbers are:
Addition, Subtraction, Multiplication, Division
$3 + 2 = 5$ $5 - 2 = 3$ $3 \times 2 = 6$ $6 \div 2 = 3$

The fifth operation with numbers is power of numbers.
A power of a number indicates repeated multiplication.

Base is a number that has to be multiplied 3
Exponent is a number that shows how many times to multiply the base 2
Power is result of multiplication of the same numbers 9

$3^2 = 3 \times 3 = 9$ — **$\text{Base}^{\text{Exponent}} = \text{Power}$**
Exponent 2 called squared — **$4^2 = 4 \times 4 = 16$**
Exponent 3 called cubed — **$2^3 = 2\cdot2\cdot2 = 4\cdot2 = 8$**

Special powers

$0^2 = 0$	$0^{10} = 0$	$0^n = 0$
$1^2 = 1$	$1^{10} = 1$	$1^n = 1$

Perfect square numbers

$1^2 = 1$	$2^2 = 4$	$3^2 = 9$	$4^2 = 16$	$5^2 = 25$
$6^2 = 36$	$7^2 = 49$	$8^2 = 64$	$9^2 = 81$	$10^2 = 100$

Perfect square numbers are 1, 4, 9, 16, 25, 36, 49, 64, 81, 100

$10^2 + 11^2 + 12^2 = 13^2 + 14^2$ $100 + 121 + 144 = 169 + 196$ $365 = 365$
$3^4 = 3\times3\times3\times3 = 81$ $3^5 = 81 \times 3 = 243$ $3^6 = 243 \times 3 = 729$
$4^4 = 4\times4\times4\times4 = 256$ $4^5 = 256 \times 4 = 1024$ $5^4 = 125 \times 5 = 625$

Basic Chapter 2 Lesson 2
Table 2
Perfect Square Numbers

Take number 20 from the left column add number 5 from the first row to get number 25
Answer is in intersection row of 20 and column of 5
$25^2 = 25 \cdot 25 = 625$

Memorize Table of Perfect Square first 2 rows and first column

a^2	0	1	2	3	4	5	6	7	8	9
	0	1	4	9	16	25	36	49	64	81
10	100	121	144	169	196	225	256	289	324	361
20	400	441	484	529	576	625	676	729	784	841
30	900	961	1024	1089	1156	1225	1296	1369	1444	1521
40	1600	1681	1764	1849	1936	2025	2116	2209	2304	2401
50	2500	2601	2704	2809	2916	3025	3136	3249	3364	3481
60	3600	3721	3844	3969	4096	4225	4356	4489	4624	4761
70	4900	5041	5184	5329	5476	5625	5776	5929	6084	6241
80	6400	6561	6724	6889	7056	7225	7396	7569	7744	7921
90	8100	8281	8464	8649	8836	9025	9216	9409	9604	9801

Basic Chapter 2 Lesson 2

Prime Factorization

Prime and Composite Numbers

Factor × Factor = Product **$3 \times 2 = 6$**

A factor of a number n is a nonzero whole number that divides evenly i nto n

All divisors of 12 are 1, 2, 3, 4, 6, 12 **$1\times12 = 2\times6 = 3\times4 = 12$**

All divisors of 24 are 1, 2, 3, 4, 6, 8, 12, 24 **$1\times24=2\times12=3\times8=4\times6=24$**

Prime Numbers

A natural number called a prime number if it has only two different divisors: one and the number itself 2, 3, 5, 7, 11, 13, 17, 19 …

Composite Numbers

A natural number that has more than two divisors called a composite number **4, 6, 8, 9, 10, 12, 14, 15, 16, 18…**

The number 1 has only one divisor: which is why 1 is not a prime or a composite. Prime numbers cannot factorize, represent like product another natural numbers.

Remember the first 10 prime numbers between 1 and 30
2, 3, 5, 7, 11, 13, 17, 19, 23, 29
To factorize composite numbers to prime factors means to represent composite numbers in the product of prime factors:
$8 = 4 \times 2 = 2 \times 2 \times 2 = 2^3$

Prime Factorization

Example 1: Find prime factorization for 100
Try to divide by prime numbers: by 2, by 3, by 5, by 7…
$100 \div 2$
$50 \div 2$
$25 \div 5$
$5 \div 5$
$100 = 2\times2\times5\times5 = 2^2\cdot5^2$
Example 2: Represent your numbes like the product of just prime numbers **$50 = 2 \times 25 = 2 \times 5 \times 5 = 2\times 5^2$**

Basic Chapter 2 Lesson 3

Tests for division

Test for division is a way to know, without doing division, if some number can divide by other number evenly or with a remainder of zero. A number is divisible by another number when there is no remainder.

Test for 2

A number is divisible by 2 if its last digit is even digits (0, 2, 4, 6, 8)
If a number can divide by 2 this number called even.
Example: 1,234

Test for 5

A number is divisible by 5 if its last digit is 0 or 5
Example: 100 can be divided by 5 $100 \div 5 = 20$

Test for 10

A number is divisible by 10 if its last digits have one or more zeros.
Example: 240 can be divided by 10 $240 \div 10 = 24$

Test for 25

A number is divisible by 25 if the last two digits to the right are 00, 25, 50, 75. Example: 100, 250, 675, 925 $100 \div 25 = 4$

Test for 4

A number is divisible by 4 if the last two digits to the right of the number are 00 or make a new number that can be divided by 4.
Example:1<u>00</u>, 7<u>28</u>, 5<u>40</u> $100 \div 4 = 25$ $28 \div 4 = 7$ $40 \div 4 = 10$

Test for 8

A number is divisible by 8 if the last three digits to the right of number are 000 or make a new number that can be divided by 8.
Example:1<u>000</u>, 2<u>160</u>, 6<u>240</u> $1000 \div 8 = 125$ $160 \div 4 = 40$ $240 \div 4 = 60$

Basic Chapter 2 Lesson 3

Test for 3

A number is divisible by 3 if the sum of its digits is divisible by 3

$342 = 300 + 40 + 2 = 3\times100 + 4\times10 + 2 = 3(99+1) + 4(9+1) + 2$

$3 \times 99+3+4\times9+4+2 = 3\times99+4\times9 + (3+4+2)$

Example: 342 $3+4+2 = 9$ $9\div3 = 3$ $342\div3 = 114$

Test for 9

A number is divisible by 9 if the sum of its digits is divisible by 9.

$468= 400 +60 + 8 = 4\times100+6\times10+8 = 4(99+1) + 6(9+1) +8$

$4 \times 99+ 4 + 6\times 9 + 6 + 8 = 4\times99 + 6\times9 + (4 + 6 + 8)$

Example 468 $4+6+8 =18$ $18\div9 = 2$ $468 \div9 = 52$

Test for 6

A number is divisible by 6 if it is divisible by both 2 and 3.

Example: 246 Last digit 6 $6\div2=3$ $2+4+6=12$ $12\div3 = 4$ $246\div6=41$

Test for 15

A number is divisible by 15 if it is divisible by both 5 and 3.

Example:465 Last digit 5 $4 + 6 + 5 = 15$ $15\div3 = 5$ $465\div15 = 31$

Test for Sum

If each addend can divide by given number the sum divisible by a Number $(a + b) \div c = a \div c + b \div c$

$(18 + 12) \div 3 = 18 \div 3 + 12 \div 3 = 6 + 4 = 10$

Test for Difference

If minuend and subtrahend can divide by given number the difference divisible

by a number

$(a - b) \div c = a \div c - b \div c$ $(18 - 12) \div 3 = 18 \div 3 - 12 \div 3 = 6 - 4 = 2$

Test for Product

If one of factors divide by given number the product divisible by a number

$(A \times B) \div C = (A \div C) \times B$ $(6 \times 5) \div 3 = (6 \div 3) \times 5 = 2 \times 5 = 10$

$(A \times B) \div C = A \times (B \div C)$ $(7 \times 6) \div 3 = 7 \times (6 \div 3) = 7 \times 2 = 14$

Basic Chapter 2 Lesson 3

Mental Mathmeans: if you can do operations mentally

If numbers can divide evenly, try to multiply or divide easily

To multiply a number by 5, divide this number by 2, and multiply that result by 10

$16 \times 5 = 16 \div 2 \times 10 = 8 \times 10 = 80$

$24 \times 5 = 24 \div 2 \times 10 = 12 \times 10 = 120$

To multiply a number by 25, divide this number by 4, and multiply that result by 100

$16 \times 25 = 16 \div 4 \times 100 = 4 \times 100 = 400$

$24 \times 25 = 24 \div 4 \times 100 = 6 \times 100 = 600$

To multiply a number by 125, divide this number by 8, and multiply result by 1000

$16 \times 125 = 16 \div 8 \times 1000 = 2 \times 1000 = 2000$

$24 \times 125 = 24 \div 8 \times 1000 = 3 \times 1000 = 3000$

To divide a number by 5, divide this number by 10, and multiply that result by 2

$50 \div 5 = 50 \div 10 \times 2 = 5 \times 2 = 10$

$80 \div 5 = 80 \div 10 \times 2 = 8 \times 2 = 16$

To divide a number by 25, divide this number by 100, and multiply that result by 4

$500 \div 25 = 500 \div 100 \times 4 = 5 \times 4 = 20$

$800 \div 25 = 800 \div 100 \times 4 = 8 \times 4 = 32$

To divide a number by 125, divide this number by 1000, and multiply that result by 8

$5000 \div 125 = 5000 \div 1000 \times 8 = 5 \times 8 = 40$

$8000 \div 125 = 8000 \div 1000 \times 8 = 8 \times 8 = 64$

Basic Chapter 2 Lesson 4

Divisors and Multiples
Greatest Common Factor

The product of the numbers 2 and 5 is 10; the numbers 2 and 5 called factors. If the number 10 divided by 2, the result is 5 and the divisor is 2. The number 2 can called a factor and can called a divisor of 10.

Greatest Common Factor or Greatest Common Divisor (in short GCF or GCD)

1) Find the GCF of numbers 12 and 18
Divisors of number 12 are 1, 2, 3, 4, 6, 12
Divisors of number 18 are 1, 2, 3, 6, 9, 18
Common divisors are 1, 2, 3, 6
Greatest common divisor is 6
GCD (12, 18) = 6 $\quad$ $12 \div 6 = 2$ $\quad$ $18 \div 6 = 3$

2) Find GCF of numbers 24, 36, and 48
Factors of number 24 are 1, 2, 3, 4, 6, 8, 12, 24
Factors of number 36 are 1, 2, 3, 4, 6, 12, 18, 36
Factors of number 48 are 1, 2, 3, 4, 6, 12, 16, 24, 48
Common factors are 1, 2, 3, 4, 6, 12
Greatest common factor is 12
GCF (24, 36, 48) = 12 $\quad$ $24 \div 12=2$ $\quad$ $36\div 12=3$ $\quad$ $48\div 12=4$

3)If the greatest number divided evenly by the smallest number, the greatest common factor of numbers is equal to the smallest of given numbers: GCF (12, 4) = 4
Factors of 12 are 1, 2, 3, 4, 6, 12
Factors of 4 are 1, 2, 4
Common factors are 1, 2, 4
Greatest common factor is 4
GCF (12, 4) = 4 $\quad$ $12 \div 4 = 3$

4) The greatest common factor of prime numbers is one:
GCF (5,7) = 1 $\quad$ $5 \div 1 = 5$ $\quad$ $7\div 1 = 7$

Basic Chapter 2 Lesson 4

Multiples

Any numbers, which is evenly divided by a given number, called the multiple of a given number: $10 \div 5 = 2$ 10 is a multiple of 5
Any number has many multiples. Multiples of 5 are 5, 10, 15, 20 ...
The least multiple for any number is the number itself.
Least Common Multiple, or in short LCM (× 1, 2, 3, 4, 5...)
The least number is the smallest number that can evenly divided by each of the given numbers.

1) Find the LCM of numbers 12 and 18
Multiples of 12 are 12, 24, 36, 48, 60, 72...
Multiples of 18 are 18, 36, 54, 72...
Common multiples are 36, 72...
Least common multiples is 36
LCM (12, 18) = 36 $36 \div 12 = 3$ $36 \div 18 = 2$

2)If the greatest of given numbers evenly divided by the smallest numbers, the greatest number is the least common multiple of the given numbers.
Find the LCM of numbers 4, 12
Multiples of 4 are 4, 8, 12, 16, 20, 24 ...
Multiples of 12 are 12, 24 ...
Common multiples are 12, 24 ...
LCM (4, 12) = 12

3)The least common multiple of prime numbers is equal to the product of the numbers.
Find the LCM of numbers 2 and 5
Multiples of 2 are 2, 4, 6, 8, 10...
Multiples of 5 are 5, 10...
Common multiples are 10, 20 ...
Least multiple is 10
LCM (2, 5) = $2 \times 5 = 10$

4) A second way to find LCM. Find the LCM of numbers 12 and 18
If the greatest number cannot be divided by the smallest number, try to multiply the big number by 2, 3, 4, 5 ... and check division
Example: $18 > 12$ 18 cannot be divided evenly by 12
$18 \times 2 = 36$ $36 \div 12 = 3$ LCM (12, 18) = 36

Basic Chapter 2 Lesson 5
Order of Five Operations

**An arithmetic expression contains numbers, sign of operations and sometimes parentheses. Result of all operations called value of expressions. Expressions with numbers can be in different form: as a sum 8 + 4, a difference 8 - 4, a product 8 × 4, or a quotient 8 ÷ 4. Operations of the first level are addition (+) and subtraction (-)
Operations of the second level are multiplication (×) and division (÷)
Operations of the third level are power (a^n), root, and logarithm.**

Order of Operations of the Same Level (+, -) or (×, ÷)

An arithmetic expression without parenthesis, that has just addition or just multiplication, add or multiply in any order:

80 + 60 + 20 = 80 + 20 + 60 = 100 + 60 = 160
25 × 6 × 4 = 25 × 4 × 6 = 100 ×600
11 + 22 + 19 + 28 = (11 + 19) + (22 + 28) = 30 + 50 = 80
4 × 5 × 25 × 2 = (4 × 25) × (5 × 2) = 100 × 10 = 1000

An expression without parenthesis that has operations with the same level of operations (+, -) or (×, ÷), work in order, from left to the right:

80 – 20 + 40 = 60 + 40 = 100 160 ÷ 4 × 2 = 40 × 2 = 80
38 – 8 + 32 – 2 = 30 + 32 – 2 = 62 – 2 = 60
64 ÷ 8 × 4 ÷ 2 = 8 × 4 ÷ 2 = 32 ÷ 2 = 16

Order of Operations of Two Levels (+, -, ×, ÷)

To work an expression without parentheses that has operations of two levels of operations, at first work in order with the second level of operation (×, ÷) from left to right, then work in order with the first level of operations (+, -,) from left to right

60 + 40 × 2 = 60 + 80 = 140
180 ÷ 9 + 40 = 20 + 40 = 60
20 × 4 – 20 = 80 – 20 = 60
80 – 150 ÷ 3 = 80 – 50 = 30
12 + 18 × 2÷2 = 12 + 36 ÷ 2 = 12 + 18 = 30
18 – 24 ÷ 6 × 2 = 18 – 4 × 2 = 18 – 8 = 10
16 × 4 ÷ 2 – 6 = 64 ÷ 2 – 6 = 32 – 6 = 26
18 ÷ 9 × 2 + 4 = 2 × 2 + 4 = 4 + 4 = 8

Order of Operations in an Expression with Parentheses

In an expression with parentheses: the first step is to work with the operations in parentheses, and the second step is to work before and after parentheses from left to right:

$(50 + 40) \times 2 = 90 \times 2 = 180$

$3 \times (90 - 70) = 3 \times 20 = 60$

$(180 - 40) \div 7 = 140 \div 7 = 20$

$200 \div (40 - 20) = 200 \div 20 = 10$

$15 \div 3 \times (5 + 4) \times 2 = 15 \div 3 \times 9 \times 2 = 5 \times 9 \times 2 = 45 \times 2 = 90$

$20 \div (4 - 2) \times 3 = 20 \div 2 \times 3 = 10 \times 3 = 30$

$6 \times (9 - 7) \div 3 = 6 \times 2 \div 3 = 12 \div 3 = 4$

$(18 - 4) \div 7 \times 5 = 14 \div 7 \times 5 = 2 \times 5 = 10$

Grouping Symbols

Expressions have three kinds of grouping symbols: first do the operations in parentheses, second do the operations - in brackets, and third do the operations - in braces: Parentheses (1) Brackets [2] Braces { 3 }

Order of Five Operations

Expressions have five operations: first do the operations of the third level (a^n); second, do the operations of the second level (× or ÷), and lastly, do the operations of the first level (+ or -), in order as they occur from left to right.

Remember (in short):

Grouping Symbols	(1) [2]	{3}
Exponents (Power)	a^n	
Multiplication or Division from left to right	× or ÷	L → R
Addition or Subtract from left to right	+ or –	L → R

Example 1

$6^2 \div 2^2 \times 3 - 3^2 \times 6 \div 3 = 36 \div 4 \times 3 - 9 \times 6 \div 3 = 9 \times 3 - 54 \div 3 = 27 - 18 = 9$

Example 2

$50 - \{200 \div [60 - (30 + 20)]\} = 50 - \{200 \div [60 - 50]\} =$

$50 - 200 \div 10 = 50 - 20 = 30$

Basic Chapter 2 Lesson 6

Laws for Whole Numbers 1

A formula is a statement of math relations between variables.
Formulas include letters that can represent any numbers.
Equality is two expressions joined by the equal symbol.

Laws of Addition

First Law of Addition (commutative)

Changing the order of two addends does not change the sum:

$A + B = B + A \qquad 2 + 3 = 5 \qquad 3 + 2 = 5 \qquad 2 + 3 = 3 + 2$

Second Law of Addition (associative)

Changing the grouping of addends does not change the sum:

$A + (B + C) = (A + B) + C \qquad 8 + (2 + 5) = (8 + 2) + 5 = 10 + 5 = 15$

Use the commutative law of addition as the first step.
Use the associative law of addition as the second step.
Add in any order. Use the two laws together.

$2 + 7 + 3 + 8 = (2 + 8) + (3 + 7) = 10 + 10 = 20$

Law of Multiplication

First Law of Multiplication (commutative)

Changing the order of two factors does not change the product:

$A\,B = B\,A \quad 2 \times 3 = 6 \qquad 3 \times 2 = 6 \qquad 2 \times 3 = 3 \times 2$

Second Law of Multiplication (associative)

Changing the grouping of factors does not change the product:

$A\,(B\,C) = (AB)\,C \qquad 5 \times (2 \times 4) = (5\times2) \times 4 = 10 \times 4 = 40$

Use the commutative law of multiplication as the first step
Use the associative law of multiplication as the second step
Multiply in any order; use the two laws together:

$5 \times 8 \times 2 = 5 \times 2 \times 8 = 10 \times 8 = 80$

$2\times5 = 10 \quad 2\times50 = 100 \qquad 4\times25 = 100 \qquad 8\times125 = 1000$

Basic Chapter 2 Lesson 6

First Distributive Law of Multiplication with Addition

To multiply a number by a sum, multiply that number by each of the addends and add the results

$A(B + C) = AB + AC$ $2(3+5) = 2\times 3+ 2\times 5= 6 +10=16$ $2(3+5)=2\times 8=16$

$(A + B)C = AC + BC$ $32\times 5=(30+2)\times 5=30\times 5+2\times 5=150+10 = 160$

Second Distributive Law of Multiplication with Subtraction

To multiply a number by a difference, multiply that number by the minuend and subtrahend, and subtract the results:

$A(B - C) = AB - AC$ $2(10 - 4) = 2 \times 10 - 2 \times 4 = 20 - 8 = 12$

$(A - B)C = AC - BC$ $(20 - 2) \times 5 = 20 \times 5 - 2 \times 5 = 100 - 10 = 90$

Division sum **$A\div C + B\div C = (A + B) \div C$**
$18 \div 3 + 12 \div 3 = (18 + 12) \div 3 = 30 \div 3 = 10$

Division difference **$A \div C - B \div C = (A - B) \div C$**
$18 \div 3 - 12 \div 3 = (18 - 12) \div 3 = 6 \div 3 = 2$

To do mental calculations means to carry out operations by laws or properties.

Memorize Formulas

1) $A + (B + C) = (A + B) + C$
2) $A \times (B \times C) = (A \times B) \times C$
3) $A \cdot (B + C) = A\cdot B + A\cdot C$
4) $A \cdot (B - C) = A\cdot B - A\cdot C$
5) $A\cdot B + A\cdot C = A \cdot (B + C)$
6) $A\cdot B - A\cdot C = A \cdot (B - C)$
7) $A \div C + B \div C = (A + B) \div C$
8) $A \div C - B \div C = (A - B) \div C$

Basic Chapter 2 Lesson 6
Review 1 A
Add numbers

$10 = 1 + 9 = 2 + 8 = 3 + 7 = 4 + 6 = 5 + 5$

1) $43 + 18 + 27$
2) $22 + 46 + 18$
3) $16 + 59 + 24$
4) $53 + 22 + 17$
5) $15 + 99 + 85$
6) $81 + 46 + 19$
7) $88 + 39 + 12$
8) $36 + 47 + 53$
9) $21 + 35 + 79$
10) $86 + 20 + 14$
11) $35 + 65 + 16$
12) $11 + 89 + 15$
13) $(95 + 90) + 5$
14) $(10 + 88) + 12$
15) $(33 + 56) + 67$
16) $(19 + 14) + 86$
17) $(92 + 3) + 8$
18) $3 + (97 + 18)$
19) $(24 + 23) + 76$
20) $(25 + 17) + 75$

Multiply numbers

$100 = 2 \times 50 = 4 \times 25 = 5 \times 20 = 10 \times 10$

21) $5 \times 25 \times 2$
22) $25 \times 5 \times 4$
23) $(20 \times 9) \times 5$
24) $(50 \times 6) \times 2$
25) $(25 \times 6) \times 4$
26) $(25 \times 7) \times 4$
27) $25 \times (8 \times 4)$
28) $25 \times (4 \times 9)$
29) $8 \times (5 \times 20)$
30) $20 \times (4 \times 5)$

Basic Chapter 2 Lesson 6
Review 1 B
Evaluate by using laws

1) $5 \cdot (10 + 6)$
2) $5 \cdot (30 + 8)$
3) $(20 + 7) \times 5$
4) $(40 + 9) \times 5$
5) $25\times (10 + 1)$
6) $25\times (10 - 4)$
7) $8\ (20 - 5)$
8) $8\ (30 + 5)$
9) $(10 - 2) \times 25$
10) $(20 - 4) \times 25$

Evaluate by using laws

$34 \times 5= (30 + 4) \times 5 = 30 \times 5 + 4 \times 5 = 150 + 20 = 170$

11) 12×5
12) 14×5
13) 16×5
14) 18×5
15) 22×5
16) 12×25
17) 14×25
18) 16×25
19) 18×25
20) 22×25

Perform the operations

21) $150 \div 5 + 200 \div 5$
22) $(200 + 35) \div 5$
23) $160 \div 4 + 120 \div 4$
24) $(360 + 12) \div 6$
25) $480 \div 8 - 240 \div 8$
26) $(100 - 15) \div 5$
27) $810 \div 9 - 360\div 9$
28) $(400 - 16) \div 8$
29) $560 \div 7 - 350 \div 7$
30) $(540 - 120) \div 6$

Basic Chapter 2 Lesson 7

Laws for operations 2

A formula is a statement of math relations between variables.
Formulas include letters that can represent any numbers.
Equality is two expressions joined by the equal symbol.
To do mental calculations means to carry out operations by laws or properties.

$24\times11=24(10+1)=24\times10+24\times1=240+24=264$ (2+4=6-middle digit)

Memorize Formulas

1) $A + (B + C) = (A + B) + C$
2) $A \times (B \times C) = (A \times B) \times C$
2) $A \cdot (B + C) = A \cdot B + A \cdot C$
4) $A \cdot (B - C) = A \cdot B - A \cdot C$
3) $A \cdot B + A \cdot C = A \cdot (B + C)$
6) $A \cdot B - A \cdot C = A \cdot (B - C)$

Review 1 A

1) $8 \times 75 + 8 \times 25$
2) $37 \times 4 + 4 \times 13$
3) $25 \times 4 - 15 \times 4$
4) $54 \times 5 - 34 \times 5$
5) $7 \times 37 + 63 \times 7$
6) $49 \times 2 + 2 \times 51$
7) $75 \times 6 - 35 \times 6$
8) $39 \times 8 - 8 \times 19$
9) $3 \times 27 + 3 \times 23$
10) $24 \times 3 + 16 \times 3$
11) $93 \times 2 - 2 \times 53$
12) $58 \times 3 + 42 \times 3$
13) $43 \times 4 + 4 \times 17$
14) $96 \times 2 - 26 \times 2$
15) $5 \times 28 - 5 \times 18$
16) $38 \times 4 - 18 \times 4$
17) $29 \times 3 + 31 \times 3$
18) $13 \times 6 + 17 \times 6$
19) $67 \times 4 - 4 \times 47$
20) $8 \times 57 - 8 \times 47$

Basic Chapter 2 Lesson 7
Evaluate using laws for whole numbers
Memorize Formulas

1) $A + (B + C) = (A + B) + C$ 2) $A \times (B \times C) = (A \times B) \times C$

3) $A \cdot (B + C) = A \cdot B + A \cdot C$ 4) $A \cdot (B - C) = A \cdot B - A \cdot C$

5) $A \cdot B + A \cdot C = A \cdot (B + C)$ 6) $A \cdot B - A \cdot C = A \cdot (B - C)$

$34 \cdot 5 = (30 + 4) \cdot 5 = 30 \cdot 5 + 4 \cdot 5 = 150 + 20 = 170$

$8 \cdot 5 + 2 \cdot 5 = 5 \cdot (8 + 2) = 5 \cdot 10 = 50$

Review 1 B

1) $80\times7 + 20\times7$
2) $70\times5 - 20\times5$
3) $70\times4 + 30\times4$
4) $60\times6 - 40\times6$
5) $40\times5 + 60\times5$
6) $90\times4 - 80\times4$
7) $50\times20 + 50\times80$
8) $30\times7 - 30\times5$
9) $30\times40 + 30\times60$
10) $50\times90 - 50\times70$
11) $80\times75 + 80\times25$
12) $25\times60 - 15\times60$
13) $20\times17 + 20\times13$
14) $75\times40 - 35\times40$
15) $30\times27 + 30\times23$
16) $93\times20 - 20\times53$
17) $14\times40 + 40\times16$
18) $70\times28 - 70\times18$
19) $29\times30 + 31\times30$
20) $77\times20 - 20\times47$

Basic Chapter 2 Lesson 8
Common Factor
Distributive laws of multiplication
Memorize Formulas

1) $A + (B + C) = (A + B) + C$
2) $A \times (B \times C) = (A \times B) \times C$
3) $A \cdot (B + C) = A \cdot B + A \cdot C$
4) $A \cdot (B - C) = A \cdot B - A \cdot C$
5) $A \cdot B + A \cdot C = A \cdot (B + C)$
6) $A \cdot B - A \cdot C = A \cdot (B - C)$

$A (B + C) = AB + AC$
$5 \times (40 + 10) = 5 \times 40 + 5 \times 10 = 200 + 50 = 250$
$AB + AC = A (B + C)$
$40 \times 5 + 10 \times 5 = 5 (40 + 10) = 5(50) = 250$
$A (B - C) = AB - AC$
$5 (40 - 10) = 5 \times 40 - 5 \times 10 = 200 - 50 = 150$
$AB - AC = A (B - C)$
$5 \times 40 - 5 \times 10 = 5 (40 - 10) = 5(30) = 150$

$AB + AC = A (B + C)$ $\quad\quad$ $AB - AC = A (B - C)$

If two products have a common factor, put the common factor before parentheses. AB + AC A – common factor Put A before parentheses. In parentheses, put results of division products by common factor

$AB \div A = B$ $\quad\quad$ $AC \div A = C$

$16 \cdot 5 + 4 \cdot 5 = 5 (16 + 4) = 5 (20) = 100$
$22 \cdot 5 - 2 \cdot 5 = 5 (22 - 2) = 5 (20) = 100$

$75 \times 5 + 25 \times 5 = 5(75 + 25) = 5(100) = 500$
$125 (5) - 25(5) = 5(125 - 25) = 5(100) = 500$

$36 \cdot 5 = (30 + 6) \cdot 5 = 30 \cdot 5 + 6 \cdot 5 = 150 + 30 = 180$
$8 \cdot 9 + 2 \cdot 9 = 9 \cdot (8 + 2) = 9 \cdot 10 = 90$

Basic Chapter 2 Lesson 8
Evaluate with laws for whole numbers
Memorize Formulas

$A + (B + C) = (A + B) + C$

$A \cdot (B + C) = A \cdot B + A \cdot C$

$A \cdot B + A \cdot C = A \cdot (B + C)$

$A \times (B \times C) = (A \times B) \times C$

$A \cdot (B - C) = A \cdot B - A \cdot C$

$A \cdot B - A \cdot C = A \cdot (B - C)$

Review 1 A

1) $80 \times 75 + 80 \times 25$
2) $37 \times 40 + 40 \times 13$
3) $25 \times 40 - 15 \times 40$
4) $54 \times 50 - 34 \times 50$
5) $70 \times 37 + 63 \times 70$
6) $49 \times 40 + 40 \times 51$
7) $85 \times 60 - 25 \times 60$
8) $39 \times 80 - 80 \times 19$
9) $30 \times 27 + 30 \times 23$
10) $24 \times 30 + 16 \times 30$

Review 1 B

11) $93 \times 20 - 20 \times 63$
12) $58 \times 30 + 42 \times 30$
13) $43 \times 40 + 40 \times 17$
14) $96 \times 20 - 26 \times 20$
15) $50 \times 28 - 50 \times 18$
16) $38 \times 45 - 18 \times 45$
17) $29 \times 30 + 31 \times 30$
18) $30 \times 60 + 70 \times 60$
19) $77 \times 40 - 40 \times 47$
20) $80 \times 57 - 80 \times 47$

Basic Chapter 2 Lesson 9
Reviews for Test 2
Review 2 A

Date **Name**

Read directions carefully. **Answers must be in indicated place.**
Show all steps for full credits. **Each problem is worth 5 points.**
Do only 1 operation per 1 step. **Solve 1 problem 5 minutes.**

Evaluate with laws for whole numbers

1) $(A + B) + C = (A + C) + B$
$(85 + 80) + 5$
2) $(A \cdot B) \cdot C = (A \cdot C) \cdot B$
$(20 \cdot 58) \cdot 5$
3) $A\,(B + C) = A \cdot B + A \cdot C$
$4 \cdot (50 + 25)$
4) $A\,(B - C) = A \cdot B - A \cdot C$
$4 \cdot (50 - 25)$
5) $A \cdot B + A \cdot C = A\,(B + C)$
$8 \cdot 125 + 8 \cdot 25$
6) $A \cdot B - A \cdot C = A\,(B - C)$
$8 \cdot 125 - 8 \cdot 25$
7) $A \div C - B \div C = (A - B) \div C$
$22592 \div 32 - 19360 \div 32$

Evaluate

8) $2^3 - 8^2 \div 4^2$

9) $58800 \div (95^2 - 25^2)$

10) $6^3 - 2^3(10 - 2^3 \div 2^2)$

Basic Chapter 2 Lesson 9
Review 2 B

Date **Name**

Read directions carefully. Answers must be in indicated place.
Show all steps for full credits. Each problem is worth 5 points.
Do only 1 operation per 1 step. Solve 1 problem 5 minutes.

Evaluate by using laws for whole numbers

1) $(A + B) + C = (A + C) + B$
   $(80 + 88) + 12$
2) $(A \cdot B) \cdot C = (A \cdot C) \cdot B$
   $(50 \cdot 67) \cdot 2$
3) $A \cdot (B + C) = A \cdot B + A \cdot C$
   $8 \cdot (125 + 50)$
4) $A \cdot (B - C) = A \cdot B - A \cdot C$
   $8 \cdot (125 - 50)$
5) $A \cdot B + A \cdot C = A \cdot (B + C)$
   $4 \cdot 75 + 4 \cdot 50$
6) $A \cdot B - A \cdot C = A \cdot (B - C)$
   $4 \cdot 75 - 4 \cdot 50$
7) $A \div C - B \div C = (A - B) \div C$
   $14528 \div 16 - 9680 \div 16$

Evaluate

8) $6^3 - 6^2 \div 2^2$

9) $52800 \div (85^2 - 25^2)$

10) $\{3 \cdot 10^2 - 2^2[2^2 + (2 + 3)^2]\} + 2^3$

Basic Chapter 2 Lesson 10
Tests for Chapter 2
Test 2 A

Date **Name**
Read directions carefully. **Answers must be in indicated place.**
Show all steps for full credits. **Each problem is worth 5 points.**
Do only 1 operation per 1 step. **Solve 1 problem 5 minutes.**

Evaluate by using laws with whole numbers

1) $(75 + 200) + 25$

1) ______

2) $(5 \cdot 84) \cdot 20$

2) _____

3) $8 \cdot (125 + 60)$

3) ______

4) $4 \cdot (125 - 50)$

4)______

5) $8 \cdot 125 + 8 \cdot 75$

5) _____

6) $4 \cdot 175 - 4 \cdot 50$

6) _____

Basic Chapter 2 Lesson 10

Test 2 A

Date **Name**

Read directions carefully. **Answers must be in indicated place.**

Show all steps for full credits. **Each problem is worth 5 points.**

Do only 1 operation per 1 step. **Solve 1 problem 5 minutes.**

Evaluate

7) $29056 \div 32 - 25824 \div 32$

7)______

8) $3^4 - 3^4 \div 3^2$

8) ______

9) $39600 \div (85^2 - 25^2)$

9)______

10) $7^3 - 2^2(3^2 \times 2^3 - 2^2)$

10) _____

Basic Chapter 2 Lesson 10
Test 2 B

Date **Name**

Read directions carefully. **Answers must be in indicated place.**

Show all steps for full credits. **Each problem is worth 5 points.**

Do only 1 operation per 1 step. **Solve 1 problem 5 minutes.**

Evaluate by using laws with whole numbers

1) $(175 + 300) + 25$

1)______

2) $(2 \cdot 96) \cdot 50$

2) ______

3) $4 \cdot (125 + 25)$

3) ______

4) $8 \cdot (125 - 70)$

4) _____

5) $4 \cdot 150 + 4 \cdot 25$

5)_____

6) $8 \cdot 225 - 8 \cdot 75$

6)___

Basic Chapter 2 Lesson 10
Test 2 B

Date | **Name**
Read directions carefully. | **Answers must be in indicated place.**
Show all steps for full credits. | **Each problem is worth 5 points.**
Do only 1 operation per 1 step. | **Solve 1 problem 5 minutes.**

Evaluate

7) $12912 \div 16 - 8064 \div 16$

7) ______

8) $3^4 - 4^3 \div 2^3$

8) ______

9) $70200 \div (95^2 - 35^2)$

9) ______

10) $\{3 \cdot 8^2 - 2^2[2^3 + (1 + 3)^2]\} + 2^3$

10) ______

Basic Answers Chapter 2

Lesson 1 1) 1200 2) 303 3) 73 4) 9 5) 104

Lesson 6 Review 1 A

1)	88	2)	86	3)	99	4)	92	5)	199
6)	146	7)	139	8)	136	9)	135	10)	120
11)	116	12)	115	13)	190	14)	110	15)	156
16)	119	17)	103	18)	118	19)	123	20)	117
21)	250	22)	500	23)	900	24)	600	25)	600
26)	700	27)	800	28)	900	29)	800	30)	400

Lesson 6 Review 1 B

1)	80	2)	190	3)	135	4)	245	5)	275
6)	150	7)	120	8)	280	9)	200	10)	400
11)	60	12)	70	13)	80	14)	90	15)	110
16)	300	17)	350	18)	400	19)	450	20)	550
21)	70	22)	47	23)	70	24)	62	25)	30
26)	17	27)	50	28)	48	29)	30	30)	70

Lesson 7 Review 1 A

1)	800	2)	200	3)	40	4)	100	5)	700
6)	200	7)	240	8)	160	9)	150	10)	120
11)	80	12)	300	13)	240	14)	140	15)	50
16)	80	17)	180	18)	180	19)	80	20)	80

Lesson 7 Review 1 B

1)	700	2)	250	3)	400	4)	120	5)	500
6)	40	7)	5000	8)	60	9)	3000	10)	1000
11)	8000	12)	600	13)	600	14)	1600	15)	1500
16)	800	17)	1200	18)	700	19)	1800	20)	600

Basic Answers Chapter 2

Lesson 8 Review 1 A

1)	8000	2)	2000	3)	400	4)	1000	5)	7000
6)	4000	7)	3600	8)	1600	9)	1500	10)	1200

Lesson 8 Review 1 B

11)	600	12)	3000	13)	2400	14)	1400	15)	500
16)	900	17)	1800	18)	6000	19)	1200	20)	800

Lesson 9 Review 2 A

1)	170	2)	5800	3)	300	4)	100	5)	1200
6)	800	7)	101	8)	4	9)	7	10)	152

Lesson 9 Review 2 B

1)	180	2)	6700	3)	1400	4)	600	5)	500
6)	100	7)	303	8)	207	9)	8	10)	192

Lesson 10 Test 2 A

1)	300	2)	8400	3)	1480	4)	300	5)	1600
6)	500	7)	101	8)	72	9)	6	10)	71

Lesson 10 Test 2 B

1)	500	2)	9600	3)	600	4)	440	5)	700
6)	1200	7)	303	8)	73	9)	9	10)	104

Basic

Chapter 3

Decimals

Students!

Students, you are smarter than a calculator!
Students, you are smarter than a computer!
Students, you need to believe in yourself.
Students, be patient and do your best.

Do not be afraid to make mistakes.
Tell yourself positive words.
I am young! I am smart!
I am strong! I can do anything!

Smart students use their brain!
Your brain is always with you!
A calculator is not always with you.
Mathematics is work for the brain.

Be open to new ideas in math!
Work hard and try your best!
Check answers and ask questions!
Practice! Practice! Practice!

Basic Chapter 3 Lessons
Decimals
Lessons 1 - 10
Basic Problems

Lesson 1 Table of Multiplication. Pretest

Lesson 2 Addition and Subtraction of decimals

Lesson 3 Multiplication and Power of Decimals. Division decimals

Lesson 4 Table 3, Mental Math # 5, 6
Teacher can Mental Math exercises put on plastic and work in class.

Lesson 5 Order of five operations with decimals

Lesson 6 Worksheet 1 A, Worksheet 1 B for Final Exam

Lesson 7 Laws for decimals

Lesson 8 Review 1 A, Review 1 B

Lesson 9 Review 3 A, Review 3 B

Lesson 10 Test 3 A, Test 3 B

Answers Chapter 3

Basic Chapter 3 Contents
Decimals

Contents have topics and exercises, which students need to solve before tests. 1)Work with lesson 1, read rules and examples, 2) Solve the problems from contents, 3) Check answers in the end of chapter 1, 4)Work with the next lessons 2, 3, 4, 5, …, 5) Solve Test 1 A, 6) Grading: Scale without minuses

A=90–100% B+=85–89% B=80–84% C+=75–79%
C=70–74% D+=65–69% D=60–64 % F=0 - 59%

7)If you solve 10 problems correctly, you get an A, 8) If have an A, you can move on to next chapter, 9) If your grade is lower than an A, work with the mistakes, then solve Test 1B, 10) When you finish all of the chapters, solve Final Exam in last chapter.

 # 1 - # 20 odd or # 1, 3, 5, 7, 9, 11, 13, 15, 17, 19

Lesson 1 Memorize Table 1 of Multiplication.
Solve the Pretest 1 # 1–# 5
Lesson 2 Addition and Subtraction of decimals

Lesson 3 Multiplication, Powers, Decimals of decimals
Lesson 4 Memorize Table 3. Solve Mental Math # 5,
Mental Math # 6 odd

Lesson 5 Order of five operations with decimals.
Evaluate # 1 - # 20 odd
Lesson 6 Solve Worksheet 1 A for Final Exam # 1 – # 20 odd
Solve Worksheet 1 B for Final Exam # 1 – # 20 odd

Lesson 7 Laws for decimals. Memorize Laws
Lesson 8 Solve the Review 1 A # 1 – # 20 odd
Solve the Review 1 B # 1 – # 20 odd

Lesson 9 Reviews
Review 3 A # 1 – # 10, Review 3 B # 1 – # 10 all
Lesson 10 Tests
Test 3 A # 1 – # 10, Test 1 B # 1 – # 10 all
Answers Chapter 3

Basic Chapter 3 Lesson 1
Table 1
Multiplication and Division

$A + 0 = A$ $0 + A = A$ $A - 0 = A$ $A - A = 0$

$A \times 1 = A$ $1 \times A = A$ $A \div 1 = A$ $A \div A = 1$

$A \times 0 = 0$ $0 \times A = 0$ $0 \div A = 0$ $A \div 0 = \emptyset$

$A + B = B + A$ $A \times B = B \times A$ $\emptyset$ no solutions

Number 5 from the left column multiply by number 4 from first row to get answer 20. $5 \times 4 = 20$ $20 \div 5 = 4$ $20 \div 4 = 5$

Memorize Table 1

×	1	2	3	<u>4</u>	5	6	7	8	9	10
1	1	2	3	4	5	6	7	8	9	10
2	2	4	6	8	10	12	14	16	18	20
3	3	6	9	12	15	18	21	24	27	30
4	4	8	12	16	20	24	28	32	36	40
<u>5</u>	5	10	15	<u>20</u>	25	30	35	40	45	50
6	6	12	18	24	30	36	42	48	54	60
7	7	14	21	28	35	42	49	56	63	70
8	8	16	24	32	40	48	56	64	72	80
9	9	18	27	36	45	54	63	72	81	90
10	10	20	30	40	50	60	70	80	90	100

Basic Chapter 3 Lesson 1
Pretest

Date **Name**

Evaluate

1) $22 \div 4.4 \times 5$

1) ________

2) $5.2 - 3.6 \div 4$

2) ________

3) $2^2 - 0.4^2 \div 0.2^2$

3) ________

4) $0.3^2 \div (0.6 - 0.3)^2$

4) ________

5) $4.8 \div (1.8^2 + 2.76)$

5) ________

Basic Chapter 3 Lesson 2

Addition and Subtraction of Decimals

Decimals

A fraction is equal parts of whole number: $1 \div 4 = \frac{1}{4}$

A fraction contains a numerator (N) and a denominator (D).

The numerator written above the bar and shows how many equal parts are to be taken.

The denominator written under the bar and shows by how many equal parts the whole should be divided.

Fractions with a denominator of one and some zeros can represent like decimals: $\frac{3}{10} = 0.3$ $\frac{7}{100} = 0.07$ $1\frac{9}{1000} = 1.009$

Decimals contain two parts: to the left of the point stand digits of whole parts, to the right of the point stand digits of fractional parts, which called decimals digits.

Money can represent in decimals:
\$ 0.5 = \$ 0.50 = 50 c **\$ 0.05 = 5 c**

After the point, the 1 st place to the right is the tenth place, the 2nd place is the hundredth place, and 3rd is the thousandth place:

0.5 **3.28** **2.125**

Basic Property of Decimal

Adding or deleting zeros after the last digit to the right of the decimal point does not change the value of the decimal:

4 = 4. = 4.0 = 4.00 **5.00 = 5**

Basic Chapter 3 Lesson 2

Rounding Decimals

Taking an approximate value for the number called rounding.
A rounded number is always rounded to a given place value.

A number can round to any digit: one, ten, hundred …

Rule for Rounding

If the digit to the right of the given place value is less than 5 (0, 1, 2, 3, 4) that digit and all digits to the right are replaced by zeros or dropped.

Example: rounding a number to the nearest ten: $283 \approx 280$ $(3 < 5)$

If the digit to the right of the given place value is greater than or equal to 5 (5, 6, 7, 8, 9) increase the digit in the given place by 1 and replace all other digits to the right by zeros or drop all digits to its right.

Example: rounding a number to the nearest ten: $238 \approx 240$ $(8 > 5)$

Comparison Decimals

To compare two decimals, make the number of decimals digits equal, add from the right the zeros, and compare digits as whole and fractional part of decimals. (Work from left to right)

$1.48 < 3.6$ $(1 < 3)$
$0.5 > 0.05$

$0.586 < 0.590$ $(8 < 9)$
$0.50 > 0.05$

Addition and Subtraction of Decimals

To add or subtract decimals, line up the decimal points.
Then add or subtract from right to left.
Write a decimal point in the sum or difference.
Add zeroes as needed.

$$\begin{array}{r} 1.25 \\ +\ 0.80 \\ \hline 2.05 \end{array} \qquad \begin{array}{r} 1.25 \\ -\ 0.80 \\ \hline 0.45 \end{array}$$

Basic Chapter 3 Lesson 3
Multiplication and Powers of decimals

Multiplication of Decimals

Multiply two decimals as if they are whole numbers.
The number of decimal digits in the product is the sum of the number of decimal digits in both factors:

1 digit	2 digits	3 digits
$\begin{array}{r} 2.5 \\ \times\ \ 4 \\ \hline 10.0 \end{array}$	$\begin{array}{r} 2.5 \\ \times 0.8 \\ \hline 2.00 \end{array}$	$\begin{array}{r} 1.25 \\ \times 0.8 \\ \hline 1.000 \end{array}$
$2.5 \times 4 = 10$	$2.5 \times 0.8 = 2$	$1.25 \times 0.8 = 1$

Multiplication of Decimal by Numbers 10, 100, 1000 …

To multiply a decimal by numbers 10, 100, or 1000 …move the decimal point to the right as many decimal places as there are zeros in the number: 1 zero – 1 place 2 zeros – 2 places 3 zeros – 3 places

$0.24 \times 10 = 2.4$ $0.24 \times 100 = 24$ $0.24 \times 1000 = 240$

Powers of Decimals

A power is multiplication of the same numbers:

$3.5^2 = 3.5 \times 3.5 = 12.25$ $0.5^3 = 0.5 \times 0.5 \times 0.5 = 0.25 \times 0.5 = 0.125$

$$\begin{array}{r} 3.5 \\ \times 3.5 \\ \hline 175 \\ +1050 \\ \hline 12.25 \end{array} \qquad \begin{array}{r} 0.5 \\ \times 0.5 \\ \hline 0.25 \end{array} \qquad \begin{array}{r} 0.25 \\ \times 0.5 \\ \hline 0.125 \end{array}$$

Basic Chapter 3 Lesson 3

Division of Decimals

Division of a Decimal by a Whole Number

To divide a decimal by a natural number, first divide the whole part by the number, then in the quotient, place the point, after that, divide the fractional part of the number.

If the digit is not enough, add zeros to the right:

$2 \div 5 = 0.4$	$0.2 \div 5 = 0.04$	$0.02 \div 5 = 0.004$

Division of Decimal by Numbers 10, 100, and 1000…

To divide a decimal by numbers 10, 100, or 1000 …move the decimal point to the left as many decimal places as there are zeros in the number:

1 zero – 1 place	**2 zeros – 2 places**	**3 zeros – 3 places**
$12.5 \div 10 = 1.25$	$12.5 \div 100 = 0.125$	$12.5 \div 1000 = 0.0125$
$5 \div 10 = 0.5$	$5 \div 100 = 0.05$	$5 \div 1000 = 0.005$

Division of Decimals

To divide two decimals, multiply dividend and divisor by 10, 100, or 1000; the divisor has to be a whole number, after that, divide the decimal by the whole number:

$2 \div 0.5 = 20 \div 5 = 4$	$0.2 \div 0.5 = 2 \div 5 = 0.4$	$0.02 \div 0.5 = 0.2 \div 5 = 0.04$
$2 \times 10 = 20$	$0.2 \times 10 = 2$	$0.02 \times 10 = 0.2$
$0.5 \times 10 = 5$	$0.5 \times 10 = 5$	$0.5 \times 10 = 5$

Basic Chapter 3 Lesson 4

Table 3 Decimals

Properties of Decimal

$4 = 4. = 4.0 = 4.00$ $\qquad$ $5.00 = 5$

Rounding of Decimals

$2.83 \approx 2.8$ $\qquad$ $2.38 \approx 2.4$

Comparison of Decimals

$1.48 < 3.6$ $\qquad$ $0.5 > 0.05$

Addition and Subtraction of Decimals

$$\begin{array}{r} 1.25 \\ +\,0.80 \\ \hline 2.05 \end{array} \qquad \begin{array}{r} 1.25 \\ -\,0.80 \\ \hline 0.45 \end{array}$$

Multiplication of Decimals

$0.24 \times 10 = 2.4$ $\qquad$ $0.24 \times 100 = 24$ $\qquad$ $0.24 \times 1000 = 240$

$2.5 \times 4 = 10$ $\qquad$ $1.25 \times 8 = 10$ $\qquad$ $0.5 \times 0.2 = 0.1$

Powers of Decimals

$3.5^2 = 3.5 \times 3.5 = 12.25$ $\qquad$ $0.5^3 = 0.5 \times 0.5 \times 0.5 = 0.25 \times 0.5 = 0.125$

Division of Decimals

$5 \div 10 = 0.5$ $\qquad$ $5 \div 100 = 0.05$ $\qquad$ $5 \div 1000 = 0.005$

$2 \div 5 = 0.4$ $\qquad$ $0.2 \div 5 = 0.04$ $\qquad$ $0.02 \div 5 = 0.004$

$2 \div 0.5 = 20 \div 5 = 4$ $\qquad$ $0.2 \div 0.5 = 2 \div 5 = 0.4$ $\qquad$ $0.02 \div 0.5 = 0.2 \div 5 = 0.04$

Order of Operations with the Same Level

$0.8 + 0.6 + 0.2 = 0.8 + 0.2 + 0.6 = 1.6$ $\qquad$ $2.5 \times 6 \times 4 = 2.5 \times 4 \times 6 = 60$

$0.8 - 0.2 + 0.4 = 0.6 + 0.4 = 1$ $\qquad$ $1.6 \div 4 \times 2 = 0.4 \times 2 = 0.8$

Order of Operations of Two Levels

$0.2 \times 4 - 0.2 = 0.8 - 0.2 = 0.6$ $\qquad$ $1.8 \div 9 + 0.4 = 0.2 + 0.4 = 0.6$

$0.6 + 0.4 \times 2 = 0.6 + 0.8 = 1.4$ $\qquad$ $0.8 - 1.5 \div 3 = 0.8 - 0.5 = 0.3$

Order of Operations in Expression with Parentheses

$(0.5 + 0.4) \times 2 = 0.9 \times 2 = 1.8$ $\qquad$ $3 \times (0.9 - 0.7) = 3 \times 0.2 = 0.6$

$(1.8 - 0.4) \div 7 = 1.4 \div 7 = 0.2$ $\qquad$ $2 \div (0.4 - 0.2) = 2 \div 0.2 = 10$

Groping Symbols: 1) Parentheses ( ) 2) Bracket [ ] 3) Brace { }

Order of Five Operations

Grouping Symbols (1) [2] {3} $\qquad$ **Exponents (Power) a^n**

Multiplication or Division from left to right $\qquad$ $\times$ **or** $\div$ $\;L \rightarrow R$

Addition or Subtraction from left to right $\qquad$ $+$ **or** $-$ $\;L \rightarrow R$

Basic Chapter 3 Lesson 4

Mental Math 5
Decimals

1) $0.2 + 0$
2) $0 + 1.5$
3) $0.2 + 0.5$
4) $0.5 + 0.2$
5) $1.8 + 0.2$
6) $0.6 - 0$
7) $1.7 - 1.7$
8) $0.5 - 0.2$
9) 0.6×0
10) 0×1.7
11) 1×1.9
12) 0.8×1
13) 0.2×5
14) 0.4×5
15) 5×0.6
16) 0.5×8
17) 1.2×5
18) 5×1.4
19) $0 \div 0.6$
20) $0.7 \div 0$
21) $0.8 \div 1$
22) $1.9 \div1.9$
23) 0.2^2
24) 0.3^2
25) 0.4^2
26) 0.5^2
27) 0.6^2
28) 0.7^2
29) 0.8^2
30) 0.9^2

Mental Math 6
Laws for Whole Numbers

1) $8\times3 + 2\times3$
2) $7\times5 - 2\times5$
3) $7\times4 + 3\times4$
4) $6\times2 - 4\times2$
5) $4\times5 + 6\times5$
6) $9\times4 - 8\times4$
7) $5\times2 + 5\times8$
8) $3\times7 - 3\times6$
9) $3\times4 + 3\times6$
10) $5\times9 - 5\times8$
11) $8\times75 + 8\times25$
12) $25\times6 - 15\times6$
13) $2\times17 + 2\times13$
14) $75\times4 - 35\times4$
15) $3\times27 + 3\times23$
16) $93\times 2 - 2\times53$
17) $14\times4 + 4\times16$
18) $6\times28 - 6\times18$
19) $29\times3 + 21\times3$
20) $67\times2 - 2\times47$

Basic Chapter 3 Lesson 5

Order of Five Operations with Decimals

If an expression has five operations, first work on operations at the third level (a^n), after that, work on operations at the second level (× or÷), last, work on operations at the first level (+ or -) in order, as they occur, from left to right.

in short

Grouping Symbols	**(1)**	**[2]**	**{3}**
Exponents or Power – third level of operations	**a^n**		

Multiplication or Division from left to right–second level of operations

× or ÷ L → R

Addition or Subtraction from left to right – first level of operations

+ or – L → R

Examples:

a) $0.2^3 \div (0.4 - 0.2)^2 = 0.008 \div (0.2)^2 = 0.008 \div 0.04 = 0.8 \div 4 = 0.2$

b) $2^2 - 0.8^2 \div 0.4^2 = 4 - 0.64 \div 0.16 = 4 - 4 = 0$

c) $3^2 - 0.3^4 \div 0.3^3 = 9 - 0.0081 \div 0.027 = 9 - 8.1 \div 27 = 9 - 0.3 = 8.7$

d) $0.8 \times 2 + (1.5^2 - 0.25) = 1.6 + (2.25 - 0.25) = 1.6 + 2 = 3.6$

e) $5\times [16.72 \div 8 - (40 - 39.01)] = 5\times [2.09 - 0.99] = 5 \times 1.1 = 5.5$

f) $3.7\div [6 - (0.3 + 0.2\times 10)]=3.7\div [6-(0.3+2)]=3.7\div [6-2.3]=3.7\div 3.7=1$

g) $30\div [5.5 - (1.5^2 + 1.75)]=30\div [5.5 - (2.25 + 1.75)]=30\div [5.5 - 4]=20$

h) $9.1\div\{1- [0.3(0.2 + 0.1)]\} = 9.1\div\{1 - [0.3\times 0.3]\}= 9.1\div\{1 - 0.09\}=10$

i) $2 \times 0.3 + \{4.9 \div [1.2 - (0.2 + 0.3)]\} = 0.6 + \{4.9 \div [1.2 - 0.5]\} =$
$0.6 + \{4.9 \div 0.7\} = 0.6 + 7 = 7.6$

j) $0.5 [1.2 - (0.6 + 0.3)] - 0.2^2 = 0.5 [1.2 - 0.9] - 0.04 =$
$0.5 \times 0.3 - 0.04 = 0.15 - 0.04 = 0.11$

Basic Chapter 3 Lesson 5

Evaluate

1) $80 \times 7.5 + 80\times 2.5$
2) $3.7 \times 40 + 40 \times 1.3$

3) $8.5 \times 40 - 4.5 \times 40$
4) $5.4 \times 50 - 3.4 \times 50$

5) $70\times 3.7 + 6.3 \times 70$
6) $4.9 \times 30 + 30 \times 5.1$

7) $7.5 \times 60 - 3.5 \times 60$
8) $6.9 \times 80 - 80 \times 1.9$

9) $30 \times 2.7 + 30 \times 2.3$
10) $2.4 \times 30 + 1.6 \times 30$

11) $9.3 \times 20 - 20 \times 5.3$
12) $5.8 \times 50 + 4.2 \times 50$

13) $4.3 \times 40 + 40 \times 2.7$
14) $9.6 \times 20 - 2.6 \times 20$

15) $50 \times 6.8 - 50 \times 1.8$
16) $7.8 \times 40 - 3.8 \times 40$

17) $2.9 \times 30 + 3.1 \times 30$
18) $3.3 \times 60 + 2.7 \times 60$

19) $9.7 \times 50 - 50 \times 2.7$
20) $80 \times 9.7 - 80 \times 5.7$

Basic Chapter 3 Lesson 6
Worksheet 1 A for Final Exam
Whole Numbers
Perform the indicated operations

1) $(4^2 - 2^2) \div 6$
2) $(10^2 - 9^2) \div 19$
3) $(7^2 - 5^2) \div 8$
4) $(11^2 - 9^2) \div 20$
5) $(8^2 - 5^2) \div 3$
6) $(13^2 - 9^2) \div 8$
7) $84 \div (8^2 - 6^2)$
8) $81 \div (14^2 - 13^2)$
9) $52 \div (7^2 - 6^2)$
10) $64 \div (9^2 - 7^2)$
11) $96 \div (8^2 - 4^2)$
12) $56 \div (4^2 - 3^2)$
13) $(9^2 - 6^2) \div (8^2 - 7^2)$
14) $(5^2 - 4^2)(3^2 - 2^2)$
15) $2^2(10 - 2^2 \div 2)$
16) $2^3 \div 2^2 \times 2 + 2^2 \times 5$
17) $5(3^2 + 2^3 \div 2^2)$
18) $(2 + 4)^2 \div 3 \times 2$
19) $2[2^2 + (1 + 2)^2]$
20) $2^2[2^2 + (2 + 2)^2]$

Worksheet 1 B for Final Exam
Decimals
Perform the indicated operations

1) $0.2^2 \div 0.2^2 - 0.2$
2) $0.3^3 \div 0.3^2 - 0.3$
3) $0.4^2 \div 0.2^2 - 2$
4) $0.4^2 \div 0.2^2 - 4$
5) $0.6^2 \div 0.2^2 - 4$
6) $2^2 - 0.8^2 \div 0.4^2$
7) $2^2 - 0.4^2 \div 0.2^2$
8) $5^2 - 0.6^2 \div 0.2^2$
9) $3^2 - 0.8^2 \div 0.4^2$
10) $7^2 - 0.9^2 \div 0.3^2$
11) $0.2^2 \div (0.4 - 0.3)^2$
12) $0.3^3 \div (0.6 - 0.3)^2$
13) $0.4^2 \div (0.6 - 0.4)^2$
14) $0.6^2 \div (0.5 - 0.2)^2$
15) $0.8^2 \div (0.7 - 0.3)^2$
16) $(0.4 - 0.2)^2 \div 0.2^2$
17) $(0.6 - 0.3)^2 \div 0.3$
18) $(0.8 - 0.4)^2 \div 0.4$
19) $(0.5 - 0.2)^2 \div 0.9$
20) $(0.7 - 0.5)^2 \div 0.4$

Basic Chapter 3 Lesson 7

Laws for Decimals

Addition

$A + 0 = A$	$0.2 + 0 = 0.2$	$0 + A = A$	$0 + 1.3 = 1.3$
$A + B = B + A$	$0.2 + 0.3 = 5$	$0.3 + 0.2 = 0.5$	$0.2 + 0.3 = 0.3 + 0.2$

$A + (B + C) = (A + B) + C$ $\quad$ $1.7 + 0.2 = 1.0 + 0.7 + 0.2 = 1.0 + 0.9 = 1.9$

Subtraction

$A - 0 = A$	$0.6 - 0 = 0.6$	$A - A = 0$	$1.7 - 1.7 = 0$
$0.2 + 0.3 = 0.5$	$0.5 - 0.2 = 0.3$	$0.5 - 0.3 = 0.2$	

Multiplication

$A \times 0 = 0$	$0.6 \times 0 = 0$	$0 \times A = 0$	$0 \times 1.7 = 0$
$A \times 1 = A$	$0.8 \times 1 = 0.8$	$1 \times A = A$	$1 \times 1.9 = 1.9$

$A\,B = B\,A$ $\quad$ $0.2 \times 0.3 = 0.06$ $\quad$ $0.3 \times 0.2 = 0.06$ $\quad$ $0.2\times0.3=0.3 \times 0.2$

$A\,(BC) = (AB)C$ $\quad$ $0.5\,(2 \times 4) = (0.5 \times 2) \times 4 = 1 \times 4 =4$

$(B + C)\,A = AB + AC$ $\quad$ $(1 + 0.4)5 =1 \times 5 + 0.4 \times 5= 5+2 = 7$

$A\,(B - C) = AB - AC$ $\quad$ $5(1 - 0.4) =5 \times 1 - 5 \times 0.4=5 - 2 = 3$

Basic Chapter 3 Lesson 7
Division

$0 \div A = 0$	**$0 \div 0.6 = 0$**	**$A \div 0 = \emptyset$**	**$0.7 \div 0 = \emptyset$** **$\emptyset$ no solutions**
$A \div 1 = A$	**$0.8 \div 1 = 0.8$**	**$A \div A = 1$**	**$1.9 \div 1.9 = 1$**

$(A + B) \div C = A \div C + B \div C$
$(1.8 + 1.2) \div 3 = 1.8 \div 3 + 1.2 \div 3 = 0.6 + 0.4 = 1$

$(A - B) \div C = A \div C - B \div C$
$(1.8 - 1.2) \div 3 = 1.8 \div 3 - 1.2 \div 3 = 0.6 - 0.4 = 0.2$

$(AB) \div C = (A \div C)\,B$ **$(0.6 \times 7) \div 0.3 = (0.6 \div 0.3) \times 7 = 2 \times 7 = 14$**
$(AB) \div C = A(B \div C)$ **$(7 \times 0.6) \div 0.3 = 7\,(0.6 \div 0.3) = 7 \times 2 = 14$**

Mental Calculations

$2.4 \times 5 = 2.4 \div 2 \times 10 = 1.2 \times 10 = 12$

$2.4 \times 25 = 2.4 \div 4 \times 100 = 0.6 \times 100 = 60$

$2.4 \times 125 = 2.4 \div 8 \times 1000 = 0.3 \times 1000 = 300$

Common Factor

$(B + C)\,A = AB + AC$ **$(0.4 + 0.1)5 = 0.4 \times 5 + 0.1 \times 5 = 2 + 0.5 = 2.5$**

$AB + AC = A(B + C)$ **$0.4 \times 5 + 0.1 \times 5 = 5\,(0.4 + 0.1) = 5(0.5) = 2.5$**

$A(B - C) = AB - AC$ **$5\,(0.4 - 0.1) = 5 \times 0.4 - 5 \times 0.1 = 2 - 0.5 = 1.5$**

$AB - AC = A(B - C)$ **$5 \times 0.4 - 5 \times 0.1 = 5\,(0.4 - 0.1) = 5(0.3) = 1.5$**

Basic Chapter 3 Lesson 8

Review 1 A

$$5 \times 0.6 + 5 \times 0.4 = 5\,(0.6 + 0.4) = 5(1) = 5$$

Simplify by laws for decimals

1) $8 \times 7.5 + 8 \times 2.5$

2) $2.5 \times 6 - 1.5 \times 6$

3) $3 \times 3.7 + 6.3 \times 3$

4) $7.5 \times 4 - 3.5 \times 4$

5) $3 \times 2.7 + 3 \times 2.3$

6) $9.3 \times 2 - 2 \times 5.3$

7) $1.4 \times 4 + 4 \times 1.6$

8) $6 \times 2.8 - 6 \times 1.8$

9) $2.9 \times 3 + 2.1 \times 3$

10) $7 \times 2.4 - 1.4 \times 7$

11) $3.7 \times 4 + 4 \times 1.3$

12) $5.4 \times 5 - 3.4 \times 5$

13) $4.9 \times 2 + 2 \times 5.1$

14) $3.9 \times 6 - 6 \times 1.9$

15) $2.4 \times 9 + 1.6 \times 9$

16) $5.8 \times 3 + 3.2 \times 3$

17) $9.6 \times 2 - 2.6 \times 2$

18) $3.8 \times 3 - 1.8 \times 3$

19) $0.3 \times 6 + 0.7 \times 6$

20) $7 \times 5.7 - 7 \times 4.7$

Basic Chapter 3 Lesson 8

Review 1 B

$$8 \times 2.5 - 8 \times 1.5 = 8\,(2.5 - 1.5) = 8(1) = 8$$

Simplify by laws for decimals

1) $2 \times 0.7 + 2 \times 0.3$

2) $4 \times 6.4 - 4 \times 5.4$

3) $0.3 \times 7.8 + 0.3 \times 2.2$

4) $4 \times 3.2 - 4 \times 2.2$

5) $3.6 \times 2.3 + 3.6 \times 7.7$

6) $5 \times 3.6 + 5 \times 6.4$

7) $7.9 \times 4.7 + 7.9 \times 5.3$

8) $3 \times 5.6 + 4.4 \times 3$

9) $3.7 \times 2.4 + 3.7 \times 7.6$

10) $2 \times 6.2 - 2 \times 5.2$

11) $5.8 \times 7.3 + 5.8 \times 2.7$

12) $7 \times 4.8 - 7 \times 3.8$

13) $6.1 \times 3.4 + 6.1 \times 6.6$

14) $9 \times 3.8 - 9 \times 2.8$

15) $6.7 \times 8.4 + 6.7 \times 1.6$

16) $3 \times 4.1 - 3 \times 3.1$

17) $4.3 \times 3 + 3.5 \times 3 - 5.8 \times 3$

18) $4.9 \times 7 - 1.8 \times 7 + 6.9 \times 7$

19) $5.6 \times 4 + 2.7 \times 4 - 8.3 \times 4$

20) $3.4 \times 5 - 1.7 \times 5 + 8.3 \times 5$

Basic Chapter 3 Lesson 9
Reviews for Chapter 3
Review 3 A

Date **Name**

Read directions carefully. **Answers must be in indicated place.**
Show all steps for full credits. **Each problem is worth 5 points.**
Do only 1 operation per 1 step. **Solve 1 problem 5 minutes.**

1) Add decimals $37.37 + 13.73$

1) ________

2) Subtract decimals $61.1 - 46.46$

2) ________

3) Multiply decimals 0.75×4.8

3) ________

4) Divide decimals $36.45 \div 0.9$

4) ________

5) Evaluate $9.5\times5.8 - 9.5\times4.8$

5)________

Basic Chapter 3 Lesson 9
Review 3 A

Date | **Name**

Read directions carefully.
Show all steps for full credits.
Do only 1 operation per 1 step.

Answers must be in indicated place.
Each problem is worth 5 points.
Solve 1 problem 5 minutes.

Evaluate

6) $33 \div 6.6 \times 5$

6) ________

7) $9.6 - 6.8 \div 4$

7) ________

8) $6^2 - 0.6^2 \div 0.3^2$

8) ________

9) $0.4^2 \div (0.6 - 0.4)^2$

9) ________

10) $12 \div (2.6^2 + 1.24)$

10) _______

Basic Chapter 3 Lesson 9
Review 3 B

Date **Name**
Read directions carefully. **Answers must be in indicated place.**
Show all steps for full credits. **Each problem is worth 5 points.**
Do only 1 operation per 1 step. **Solve 1 problem 5 minutes.**

1) Add decimals $46.46 + 14.64$

1)________

2) Subtract decimals $51.1 - 37.37$

2)________

3) Multiply decimals 0.75×2.4

3) ________

4) Divide decimals $45.54 \div 0.9$

4) ________

5) Evaluate $8.6\times4.7 - 8.6\times3.7$

5) ________

Basic Chapter 3 Lesson 9
Review 3 B

Date **Name**

Read directions carefully. **Answers must be in indicated place.**
Show all steps for full credits. **Each problem is worth 5 points.**
Do only 1 operation per 1 step. **Solve 1 problem 5 minutes.**

Evaluate

6) $33 \div 5.5 \times 6$

6) ________

7) $4.8 - 3.4 \div 2$

7) ________

8) $4^2 - 0.4^2 \div 0.2^2$

8) ________

9) $0.6^2 \div (0.5 - 0.3)^2$

9) ________

10) $9.9 \div (2.8^2 + 1.16)$

10) ________

Basic Chapter 3 Lesson 10
Tests for Chapter 3
Test 3 A

Date **Name**

Read directions carefully. **Answers must be in indicated place.**
Show all steps for full credits. **Each problem is worth 5 points.**
Do only 1 operation per 1 step. **Solve 1 problem 5 minutes.**

1) Add decimals $19.19 + 11.91$

1) ________

2) Subtract decimals $41.1 - 28.28$

2)________

3) Multiply decimals 0.25×1.6

3) ________

4) Divide decimals $18.3 \div 0.6$

4)__________

5) Evaluate $7.8\times6.2 - 7.8\times5.2$

5) _______

Basic Chapter 3 Lesson 10
Test 3 A

Date **Name**

Read directions carefully. **Answers must be in indicated place.**
Show all steps for full credits. **Each problem is worth 5 points.**
Do only 1 operation per 1 step. **Solve 1 problem 5 minutes.**

Evaluate

6) $22 \div 5.5 \times 4$

6) ________

7) $4.8 - 2.4 \div 2$

7) ________

8) $3^2 - 0.9^2 \div 0.3^2$

8) ________

9) $0.2^2 \div (0.4 - 0.2)^2$

9) ________

10) $4.4 \div (1.6^2 + 1.44)$

10)________

Basic Chapter 3 Lesson 10
Test 3 B

Date | **Name**
Read directions carefully. | **Answers must be in indicated place.**
Show all steps for full credits. | **Each problem is worth 5 points.**
Do only 1 operation per 1 step. | **Solve 1 problem 5 minutes.**

1) Add decimals $28.28 + 12.82$

1) ________

2) Subtract decimals $31.1 - 19.19$

2) ________

3) Multiply decimals 0.25×3.2

3) ________

4) Divide decimals $24.3 \div 0.6$

4) ________

5) Evaluate $6.8 \times 3.9 - 6.8 \times 2.9$

5) ________

Basic Chapter 3 Lesson 10
Test 3 B

Date **Name**

Read directions carefully. **Answers must be in indicated place.**
Show all steps for full credits. **Each problem is worth 5 points.**
Do only 1 operation per 1 step. **Solve 1 problem 5 minutes.**

Evaluate

6) $22 \div 4.4 \times 5$

6) ________

7) $5.2 - 3.6 \div 4$

7) ________

8) $2^2 - 0.4^2 \div 0.2^2$

8) ________

9) $0.3^2 \div (0.6 - 0.3)^2$

9) ________

10) $4.8 \div (1.8^2 + 2.76)$

10) ________

Basic Answers Chapter 3

Lesson 1

1)	25	2)	4.3	3)	0	4)	1	5)	0.8

Lesson 4 Mental Math 5

1)	0.2	2)	1.5	3)	0.7	4)	0.7	5)	2
6)	0.6	7)	0	8)	0.3	9)	0	10)	0
11)	1.9	12)	0.8	13)	1	14)	2	15)	3
16)	4	17)	6	18)	7	19)	0	20)	None
21)	0.8	22)	1	23)	0.04	24)	0.09	25)	0.16
26)	0.25	27)	0.36	28)	0.49	29)	0.64	30)	0.81

Lesson 4 Mental Math 6

1)	30	2)	25	3)	40	4)	4	5)	50
6)	4	7)	50	8)	3	9)	30	10)	5
11)	800	12)	60	13)	60	14)	160	15)	150
16)	80	17)	120	18)	60	19)	150	20)	40

Lesson 5 Evaluate

1)	800	2)	200	3)	160	4)	100	5)	700
6)	300	7)	240	8)	400	9)	150	10)	120
11)	80	12)	500	13)	280	14)	140	15)	250
16)	160	17)	180	18)	360	19)	350	20)	320

Lesson 6 Worksheet 1 A

1)	2	2)	1	3)	3	4)	2	5)	13
6)	11	7)	3	8)	3	9)	4	10)	2
11)	2	12)	8	13)	3	14)	45	15)	32
16)	24	17)	55	18)	24	19)	26	20)	80

Lesson 6 Worksheet 1 B

1)	0.8	2)	0	3)	2	4)	0	5)	5
6)	0	7)	0	8)	16	9)	5	10)	40
11)	4	12)	0.3	13)	4	14)	4	15)	4
16)	1	17)	0.3	18)	0.4	19)	0.1	20)	0.1

Basic Answers Chapter 3

Lesson 8 Review 1 A

1) 80	2) 6	3) 30	4) 16	5) 15
6) 8	7) 12	8) 6	9) 15	10) 7
11) 20	12) 10	13) 20	14) 12	15) 36
16) 27	17) 14	18) 6	19) 6	20) 7

Lesson 8 Review 1 B

1) 2	2) 4	3) 3	4) 4	5) 36
6) 50	7) 79	8) 30	9) 37	10) 2
11) 58	12) 7	13) 61	14) 9	15) 67
16) 3	17) 6	18) 70	19) 0	20) 50

Lesson 9 Review 3 A

1) 51.1	2) 14.64	3) 3.6	4) 40.5	5) 9.5
6) 25	7) 7.9	8) 32	9) 4	10) 1.5

Lesson 9 Review 3 B

1) 61.1	2) 13.73	3) 1.8	4) 50.6	5) 8.6
6) 36	7) 3.1	8) 12	9) 9	10) 1.1

Lesson 10 Test 3 A

1) 31.1	2) 12.82	3) 0.4	4) 30.5	5) 7.8
6) 16	7) 3.6	8) 0	9) 1	10) 1.1

Lesson 10 Test 3 B

1) 41.1	2) 11.91	3) 0.8	4) 40.5	5) 6.8
6) 25	7) 4.3	8) 0	9) 1	10) 0.8

Basic

Chapter 4

Fractions and Mixed Numbers

Students!

Students, you are smarter than a calculator!
Students, you are smarter than a computer!
Students, you need to believe in yourself.
Students, be patient and do your best.

Do not be afraid to make mistakes.
Tell yourself positive words.
I am young! I am smart!
I am strong! I can do anything!

Smart students use their brain!
Your brain is always with you!
A calculator is not always with you.
Mathematics is work for the brain.

Be open to new ideas in math!
Work hard and try your best!
Check answers and ask questions!
Practice! Practice! Practice!

Basic Chapter 4 Lessons
Fractions and Mixed Numbers
Lessons 1 – 10
Basic Problems

Lesson 1 Table of Multiplication. Pretest

Lesson 2 Fractions and Mixed Numbers. Table 4

Lesson 3 Comparison of Fractions and Mixed Numbers

Lesson 4 Addition and Subtraction of Fractions and Mixed Numbers

Lesson 5 Multiplication, Division and Power of Fractions and Mixed Numbers

Lesson 6 Table 5, Table 6, Mental Math # 7, # 8

Lesson 7 Order of Operations. Examples.

Lesson 8 Laws for Fractions and Mixed Numbers. Worksheet 2 A, 2 B

Lesson 9 Review 4 A, Review 4 B

Lesson 10 Test 4 A, Test 4 B

Answers Chapter 4

Basic Chapter 4 Contents
Fractions and Mixed Numbers

Contents have topics and exercises, which students need to solve before tests.1)Work with lesson 1, read rules and examples, 2) Solve the problems from contents, 3) Check answers in the end of chapter 1, 4) Work with the next lessons 2, 3, 4, 5, …, 5) Solve Test 1 A, 6) Grading Scale without minuses

A=90–100% B+=85–89% B=80–84% C+=75–79%
C=70–74% D+=65–69% D=60–64 % F=0 - 59%

7)If you solve 10 problems correctly, you get an A, 8) If have an A, you can move on to next chapter, 9) If your grade is lower than an A, work with the mistakes, then solve Test 1B, 10) When you finish all of the chapters, solve Final Exam in last chapter.

 # 1 - # 20 odd or # 1, 3, 5, 7, 9, 11, 13, 15, 17, 19

Lesson 1 Memorize table of multiplication. Solve the Pretest 1 # 1–#5
Lesson 2 Fractions and Mixed Numbers. Work with Table 4

Lesson 3 Comparison of Fractions and Mixed Numbers
Lesson 4 Addition and Subtraction of Fractions and Mixed Numbers

Lesson 5 Multiplication, Division and Power of Mixed Numbers
Lesson 6 Table 5 and Table 6, Solve Mental Math # 7, # 8 odd

Lesson 7 Order of five operations. Solve Review 1 A # 1 – # 20 odd
Solve Review 1 B # 1 – # 20 odd. Solve Review 1 C # 1 - # 10 all

Lesson 8 Laws for Fractions and Mixed Numbers
Solve the Review 1 A # 1 - # 20 odd. Solve the Review 1 B # 1 – # 20 odd
Solve Worksheet 2 A # 1 – # 10 all. Solve Worksheet 2 B # 1 – # 10 all

Lesson 9
Solve Review 4 A # 1 – # 10 all. Solve Review 4 B # 1 – # 10 all

Lesson 10
Solve Test 4 A # 1 – # 10 all. Solve Test 4 B # 1 – # 10 all

Answers Chapter 4

Basic Chapter 4 Lesson 1
Table 1
Multiplication and Division

$A + 0 = A$ $0 + A = A$ $A - 0 = A$ $A - A = 0$

$A \times 1 = A$ $1 \times A = A$ $A \div 1 = A$ $A \div A = 1$

$A \times 0 = 0$ $0 \times A = 0$ $0 \div A = 0$ $A \div 0 = \emptyset$

$A + B = B + A$ $A \times B = B \times A$ $\emptyset$ no solutions

Number 5 from the left column multiply by number 4 from first row to get answer 20. $5 \times 4 = 20$ $20 \div 5 = 4$ $20 \div 4 = 5$

Memorize Table 1

×	1	2	3	<u>4</u>	5	6	7	8	9	10
1	1	2	3	4	5	6	7	8	9	10
2	2	4	6	8	10	12	14	16	18	20
3	3	6	9	12	15	18	21	24	27	30
4	4	8	12	16	20	24	28	32	36	40
<u>5</u>	5	10	15	<u>20</u>	25	30	35	40	45	50
6	6	12	18	24	30	36	42	48	54	60
7	7	14	21	28	35	42	49	56	63	70
8	8	16	24	32	40	48	56	64	72	80
9	9	18	27	36	45	54	63	72	81	90
10	10	20	30	40	50	60	70	80	90	100

Basic Chapter 4 Lesson 1
Pretest Chapter 4

Date **Name**

1) Divide numbers

$$\frac{3}{8} \div \frac{3}{4} \div 1\frac{1}{4}$$

1) ________

2) Evaluate

$$(1\frac{4}{5}) \div \left(1\frac{1}{2}\right)^2$$

2) ________

3) Evaluate

$$\left(\frac{1}{5} - \frac{1}{10}\right)^2 \div \frac{1}{10}$$

3) ________

4) Evaluate

$$1\frac{1}{2} \div (5\ \frac{1}{2}\,4\ \frac{1}{4})$$

4) ________

5) Evaluate

$$1\frac{3}{4} \times 12 - 1\frac{1}{4} \times 12$$

5) ________

Basic Chapter 4 Lesson 2

Fractions and Mixed Numbers

A fraction is equal parts of whole number: $1 \div 4 = \frac{1}{4}$

A fraction contains a numerator (N) and a denominator (D).

The numerator written above the bar and shows how many equal parts are to be taken.

The denominator written under the bar and shows by how many equal parts the whole should be divided.

A fraction is a quotient of division, where the numerator of the fraction is the dividend, and the denominator of the fraction is the divisor.

The bar of the fraction is another sign of division.

Fractional expressions written with a bar:

$3 \div 4 = \frac{3}{4}$ $\quad\quad$ $N \div D = \frac{N}{D}$

The fraction called proper if the numerator of the fraction is less than the denominator of the fraction.

A proper fraction is always less than one:

$N < D$ $\quad\quad$ $\frac{N}{D} < 1$ $\quad\quad$ $3 < 4$ $\quad\quad$ $\frac{3}{4} < 1$

The fraction called improper if the numerator of the fraction equal to or more than the denominator of the fraction.

If the numerator is equal to the denominator, then the fraction equals one. $\quad$ $N = D$ $\quad\quad$ $\frac{4}{4} = 1$

Basic Chapter 4 Lesson 2

If the numerator is greater than the denominator, then the fraction is greater than one:

N > D $\frac{5}{4} > 1$ $5 > 4$

A mixed number contains a whole number and a fraction together:

$2+\frac{1}{2}=2\frac{1}{2}$ $\frac{1}{4}+1=1\frac{1}{4}$

To change the mixed number to the improper fraction, multiply the denominator by the whole number and add the numerator.
Put this result in as the new numerator and the denominator remains the same: $1\frac{3}{4}=\frac{4\cdot 1+3}{4}=\frac{7}{4}$ $2\frac{1}{3}=\frac{3\cdot 2+1}{3}=\frac{7}{3}$

To change the improper fraction to the mixed number, divide the numerator by the denominator.

The partial quotient is the whole number, the remainder is the numerator, and the denominator remains the same:

$\frac{7}{4}=1\frac{3}{4}$ $\frac{24}{6}=4$

Main Property of Fraction

If the numerator and denominator of a fraction multiplied or divided by the same natural number, to the new fraction will equal the first fraction:

$\frac{1}{2}=\frac{2}{4}$ $\frac{2}{4}=\frac{1}{2}$

Two equal fractions can have different notations of the same number:

$\frac{a}{b}=\frac{a\cdot n}{b\cdot n}$ $\frac{1}{2}=\frac{1\cdot 2}{2\cdot 2}=\frac{1}{4}$

Basic Chapter 4 Lesson 2

Find the missing numerator:

$\frac{4}{5} = \frac{?}{25}$ $25 \div 5 = 5$ $4 \times 5 = 20$ $\frac{4}{5} = \frac{4 \times 5}{5 \times 5} = \frac{20}{25}$

To reduce the fraction means to divide the numerator and denominator of the fraction by the same natural number:

$\frac{a \div n}{b \div n} = \frac{a}{b}$ $\frac{2}{4} = \frac{2 \div 2}{4 \div 2} = \frac{1}{2}$

If the GCF of the numerator and denominator equal one, the fraction cannot reduced.

If the GCF (5, 6) = 1, fraction $\frac{5}{6}$ cannot be reduced.

Greatest Common Factor (GCF)

The greatest number by which a fraction can reduced is the greatest common factor of the numerator and denominator:

GC F (12, 18) = 6

$\frac{12}{18} = \frac{12 \div 6}{18 \div 6} = \frac{2}{3}$ or $\frac{12}{18} = \frac{12 \div 2}{18 \div 2} = \frac{6}{9} = \frac{6 \div 3}{9 \div 3} = \frac{2}{3}$

To easily reduce a fraction, divide by small numbers 2, 3, 4, 5, …

$\frac{60}{75} = \frac{60 \div 3}{75 \div 3} = \frac{20}{25} = \frac{20 \div 5}{25 \div 5} = \frac{4}{5}$

$\frac{360}{600} = \frac{360 \div 10}{600 \div 10} = \frac{36}{60} = \frac{36 \div 2}{60 \div 2} = \frac{18}{30} = \frac{18 \div 2}{30 \div 2} = \frac{9}{15} = \frac{9 \div 3}{15 \div 3} = \frac{3}{5}$

Reduce a fraction by using prime factorization:

$\frac{360}{600} = \frac{2 \times 2 \times 2 \times 3 \times 3 \times 5}{2 \times 2 \times 2 \times 3 \times 5 \times 5} = \frac{3}{5}$

To work with fractions let us refresh our memory of GCF and LCM from Table 4

Basic Chapter 4 Lesson 2

Table 4

Greatest Common Factor (in short GCF)

The Greatest Common Factor is the greatest number by which all given numbers can be divided.

1) Find the GCF of numbers 12 and 18 (÷ 1, 2, 3, 4, …)

Divisors of number 12 are 1, 2, 3, 4, 6, 12

Divisors of number 18 are 1, 2, 3, 6, 9, 18

Common divisors are 1, 2, 3, 6 Greatest common divisor is 6

GCD (12, 18) = 6 $12 \div 6 = 2$ $18 \div 6 = 3$

2) If greatest of the numbers can divided evenly by the smallest number, then the greatest common factor of numbers is equal to the smallest of given numbers: Find the GCF of numbers 12 and 4

Factors of 12 are 1, 2, 3, 4, 6, 12 Factors of 4 are 1, 2, 4

Common factors are 1, 2, 4 Greatest common factor is 4

GCF (12, 4) = 4 $12 \div 4 = 3$

3) The greatest common factor of prime numbers equal one:

GCF (5,7) = 1

Least Common Multiples (in short LCM)

The least number is the smallest number that can be evenly divided by each of the given numbers

1) Find the LCM of numbers 12 and 18 (×1, 2, 3, 4, …)

Multiples of 12 are 12, 24, 36, 48, 60, 72…

Multiples of 18 are 18, 36, 54, 72…

Common multiples are 36, 72… Least common multiples is 36

LCM (12, 18) = 36 $36 \div 12 = 3$ $36 \div 18 = 2$

2)If the greatest of given numbers evenly divided by the smallest number,

the LCM of the numbers is equal to the greatest of the given numbers:

Find the LCM of numbers 4, 12

Multiples of 4 are 4, 8, 12, 16, 20, 24… Multiples of 12 are 12, 24…

Common multiples are 12, 24… LCM (4, 12) = 12

3) The least common multiple of prime numbers is equal to the product of the numbers. Find the LCM of numbers 2 and 5

Multiples of 2 are 2, 4, 6, 8, 10… Multiples of 5 are 5, 10…

Common multiples are 10, 20 … LCM (2, 5) = $2 \times 5 = 10$

Basic Chapter 4 Lesson 3

Comparison of Fractions and Mixed numbers

To compare two fractions with the same denominators, compare the numerators.

The greater fraction has the greater numerator.
The smaller fraction has the smaller numerator:

$\frac{4}{5} > \frac{2}{5}$ $\quad 4 > 2 \quad$ $\frac{1}{7} < \frac{6}{7}$ $\quad 1 < 6$

To compare two fractions with the same numerators, compare the denominators.

The greater fraction has the smaller denominator.
The smaller fraction has the greater denominator:

$\frac{1}{2} > \frac{1}{3}$ $\quad 2 < 3 \quad$ $\frac{1}{4} < \frac{1}{3}$ $\quad 4 > 3$

Find the Least Common Denominator (LCD) of fractions $\frac{1}{4}$ and $\frac{3}{8}$

If the greatest denominator can be divided by the smallest denominator, the greatest denominator is LCD $\quad 8 \div 4 = 2 \qquad$ **LCD (4, 8) = 8**

The number needed to multiply the numerator and denominator of a fraction to get a new fraction called the factor.

To find the factor of each fraction, divide the least common denominator (LCD) by each denominator: $\quad 8 \div 4 = 2 \qquad 8 \div 8 = 1$

Multiply the numerator and denominator of each fraction by the factor to get fractions with the same denominators:

$\frac{1}{4} = \frac{1 \cdot 2}{4 \cdot 2} = \frac{2}{8}$ $\qquad \frac{3}{8} = \frac{3 \times 1}{8 \times 1} = \frac{3}{8}$ $\qquad \frac{2}{8}$ **and** $\frac{3}{8}$

Basic Chapter 4 Lesson 3

Compare Fractions and Mixed Numbers with Different Denominators

To compare two fractions with different denominators, find the least common denominator and compare numerators: $1\frac{5}{6}$ **and** $1\frac{1}{8}$

To find least common denominators of fractions (LCD), find the least common multiple of each denominator: LCM (6, 8).
If the greatest denominator can divide by the smallest denominator, the greatest denominator is LCD
If the greater denominator cannot be divided by the smaller denominator, multiply the greater denominator by 2, 3, 4, … until the new denominator can divided evenly, (remainder zero) by the smaller denominator.

Example: Compare Mixed Numbers: $1\frac{5}{6}$ **and** $1\frac{1}{8}$

Find the LCM (6, 8) **8 cannot divided evenly by 6**

$8 \times 2 = 16$ **16 cannot be divided evenly by 6**

$8 \times 3 = 24$ **24 can be divided evenly by 8 and by 6** **LCD (6, 8) = 24**

$24 \div 6 = 4$ $24 \div 8 = 3$ $\frac{5}{6} = \frac{5 \cdot 4}{6 \cdot 4} = \frac{20}{24}$ $\frac{1}{8} = \frac{1 \cdot 3}{8 \cdot 3} = \frac{3}{24}$

$\frac{20}{24} > \frac{3}{24}$ $\frac{5}{6} > \frac{1}{8}$ $1\frac{5}{6} > 1\frac{1}{8}$

Second way $\frac{5}{6} \ ? \ \frac{1}{8}$

Find cross product 5×8 and 6×1 40 and 6

But $40 > 6$ $\frac{5}{6} > \frac{1}{8}$

Basic Chapter 4 Lesson 4
Addition and Subtraction of Fractions and Mixed Numbers
Addition of Fractions and Mixed numbers
The Same Denominators

To add fractions with the same denominators, add their numerators; the denominator remains the same

$$\frac{a}{c}+\frac{b}{c}=\frac{a+b}{c}$$

$$1=\frac{a}{a}=\frac{2}{2}=\frac{3}{3}=\frac{4}{4}=\frac{5}{5}$$

a) $\frac{4}{8}+\frac{3}{8}+\frac{1}{8}=\frac{4+3+1}{8}=\frac{8}{8}=1$

b) $\frac{1}{20}+\frac{3}{20}+\frac{7}{20}+\frac{9}{20}=\frac{1+3+7+9}{20}=\frac{20}{20}=1$

Different denominators

To add fractions with different denominators, find the least common denominator, multiply the fractions by a factor and add the numerators; the denominator remains the same.

LCD (6, 12) = 12 **LCD (5, 8, 20) = 40**

c) $\frac{1}{6}+\frac{7}{12}=\frac{1\cdot 2}{6\cdot 2}+\frac{7}{12}=\frac{2}{12}+\frac{7}{12}=\frac{9}{12}=\frac{9\div 3}{12\div 3}=\frac{3}{4}$

d) $\frac{2}{5}+\frac{3}{8}+\frac{7}{20}=\frac{2\cdot 8}{5\cdot 8}+\frac{3\cdot 5}{8\cdot 5}+\frac{7\cdot 2}{20\cdot 2}=\frac{16}{40}+\frac{15}{40}+\frac{14}{40}=\frac{45}{40}=\frac{45\div 5}{40\div 5}=\frac{9}{8}=1\frac{1}{8}$

Mixed numbers

To add mixed numbers with different denominators, find the least common denominator, add whole numbers and add numerators, the denominator remains the same. In addition, mixed numbers can change to improper fractions and then fractions can add.

Remember: $1=\frac{a}{a}=\frac{2}{2}=\frac{3}{3}=\frac{4}{4}=\frac{5}{5}$ $\quad 1=\frac{10}{10}$ $\quad \frac{14}{12}=1\frac{2}{12}$

Basic Chapter 4 Lesson 4

The Same Denominators

a) $1\frac{1}{8}+2\frac{3}{8}=3\frac{4}{8}=3\frac{1}{2}$

b) $1\frac{1}{9}+2\frac{2}{9}+4\frac{3}{9}=7\frac{6}{9}=7\frac{2}{3}$

c) $2\frac{3}{10}+1\frac{7}{10}=3\frac{10}{10}=3+\frac{10}{10}=3+1=4$

d) $1\frac{5}{12}+2\frac{9}{12}=3\frac{14}{12}=3+\frac{14}{12}=3+1\frac{2}{12}=4\frac{2}{12}=4\frac{1}{6}$

Different Denominators

e) $1\frac{1}{3}+2\frac{3}{9}=1\frac{1}{3}+2\frac{3\div 3}{9\div 3}=1\frac{1}{3}+2\frac{1}{3}=3\frac{2}{3}$

f) $1\frac{2}{6}+2\frac{1}{9}=1\frac{2\div 2}{6\div 2}+2\frac{1}{9}=1\frac{1}{3}+2\frac{1}{9}=1\frac{1\times 3}{3\times 3}+2\frac{1}{9}=1\frac{3}{9}+2\frac{1}{9}=3\frac{4}{9}$

g) $1\frac{1}{2}+1\frac{1}{3}+2\frac{1}{6}=1\frac{1\cdot 3}{2\cdot 3}+1\frac{1\cdot 2}{3\cdot 2}+2\frac{1}{6}=1\frac{3}{6}+1\frac{2}{6}+2\frac{1}{6}=4\frac{6}{6}=5$

LCD (2, 3, 6) = 6

Change mixed numbers to improper fractions:

h) $1\frac{1}{2}+1\frac{1}{3}+2\frac{1}{6}=\frac{3}{2}+\frac{4}{3}+\frac{13}{6}=\frac{3\cdot 3}{2\cdot 3}+\frac{4\cdot 2}{3\cdot 2}+\frac{13}{6}=\frac{9}{6}+\frac{8}{6}+\frac{13}{6}=\frac{30}{6}=5$

Basic Chapter 4 Lesson 4

Subtraction of Fractions and Mixed Numbers

To subtract fractions with the same denominator, subtract the numerators; the denominator remains the same.
Always first, try to reduce fractions:

$$\frac{a}{c}-\frac{b}{c}=\frac{a-b}{c}$$

a) $\frac{5}{8}-\frac{3}{8}=\frac{5-3}{8}=\frac{2}{8}=\frac{1}{4}$

To subtract fractions with different denominators, expand fractions to the least common denominator and subtract numerators; the denominator remains the same.LCD (10, 4) = 20 LCD (12, 18) = 36

b) $\frac{6}{20}-\frac{2}{8}=\frac{3}{10}-\frac{1}{4}=\frac{6}{20}-\frac{5}{20}=\frac{1}{20}$

c) $\frac{11}{12}-\frac{3}{18}=\frac{11}{12}-\frac{1}{6}=\frac{11-2}{12}=\frac{9}{12}=\frac{3}{4}$

To subtract mixed numbers, expand fractions to the least common denominator, subtract whole numbers and subtract numerators; the denominator remains the same.

Also, mixed numbers can be changed to improper fractions and then subtract fractions.

Remember: $4=3\frac{8}{8}$ $\quad 8=7+1=7+\frac{6}{6}=7\frac{6}{6}$

d) $6\frac{7}{8}-5\frac{3}{8}=1\frac{4}{8}=1\frac{1}{2}$

e) $1-\frac{2}{3}=\frac{3}{3}-\frac{2}{3}=\frac{1}{3}$

f) $8-3\frac{5}{6}=7\frac{6}{6}-3\frac{5}{6}=4\frac{1}{6}$

g) $3\frac{1}{2}-2=1\frac{1}{2}$

h) $4-\frac{5}{8}=3\frac{8}{8}-\frac{5}{8}=3\frac{3}{8}$

i) $3\frac{3}{4}-2\frac{1}{2}=\frac{15}{4}-\frac{5}{2}=\frac{15}{4}-\frac{10}{4}=\frac{5}{4}=1\frac{1}{4}$

Basic Chapter 4 Lesson 4

Special Case of Subtraction

If the numerator of the minuend is less than the numerator of the subtrahend, the whole numbers of the minuend need to decrease by 1.

To get the new numerator of the minuend, add the old numerator to the denominator.

Subtract 1 the whole number and add denominator to old numerator; the denominator remains the same.

a) $6\frac{2}{7} - 2\frac{5}{7} = 5\frac{9}{7} - 2\frac{5}{7} = 3\frac{4}{7}$ $\quad 2 < 5$

$(6\frac{2}{7} = 5 + 1 + \frac{2}{7} = 5 + 1\frac{2}{7} = 5 + \frac{9}{7} = 5\frac{9}{7})$

b) $5\frac{1}{3} - 2\frac{5}{6} = 5\frac{2}{6} - 2\frac{5}{6} = 4\frac{8}{6} - 2\frac{5}{6} = 2\frac{3}{6} = 2\frac{1}{2}$ $\quad 2 < 5$

$(5\frac{2}{6} = 4\frac{8}{6} \quad 5 - 1 = 4 \quad 2 + 6 = 8)$

c) $6\frac{1}{4} - 2\frac{2}{3} = 6\frac{3}{12} - 2\frac{8}{12} = 5\frac{15}{12} - 2\frac{8}{12} = 3\frac{7}{12}$ $\quad 3 < 8$

$(6\frac{3}{12} = 5\frac{15}{12} \quad 6 - 1 = 5 \quad 3 + 12 = 15)$

d) $6\frac{1}{4} - 2\frac{2}{3} = \frac{25}{4} - \frac{8}{3} = \frac{75}{12} - \frac{32}{12} = \frac{43}{12} = 3\frac{7}{12}$

Basic Chapter 4 Lesson 5

Multiplication, Powers, Division of Fractions and Mixed Numbers

Multiplication of the Fractions

To multiply the fractions, product of numerators put in new numerator and product of denominators put in new denominator.

Always first try to reduce fractions $\frac{a}{b}\times\frac{c}{d}=\frac{a\times c}{b\times d}=\frac{ac}{bd}$

a) $\frac{5}{6}\times\frac{3}{10}=\frac{5\times3}{6\times10}=\frac{1\times1}{2\times2}=\frac{1}{4}$

b) $\frac{3}{5}\times\frac{10}{11}\times\frac{22}{30}=\frac{3\times10\times22}{5\times11\times30}=\frac{1\times2\times2}{1\times1\times10}=\frac{2}{5}$

Multiplication of the Fraction and Natural Number

To multiply the fractions by natural number, multiply numerator by natural number, denominator remains the same.

$a=\frac{a}{1}$ $\quad 8=\frac{8}{1}$

$\frac{a}{b}\cdot c=\frac{a}{b}\cdot\frac{c}{1}=\frac{a\cdot c}{b}$ $\quad a\cdot\frac{b}{c}=\frac{a}{1}\cdot\frac{b}{c}=\frac{ab}{c}$

c) $\frac{3}{4}\times8=\frac{3}{4}\times\frac{8}{1}=\frac{3}{1}\times\frac{2}{1}=\frac{6}{1}=6$

d) $8\times\frac{5}{2}=8\div2\times5=4\times5=20$

Multiplication of Mixed Numbers

To multiply the mixed numbers, change the mixed numbers to improper fractions and multiply fractions.

e) $3\frac{1}{3}\cdot2\frac{2}{5}=\frac{10}{3}\cdot\frac{12}{5}=\frac{2\cdot4}{1\cdot1}=8$

f) $\frac{5}{8}\cdot4\frac{4}{5}\cdot\frac{1}{6}=\frac{5}{8}\cdot\frac{24}{5}\cdot\frac{1}{6}=\frac{3}{6}=\frac{1}{2}$

Power of Mixed Number

To find the power of a mixed numbers, change the mixed number to an improper fraction, then find the power of the numerator and denominator:

$\left(\frac{a}{b}\right)^n=\frac{a^n}{b^n}$ $\quad \left(\frac{1}{2}\right)^2=\frac{1^2}{2^2}=\frac{1}{4}$

g) $\left(\frac{3}{4}\right)^3=\frac{3^3}{4^3}=\frac{27}{64}$

h) $\left(1\frac{1}{2}\right)^2=\left(\frac{3}{2}\right)^2=\frac{3^2}{2^2}=\frac{9}{4}=2\frac{1}{4}$

Basic Chapter 4 Lesson 5

Division of Fractions and Mixed numbers

Reciprocals Fractions

The reciprocal of a fraction is the fraction with the numerator and denominator interchanged. Product of two reciprocals fractions is equal one.

$a \cdot \frac{1}{a} = 1$ $\qquad \frac{a}{b} \cdot \frac{b}{a} = 1$ $\qquad$ a) $5 \cdot \frac{1}{5} = \frac{5}{1} \cdot \frac{1}{5} = 1$

b) $\frac{1}{3} \cdot 3 = \frac{1}{3} \cdot \frac{3}{1} = 1$ $\qquad$ c) $\frac{2}{3} \cdot \frac{3}{2} = \frac{2 \cdot 3}{3 \cdot 2} = 1$ $\qquad$ d) $1\frac{3}{4} \cdot \frac{4}{7} = \frac{7}{4} \cdot \frac{4}{7} = 1$

Division of Fractions

To divide fractions, dividend multiplies by number, reciprocal of divisor. In multiplication, always first try to reduce the fraction

$\frac{a}{b} \div \frac{c}{d} = \frac{a}{b} \cdot \frac{d}{c} = \frac{ad}{dc}$ $\qquad$ e) $\frac{3}{4} \div \frac{3}{8} = \frac{3}{4} \cdot \frac{8}{3} = \frac{1 \times 2}{1 \times 1} = \frac{2}{1} = 2$

$\frac{a}{b} \div c = \frac{a}{b} \cdot \frac{1}{c} = \frac{a}{bc}$ $\qquad$ f) $\frac{3}{4} \div 9 = \frac{3}{4} \div \frac{9}{1} = \frac{3}{4} \cdot \frac{1}{9} = \frac{1}{12}$

$a \div \frac{b}{c} = \frac{a}{1} \cdot \frac{c}{b} = \frac{ac}{b}$ $\qquad$ g) $9 \div \frac{3}{4} = 9 \times \frac{4}{3} = 9 \div 3 \times 4 = 3 \times 4 = 12$

Division of Mixed Numbers

To divided mixed numbers, change mixed numbers to improper fractions and divide the fractions

h) $1\frac{1}{6} \div 4\frac{2}{3} = \frac{7}{6} \div \frac{14}{3} = \frac{7}{6} \times \frac{3}{14} = \frac{1}{2} \times \frac{1}{2} = \frac{1}{4}$

i) $3\frac{1}{5} \div 2\frac{2}{5} = \frac{16}{5} \div \frac{12}{5} = \frac{16}{5} \times \frac{5}{12} = \frac{4}{1} \times \frac{1}{3} = \frac{4}{3} = 1\frac{1}{3}$

j) $8 \div 4\frac{4}{5} = \frac{8}{1} \div \frac{24}{5} = \frac{8}{1} \times \frac{5}{24} = \frac{1}{1} \times \frac{5}{3} = \frac{5}{3} = 1\frac{2}{3}$

k) $4\frac{1}{3} \div 13 = \frac{13}{3} \div \frac{13}{1} = \frac{13}{3} \times \frac{1}{13} = \frac{1}{3}$

Basic Chapter 4 Lesson 6

Table 5 Fractions

Properties of Fraction

$$\frac{a}{b}=\frac{a\cdot n}{b\cdot n}$$

$$\frac{a\div n}{b\div n}=\frac{a}{b}$$

$$\frac{2}{3}=\frac{2\times 6}{3\times 6}=\frac{12}{18}$$

$$\text{a)}\ \frac{60}{75}=\frac{60\div 3}{75\div 3}=\frac{20}{25}=\frac{20\div 5}{25\div 5}=\frac{4}{5}$$

Comparison of Fractions

$$\frac{6}{11}>\frac{7}{22}$$

$$\frac{6}{11}=\frac{6\cdot 2}{11\cdot 2}=\frac{12}{22}$$

$$\frac{12}{22}>\frac{7}{22}$$

Addition of Fractions

$$\frac{a}{c}+\frac{b}{c}=\frac{a+b}{c}$$

$$\frac{2}{5}+\frac{3}{8}+\frac{7}{20}=\frac{2\cdot 8}{5\cdot 8}+\frac{3\cdot 5}{8\cdot 5}+\frac{7\cdot 2}{20\cdot 2}=\frac{16}{40}+\frac{15}{40}+\frac{14}{40}=\frac{45}{40}=1\frac{1}{8}$$

Subtraction of Fractions

$$\frac{a}{c}-\frac{b}{c}=\frac{a-b}{c}$$

$$\frac{3}{10}-\frac{1}{4}=\frac{6}{20}-\frac{5}{20}=\frac{1}{20}$$

$$\frac{1}{12}-\frac{1}{18}=\frac{3}{36}-\frac{2}{36}=\frac{3-2}{36}=\frac{1}{36}$$

Multiplication of Fractions

$$\frac{a}{b}\times\frac{c}{d}=\frac{a\times c}{b\times d}=\frac{ac}{bd}$$

$$\frac{4}{6}\times\frac{9}{8}=\frac{4\times 9}{6\times 8}=\frac{1\times 3}{2\times 2}=\frac{3}{4}$$

$$\frac{a}{b}\cdot c=\frac{a}{b}\cdot\frac{c}{1}=\frac{a\cdot c}{b}$$

$$\frac{3}{8}\times 4=\frac{3}{8}\times\frac{4}{1}=\frac{3\times 1}{2\times 1}=\frac{3}{2}=1\frac{1}{2}$$

$$a\cdot\frac{b}{c}=\frac{a}{1}\cdot\frac{b}{c}=\frac{ab}{c}$$

$$16\times\frac{5}{8}=\frac{16}{1}\times\frac{5}{8}=\frac{2\times 5}{1}=10$$

Division of Fractions

$$\frac{a}{b}\div\frac{c}{d}=\frac{a}{b}\cdot\frac{d}{c}=\frac{ad}{dc}$$

$$\frac{5}{6}\div\frac{10}{3}=\frac{5}{6}\times\frac{3}{10}=\frac{1\times 1}{2\times 2}=\frac{1}{4}$$

$$\frac{a}{b}\div c=\frac{a}{b}\cdot\frac{1}{c}=\frac{a}{bc}$$

$$\frac{3}{4}\div 6=\frac{3}{4}\div\frac{6}{1}=\frac{3}{4}\times\frac{1}{6}=\frac{1}{8}$$

$$a\div\frac{b}{c}=\frac{a}{1}\cdot\frac{c}{b}=\frac{ac}{b}$$

$$6\div\frac{3}{4}=\frac{6}{1}\times\frac{4}{3}=\frac{8}{1}=8$$

Power of Fractions

$$\left(\frac{a}{b}\right)^{n}=\frac{a^{n}}{b^{n}}$$

$$\left(\frac{1}{3}\right)^{2}=\frac{1^{2}}{3^{2}}=\frac{1}{9}$$

Basic Chapter 4 Lesson 6

Table 6 Mixed Numbers

Mixed Number $2+\frac{1}{2}=2\frac{1}{2}$ $\frac{1}{4}+1=1\frac{1}{4}$ $1\frac{3}{4}=\frac{4\cdot 1+3}{4}=\frac{7}{4}$

Comparison of Mixed Numbers $1\frac{1}{3}<1\frac{5}{6}$ $1\frac{2}{6}<1\frac{5}{6}$

Addition of Mixed Numbers

$1\frac{1}{8}+2\frac{3}{8}=3\frac{4}{8}=3\frac{1}{2}$ $1\frac{1}{3}+2\frac{3}{9}=1\frac{1}{3}+2\frac{1}{3}=3\frac{2}{3}$

$1\frac{2}{6}+2\frac{1}{9}=1\frac{1}{3}+2\frac{1}{9}=1\frac{3}{9}+2\frac{1}{9}=3\frac{4}{9}$ $1\frac{1}{2}+1\frac{1}{3}+2\frac{1}{6}=1\frac{3}{6}+1\frac{2}{6}+2\frac{1}{6}=4\frac{6}{6}=5$

Subtraction of Mixed Numbers

$6\frac{7}{8}-5\frac{3}{8}=1\frac{4}{8}=1\frac{1}{2}$ $4\frac{3}{4}-2\frac{1}{4}=2\frac{2}{4}=2\frac{1}{2}$ $3\frac{1}{2}-2=1\frac{1}{2}$

$1-\frac{2}{3}=\frac{3}{3}-\frac{2}{3}=\frac{1}{3}$ $4-\frac{5}{8}=3\frac{8}{8}-\frac{5}{8}=3\frac{3}{8}$ $8-3\frac{5}{6}=7\frac{6}{6}-3\frac{5}{6}=4\frac{1}{6}$

$6\frac{2}{7}-2\frac{5}{7}=5\frac{9}{7}-2\frac{5}{7}=3\frac{4}{7}$ $5\frac{1}{3}-2\frac{5}{6}=5\frac{2}{6}-2\frac{5}{6}=4\frac{8}{6}-2\frac{5}{6}=2\frac{3}{6}=2\frac{1}{2}$

Multiplication of Mixed Numbers

$3\frac{1}{3}\cdot 2\frac{2}{5}=\frac{10}{3}\cdot\frac{12}{5}=\frac{2\cdot 4}{1\cdot 1}=8$ $\frac{5}{8}\cdot 4\frac{4}{5}\cdot\frac{1}{6}=\frac{5}{8}\cdot\frac{24}{5}\cdot\frac{1}{6}=\frac{3}{6}=\frac{1}{2}$

Division of Mixed Numbers

$1\frac{1}{6}\div 4\frac{2}{3}=\frac{7}{6}\div\frac{14}{3}=\frac{7}{6}\times\frac{3}{14}=\frac{1}{4}$ $3\frac{1}{5}\div 2\frac{2}{5}=\frac{16}{5}\times\frac{5}{12}=\frac{4}{3}=1\frac{1}{3}$

$8\div 4\frac{4}{5}=\frac{8}{1}\div\frac{24}{5}=\frac{8}{1}\times\frac{5}{24}=\frac{5}{3}=1\frac{2}{3}$ $4\frac{1}{3}\div 13=\frac{13}{3}\div\frac{13}{1}=\frac{13}{3}\times\frac{1}{13}=\frac{1}{3}$

Power of Mixed Numbers

$\left(\frac{3}{4}\right)^3=\frac{3^3}{4^3}=\frac{27}{64}$ $\left(1\frac{1}{2}\right)^2=\left(\frac{3}{2}\right)^2=\frac{3^2}{2^2}=\frac{9}{4}=2\frac{1}{4}$

Mental Math 7

Fractions

1) $\frac{1}{2} \times \frac{2}{5}$

2) $\frac{1}{3} \times \frac{3}{4}$

3) $\frac{5}{8} \times 8$

4) $4 \times \frac{1}{4}$

5) $\frac{1}{2} \div \frac{1}{2}$

6) $\frac{1}{3} \div \frac{1}{6}$

7) $\frac{1}{4} \div \frac{1}{12}$

8) $2 \div \frac{1}{2}$

9) $(\frac{1}{2})^2$

10) $(\frac{1}{5})^2$

11) $(\frac{2}{5})^2$

12) $(\frac{3}{5})^2$

13) $\frac{1}{4} + \frac{3}{4}$

14) $\frac{1}{2} + \frac{3}{2}$

15) $\frac{7}{8} - \frac{5}{8}$

16) $\frac{7}{8} - \frac{3}{8}$

17) $\frac{1}{2} + \frac{1}{4}$

18) $\frac{3}{4} + \frac{1}{2}$

19) $\frac{1}{2} - \frac{1}{4}$

20) $\frac{3}{4} - \frac{1}{2}$

Mental Math 8

Mixed numbers

1) $1+\frac{1}{2}$

2) $\frac{1}{4}+2$

3) $1-\frac{1}{2}$

4) $2\frac{1}{2}-1$

5) $2\frac{3}{8}+1\frac{1}{8}$

6) $3\frac{3}{4}+2\frac{1}{4}$

7) $4\frac{3}{4}-2\frac{1}{4}$

8) $3\frac{5}{8}-2\frac{1}{8}$

9) $3\frac{1}{2}+2\frac{1}{2}$

10) $4\frac{1}{2}+2\frac{1}{2}$

11) $2\frac{3}{4}-1\frac{1}{4}$

12) $4\frac{5}{6}-1\frac{2}{6}$

13) $2\frac{3}{8}-1\frac{1}{8}$

14) $3\frac{3}{8}-2\frac{1}{8}$

15) $4\frac{7}{8}-3\frac{3}{8}$

16) $5\frac{5}{8}-4\frac{1}{8}$

17) $2-1\frac{1}{2}$

18) $2-1\frac{3}{4}$

19) $3-1\frac{1}{4}$

20) $3-1\frac{1}{5}$

Basic Chapter 4 Lesson 7

Order of Operations

An expression has five operations:
first, work on operations at the third level (a^n)
then, work on operations at second level (× or÷)
last, work on operations at the first level (+, -),
in order, as they occur, from left to right.

In short

Grouping Symbols (1) [2] {3}
Exponents or Power – third level of operations a^n
Multiplication or Division from left to right–second level
× or ÷ L → R
Addition or Subtraction from left to right–first level
+ or – L → R

Examples

$$a)\ 9-2\frac{2}{5}\div\frac{8}{25}=9-\frac{12}{5}\div\frac{8}{25}=9-\frac{12}{5}\times\frac{25}{8}=9-\frac{3}{1}\times\frac{5}{2}=9-\frac{15}{2}=8\frac{2}{2}-7\frac{1}{2}=1\frac{1}{2}$$

$$b)\ \frac{14}{15}\div\left(2\frac{1}{3}-1\frac{1}{6}\right)=\frac{14}{15}\div\left(2\frac{2}{6}-1\frac{1}{6}\right)=\frac{14}{15}\div1\frac{1}{6}=\frac{14}{15}\div\frac{7}{6}=\frac{14}{15}\times\frac{6}{7}=\frac{2}{5}\times\frac{2}{1}=\frac{4}{5}$$

$$c)\ \frac{14}{17}\div\frac{7}{17}-1\frac{3}{5}=\frac{14}{17}\times\frac{17}{7}-1\frac{3}{5}=\frac{2}{1}\times\frac{1}{1}-1\frac{3}{5}=2-1\frac{3}{5}=1\frac{5}{5}-1\frac{3}{5}=\frac{2}{5}$$

$$d)\ \left(2\frac{1}{4}-1\frac{1}{2}\right)\div\frac{15}{16}=\left(1\frac{5}{4}-1\frac{2}{4}\right)\div\frac{15}{16}=\frac{3}{4}\times\frac{16}{15}=\frac{1}{1}\times\frac{4}{5}=\frac{4}{5}$$

Composite Fractions

$$e)\ \frac{\left(2+\frac{1}{3}\right)}{\left(2-\frac{5}{6}\right)}=\frac{2\frac{1}{3}}{\left(1\frac{6}{6}-\frac{5}{6}\right)}=\frac{2\frac{1}{3}}{1\frac{1}{6}}=2\frac{1}{3}\div1\frac{1}{6}=\frac{7}{3}\div\frac{7}{6}=\frac{7}{3}\times\frac{6}{7}=\frac{1}{1}\times\frac{2}{1}=\frac{2}{1}=2$$

or

$$d)\ \frac{\left(2+\frac{1}{3}\right)}{\left(2-\frac{5}{6}\right)}=\frac{\left(2+\frac{1}{3}\right)\times6}{\left(2-\frac{5}{6}\right)\times6}=\frac{2\times6+\frac{1}{3}\times\frac{6}{1}}{2\times6-\frac{5}{6}\times\frac{6}{1}}=\frac{12+2}{12-5}=\frac{14}{7}=2$$

LCD(3, 6) = 6

Basic Chapter 4 Lesson 7
Review 1 A
Perform the operations

1) $7 - 1\frac{1}{3} \div \frac{4}{15}$

2) $6 - 2\frac{1}{10} \div \frac{7}{15}$

3) $4 - 2\frac{1}{2} \div 1\frac{2}{3}$

4) $3 - 1\frac{1}{9} \div \frac{5}{12}$

5) $4\frac{3}{4} - 1\frac{2}{5} \div \frac{2}{5}$

6) $3\frac{3}{11} \div \frac{9}{11} - 2\frac{3}{8}$

7) $4\frac{1}{12} \div \frac{7}{12} - 5\frac{2}{9}$

8) $10\frac{1}{8} \div 2\frac{1}{4} - 2\frac{1}{3}$

9) $12 \div 3\frac{3}{5} - 1\frac{2}{3}$

10) $6\frac{2}{5} \div \frac{12}{35} - 16$

11) $(1\frac{5}{8} + 1\frac{1}{12}) \div 1\frac{1}{12}$

12) $(1\frac{1}{6} + 3\frac{3}{4}) \div 3\frac{5}{18}$

13) $(2\frac{1}{2} + 1\frac{1}{3}) \div \frac{1}{6}$

14) $(2\frac{2}{3} - 2\frac{2}{9}) \div 2$

15) $(1\frac{1}{2} - \frac{1}{4}) \div 3\frac{3}{4}$

16) $2\frac{7}{8} \div \left(\frac{3}{4} + 1\frac{1}{6}\right)$

17) $2\frac{1}{3} \div (4\frac{1}{6} - 3\frac{7}{9})$

18) $5\frac{3}{8} \div (5\frac{5}{16} - 2\frac{5}{8})$

19) $3\frac{1}{8} \div (3\frac{4}{6} - 1\frac{1}{6})$

20) $1\frac{8}{9} \div (3\frac{1}{6} - 1\frac{3}{4})$

Basic Chapter 4 Lesson 7
Review 1 B

Solve all problems in fractions, but just final answers put in decimals

1) $8-\frac{4}{5}\div 1\frac{1}{15}$

2) $\frac{14}{15}\div\frac{7}{15}-1\frac{2}{5}$

3) $\left(3\frac{1}{4}-2\frac{1}{2}\right)\div 1\frac{1}{2}$

4) $\frac{14}{15}\div\left(3\frac{1}{3}-2\frac{1}{6}\right)$

5) $\left(1+\frac{1}{2}\right)\div\left(1-\frac{1}{2}\right)$

6) $\left(1-\frac{1}{4}\right)\div\left(1+\frac{1}{4}\right)$

7) $\dfrac{2-\frac{1}{2}}{2+\frac{1}{2}}$

8) $\dfrac{2+\frac{3}{4}}{2-\frac{3}{4}}$

9) $\dfrac{3-\frac{1}{3}}{3+\frac{1}{3}}$

10) $\dfrac{7+\frac{2}{6}}{6-\frac{2}{3}}$

11) $\dfrac{2+\frac{1}{4}}{2-\frac{1}{2}}$

12) $\dfrac{2-\frac{5}{6}}{2+\frac{1}{3}}$

13) $\dfrac{1-\frac{1}{4}}{1+\frac{7}{8}}$

14) $\dfrac{4-\frac{1}{4}}{3+\frac{1}{8}}$

15) $\dfrac{2+\frac{1}{3}}{2-\frac{5}{6}}$

16) $\dfrac{2-\frac{1}{3}}{2+\frac{1}{3}}$

17) $\dfrac{3+\frac{1}{3}}{3-\frac{1}{3}}$

18) $\dfrac{1+\frac{7}{8}}{1-\frac{1}{4}}$

19) $\dfrac{3-\frac{1}{2}}{6+\frac{1}{4}}$

20) $\dfrac{6+\frac{1}{4}}{3-\frac{1}{2}}$

Basic Chapter 4 Lesson 7

Review 1 C

1) $1\frac{3}{5}\times2\frac{1}{8}+1\frac{3}{5}\times2\frac{7}{8}$

2) $4\frac{2}{3}\times\frac{3}{4}-3\frac{1}{3}\times\frac{3}{4}$

3) $4\frac{3}{5}\times\frac{5}{6}-\frac{5}{6}\times1\frac{3}{5}$

4) $4\frac{1}{2}\times\frac{1}{6}+\frac{1}{6}\times1\frac{1}{2}$

5) $3\frac{3}{4}\times1\frac{1}{6}-2\frac{1}{4}\times1\frac{1}{6}$

6) $(\frac{1}{3}-\frac{1}{5})\div\frac{1}{15}$

7) $(1\frac{3}{4}-1\frac{1}{2})\div\frac{1}{8}$

8) $(1\frac{1}{2}-1\frac{1}{6})\div\frac{1}{6}$

9) $(2\frac{1}{2}+1\frac{1}{3})\div\frac{1}{6}$

10) $(1\frac{1}{3}+1\frac{3}{4})\div\frac{1}{12}$

Basic Chapter 4 Lesson 8

Laws for Fractions and Mixed Numbers

First Law of Addition (commutative)

Changing the order of addends does not change the sum.

Add in any order: $A + B = B + A \quad 2 + 3 = 5 \quad 3 + 2 = 5 \quad 2 + 3 = 3 + 2$

Second Law of Addition (associative)

Changing the grouping of addends does not change the sum:

$A + (B + C) = (A + B) + C \qquad 8 + (2 + 5) = (8 + 2) + 5 = 10 + 5 = 15$

First Law of Multiplication (commutative)

Changing the order of factors does not change the product.

Multiply in any order: $A\,B = B\,A \quad 2 \times 3 = 6 \quad 3 \times 2 = 6 \quad 2 \times 3 = 3 \times 2$

Second Law of Multiplication (associative)

Changing the grouping of factors does not change the product:

$A\,(B\,C) = (AB)\,C \qquad 5 \times (2 \times 4) = (5 \times 2) \times 4 = 10 \times 4 = 40$

Test for Sum

If each addend can divide by number the sum divisible by a number

$(A + B) \div C = A \div C + B \div C \quad (18 + 12) \div 3 = 18 \div 3 + 12 \div 3 = 6 + 4 = 10$

Test for Difference

If the minuend and subtrahend can divide by number the difference divisible by a number $(A - B) \div C = A \div C - B \div C$

$(18 - 12) \div 3 = 18 \div 3 - 12 \div 3 = 6 - 4 = 2$

Distributive Laws of Multiplication

To multiply a number by the sum or difference, multiply the number by each number in parentheses, then add or subtract the products:

$A\,(B + C) = AB + AC \qquad A\,(B - C) = AB - AC$

$(B + C)A = AB + AC \qquad (B - C)A = AB - AC$

$14 \cdot 5 = (10 + 4) \cdot 5 = 10 \cdot 5 + 4 \cdot 5 = 50 + 20 = 70$

$18 \cdot 5 = (20 - 2) \cdot 5 = 20 \cdot 5 - 2 \cdot 5 = 100 - 10 = 90$

Common Factor (Put A before parentheses)

$AB + AC = A\,(B + C) \qquad AB - AC = A\,(B - C)$

If two products have a common factor, put the common factor before parentheses. $AB + AC$ **A – common factor**

In parentheses, put results of divisions products by common factor

$AB \div A = B \qquad AC \div A = C$

$16 \cdot 5 + 4 \cdot 5 = 5\,(16 + 4) = 5\,(20) = 100$

$22 \cdot 5 - 2 \cdot 5 = 5\,(22 - 2) = 5\,(20) = 100$

Basic Chapter 4 Lesson 8

Review 1 A

Evaluate by using laws

1) $\frac{7}{18}+\frac{1}{3}+\frac{5}{18}$

2) $\frac{3}{14}+\frac{1}{2}+\frac{4}{14}$

3) $\frac{3}{16}+\frac{3}{4}+\frac{1}{16}$

4) $\frac{1}{15}+\frac{4}{5}+\frac{2}{15}$

5) $(\frac{1}{4}+\frac{9}{16})+\frac{7}{16}$

6) $\frac{7}{8}+\frac{3}{4}+\frac{1}{8}$

7) $\frac{3}{7}+\frac{5}{9}+\frac{4}{7}+\frac{4}{9}$

8) $(\frac{1}{7}+\frac{11}{18})+(\frac{6}{7}+\frac{7}{18})$

9) $\frac{5}{8}+(\frac{7}{24}+\frac{3}{8})+\frac{5}{24}$

10) $(\frac{8}{15}+\frac{1}{5})+(\frac{4}{15}+\frac{4}{5})$

11) $9(1-\frac{4}{9})$

12) $(\frac{2}{3}-\frac{3}{5})\times 15$

13) $9(3-\frac{5}{9})$

14) $(1+\frac{5}{6})\times 6$

15) $7(2+\frac{3}{7})$

16) $(1-\frac{1}{6})\times 6$

17) $6(1\frac{1}{2}-\frac{2}{3})$

18) $(\frac{5}{7}+\frac{4}{7})\times 7$

19) $7(\frac{5}{7}-\frac{4}{7})$

20) $(3+\frac{2}{3})\times 3$

Basic Chapter 4 Lesson 8
Review 1 B
Evaluate by using laws

1) $3(2+\frac{1}{6})$

2) $(2\frac{1}{4}-\frac{3}{4})\times 4$

3) $3(1\frac{5}{6}-1\frac{1}{3})$

4) $(3\frac{1}{2}-\frac{1}{4})\times 4$

5) $3(1\frac{2}{3}-\frac{5}{6})$

6) $(1\frac{2}{7}-\frac{1}{3})\times 21$

7) $6(2\frac{1}{2}+1\frac{1}{3})$

8) $(4+\frac{5}{8})\times 2$

9) $3(1\frac{1}{3}-\frac{5}{12})$

10) $(1\frac{1}{4}-\frac{5}{8})\times 4$

Simplify by finding common factor

11) $8\times 1\frac{7}{8}-8\times\frac{3}{8}$

12) $3\frac{2}{9}\times\frac{3}{5}+\frac{3}{5}\times 6\frac{7}{9}$

13) $7\times 1\frac{3}{14}-7\times\frac{1}{14}$

14) $4\times 2\frac{7}{8}-4\times 2\frac{1}{8}$

15) $2\frac{6}{7}\times\frac{7}{10}+2\frac{1}{7}\times\frac{7}{10}$

16) $5\frac{5}{6}\times\frac{3}{4}-1\frac{5}{6}\times\frac{3}{4}$

17) $4\frac{4}{5}\times\frac{5}{8}+\frac{5}{8}\times 3\frac{1}{5}$

18) $3\frac{2}{3}\times 1\frac{1}{2}-1\frac{2}{3}\times 1\frac{1}{2}$

19) $\frac{5}{6}\times 5\frac{1}{5}-\frac{5}{6}\times 2\frac{1}{5}$

20) $3\frac{1}{3}\times 1\frac{1}{5}+1\frac{1}{5}\times 1\frac{2}{3}$

Basic Chapter 4 Lesson 8
Worksheet 2 A for Final Exam

Laws of Fractions and Mixed Numbers

Evaluate by using laws

1) $1\frac{3}{5}\times 2\frac{1}{2}+1\frac{3}{5}\times 7\frac{1}{2}$

2) $4\frac{2}{3}\times \frac{3}{4}-3\frac{1}{3}\times \frac{3}{4}$

3) $4\frac{3}{5}\times \frac{5}{6}-\frac{5}{6}\times 1\frac{3}{5}$

4) $4\frac{1}{2}\times \frac{1}{6}+\frac{1}{6}\times 1\frac{1}{2}$

5) $3\frac{3}{4}\times 1\frac{1}{6}-2\frac{1}{4}\times 1\frac{1}{6}$

6) $\frac{3}{5}\times 15-\frac{1}{5}\times 15$

7) $1\frac{3}{4}\times 8-1\frac{1}{2}\times 8$

8) $1\frac{1}{2}\times 6-1\frac{1}{6}\times 6$

9) $2\frac{1}{2}\times 6-1\frac{1}{2}\times 6$

10) $1\frac{3}{4}\times 12-1\frac{1}{4}\times 12$

Worksheet 2 B for Final Exam

1) $(1\frac{3}{4}-1\frac{1}{2})\div \frac{1}{8}$

2) $(1\frac{3}{5}+1\frac{2}{5})\div \frac{1}{2}$

3) $(1\frac{3}{5}-1\frac{2}{5})\div \frac{2}{5}$

4) $(4\frac{3}{5}+1\frac{2}{5})\div \frac{1}{5}$

5) $(2\frac{1}{2}+1\frac{1}{2})\div \frac{1}{6}$

6) $(3\frac{2}{3}-1\frac{2}{3})\div \frac{2}{3}$

7) $(1\frac{4}{5}+1\frac{1}{5})\div \frac{1}{3}$

8) $(4\frac{3}{5}-1\frac{3}{5})\div \frac{1}{5}$

9) $(1\frac{1}{3}+1\frac{2}{3})\div \frac{1}{4}$

10) $(4\frac{1}{5}-2\frac{1}{5})\div \frac{1}{3}$

Basic Chapter 4 Lesson 9
Reviews for Test 4
Review 4 A

Date **Name**

Read directions carefully. **Answers must be in indicated place.**
Show all steps for full credits. **Each problem is worth 5 points.**
Do only 1 operation per 1 step. **Solve 1 problem for 5 minutes.**

1) Reduce fraction

$$\frac{60}{100}$$

1) ________

2) Add mixed numbers

$$2\frac{5}{12}+1\frac{5}{6}$$

2) ________

3) Subtract mixed numbers

$$4-2\frac{3}{4}$$

3) ________

4) Multiply mixed numbers

$$2\frac{2}{5}\times 5\frac{5}{6}$$

4) ________

5) Divide mixed numbers

$$4\frac{5}{8}\div 6\frac{1}{6}$$

5) ________

Basic Chapter 4 Lesson 9
Review 4 A

Date **Name**

Read directions carefully. **Answers must be in indicated place.**

Show all steps for full credits. **Each problem is worth 5 points.**

Do only 1 operation per 1 step. **Solve 1 problem for 5 minutes.**

6) Divide numbers

$$\frac{2}{3} \div 1\frac{1}{3} \div \frac{5}{8}$$

6) ________

7) Evaluate

$$\left(1\frac{1}{2}\right)^3 \div \left(1\frac{1}{2}\right)$$

7) ________

8) Evaluate

$$\left(\frac{1}{2} - \frac{1}{3}\right)^2 \div \frac{1}{9}$$

8) ________

9) Evaluate

$$6\frac{1}{2} \div \left(6\frac{1}{3} - 4\frac{1}{6}\right)$$

9) ________

10) Evaluate

$$1\frac{3}{4} \times 8 - 1\frac{1}{2} \times 8$$

10) ________

Basic Chapter 4 Lesson 9
Review 4 B

Date | **Name**
Read directions carefully. | **Answers must be in indicated place.**
Show all steps for full credits. | **Each problem is worth 5 points.**
Do only 1 operation per 1 step. | **Solve 1 problem for 5 minutes.**

1) Reduce fraction

$$\frac{40}{100}$$

1) ________

2) Add mixed numbers

$$1\frac{1}{6} + 2\frac{1}{12}$$

2)________

3) Subtract mixed numbers

$$4 - 1\frac{1}{5}$$

3)________

4) Multiply mixed numbers

$$1\frac{5}{6} \times 2\frac{2}{11}$$

4)________

5) Divide mixed numbers

$$3\frac{3}{8} \div 5\frac{5}{8}$$

5)________

Basic Chapter 4 Lesson 9
Review 4 B

Date **Name**

Read directions carefully. **Answers must be in indicated place.**
Show all steps for full credits. **Each problem is worth 5 points.**
Do only 1 operation per 1 step. **Solve 1 problem for 5 minutes.**

6) Divide numbers

$\frac{1}{6} \div 1\frac{1}{3} \div \frac{5}{8}$

6)_______

7) Evaluate

$(1\frac{1}{2})^2 \div \left(1\frac{1}{4}\right)$

7) ________

8) Evaluate

$\left(\frac{1}{3} - \frac{1}{6}\right)^2 \div \frac{1}{18}$

8) ________

9) Evaluate

$4\frac{1}{2} \div (6\frac{1}{2} - 4\frac{1}{4})$

9) ________

10) Evaluate

$1\frac{1}{2} \times 6 - 1\frac{1}{6} \times 6$

10) ________

Basic Chapter 4 Lesson 10
Tests for Chapter 4
Test 4 A

Date **Name**

Read directions carefully. **Answers must be in indicated place.**
Show all steps for full credits. **Each problem is worth 5 points.**
Do only 1 operation per 1 step. **Solve 1 problem for 5 minutes.**

1) Reduce fraction

$\frac{50}{100}$

1)_______

2) Add mixed numbers

$1\frac{6}{15}+2\frac{1}{5}$

2)_________

3) Subtract mixed numbers

$5-1\frac{1}{5}$

3) _________

4) Multiply mixed numbers

$3\frac{3}{4}\times 5\frac{1}{3}$

4)__________

5) Divide mixed numbers

$2\frac{7}{8}\div 3\frac{5}{6}$

5) ________

Basic Chapter 4 Lesson 10
Test 4 A

Date **Name**

Read directions carefully. **Answers must be in indicated place.**
Show all steps for full credits. **Each problem is worth 5 points.**
Do only 1 operation per 1 step. **Solve 1 problem for 5 minutes.**

6) Divide numbers

$$1\frac{1}{4} \div \frac{3}{4} \div 1\frac{2}{3}$$

6) ________

7) Evaluate

$$\left(1\frac{1}{2}\right)^3 \div \left(1\frac{1}{2}\right)^2$$

7)________

8) Evaluate

$$\left(\frac{1}{4}-\frac{1}{8}\right)^2 \div \frac{1}{32}$$

8) ________

9) Evaluate

$$2\frac{1}{2} \div \left(3\frac{1}{2}-2\frac{1}{4}\right)$$

9)________

10) Evaluate

$$2\frac{1}{2}\times 16-1\frac{1}{4}\times 16$$

10) ________

Basic Chapter 4 Lesson 10
Test 4 B

Date **Name**
Read directions carefully. **Answers must be in indicated place.**
Show all steps for full credits. **Each problem is worth 5 points.**
Do only 1 operation per 1 step. **Solve 1 problem for 5 minutes.**

1) Reduce fraction

$$\frac{20}{100}$$

1) ________

2) Add mixed numbers

$$1\frac{6}{15}+1\frac{2}{5}$$

2) ________

3) Subtract mixed numbers

$$3-1\frac{1}{4}$$

3)________

4) Multiply mixed numbers

$$4\frac{2}{5}\times 1\frac{4}{11}$$

4) ________

5) Divide mixed numbers

$$3\frac{3}{4}\div 6\frac{1}{4}$$

5)________

Basic Chapter 4 Lesson 10
Test 4 B

Date **Name**

Read directions carefully. **Answers must be in indicated place.**

Show all steps for full credits. **Each problem is worth 5 points.**

Do only 1 operation per 1 step. **Solve 1 problem for 5 minutes.**

6) Divide numbers

$$\frac{3}{8} \div \frac{3}{4} \div 1\frac{1}{4}$$

6) ________

7) Evaluate

$$(1\frac{4}{5}) \div \left(1\frac{1}{2}\right)^2$$

7) ________

8) Evaluate

$$\left(\frac{1}{5} - \frac{1}{10}\right)^2 \div \frac{1}{10}$$

8) ________

9) Evaluate

$$1\frac{1}{2} \div (5\frac{1}{2} - 4\frac{1}{4})$$

9) ________

10) Evaluate

$$1\frac{3}{4} \times 12 - 1\frac{1}{4} \times 12$$

10) ________

Basic Answers of Chapter 4

Lesson 1 Pretest

1)	$\frac{2}{5}$	2)	$\frac{4}{5}$	3)	$\frac{1}{10}$	4)	$1\frac{1}{5}$	5)	6

Lesson 6 Mental Math 7

1)	$\frac{1}{5}$	2)	$\frac{1}{4}$	3)	5	4)	1	5)	1
6)	2	7)	3	8)	4	9)	$\frac{1}{4}$	10)	$\frac{1}{25}$
11)	$\frac{4}{25}$	12)	$\frac{9}{25}$	13)	1	14)	2	15)	$\frac{1}{4}$
16)	$\frac{1}{2}$	17)	$\frac{3}{4}$	18)	$1\frac{1}{4}$	19)	$\frac{1}{4}$	20)	$\frac{1}{4}$

Lesson 6 Mental Math 8

1)	$1\frac{1}{2}$	2)	$2\frac{1}{4}$	3)	$\frac{1}{2}$	4)	$1\frac{1}{2}$	5)	$3\frac{1}{2}$
6)	6	7)	$2\frac{1}{2}$	8)	$1\frac{1}{2}$	9)	6	10)	7
11)	$1\frac{1}{2}$	12)	$3\frac{1}{2}$	13)	$1\frac{1}{4}$	14)	$1\frac{1}{4}$	15)	$1\frac{1}{2}$
16)	$1\frac{1}{2}$	17)	$\frac{1}{2}$	18)	$\frac{1}{4}$	19)	$1\frac{3}{4}$	20)	$1\frac{4}{5}$

Basic Answers of Chapter 4

Lesson 7 Review 1 A

1)	2	2)	$1\frac{1}{2}$	3)	$2\frac{1}{2}$	4)	$\frac{1}{3}$	5)	$1\frac{1}{4}$
6)	$1\frac{5}{8}$	7)	$1\frac{7}{9}$	8)	$2\frac{1}{6}$	9)	$1\frac{2}{3}$	10)	$2\frac{2}{3}$
11)	$2\frac{1}{2}$	12)	$1\frac{1}{2}$	13)	23	14)	$\frac{2}{9}$	15)	$\frac{1}{3}$
16)	$1\frac{1}{2}$	17)	6	18)	2	19)	$1\frac{1}{4}$	20)	$1\frac{1}{3}$

Lesson 7 Review 1 B

21)	$7\frac{1}{4}$	22)	$\frac{3}{5}$	23)	$\frac{1}{2}$	24)	$\frac{4}{5}$	25)	3
26)	$\frac{3}{5}$	27)	$\frac{3}{5}$	28)	$2\frac{1}{5}$	29)	$\frac{4}{5}$	30)	$1\frac{3}{8}$
31)	$1\frac{1}{2}$	32)	$\frac{1}{2}$	33)	$\frac{2}{5}$	34)	$1\frac{1}{5}$	35)	2
36)	$\frac{5}{7}$	37)	$1\frac{1}{4}$	38)	$2\frac{1}{2}$	39)	$\frac{2}{5}$	40)	$2\frac{1}{2}$

Lesson 7 Review 1 C

1)	8	2)	1	3)	$2\frac{1}{2}$	4)	1	5)	$1\frac{3}{4}$
6)	2	7)	2	8)	2	9)	23	10)	37

Basic Answers of Chapter 4

Lesson 8 Review 1 A

1)	1	2)	1	3)	1	4)	1	5)	$1\frac{1}{4}$
6)	$1\frac{3}{4}$	7)	2	8)	2	9)	$1\frac{1}{2}$	10)	$1\frac{4}{5}$
11)	5	12)	1	13)	22	14)	11	15)	17
16)	5	17)	5	18)	9	19)	1	20)	11

Lesson 8 Review 1 B

21)	$6\frac{1}{2}$	22)	6	23)	$1\frac{1}{2}$	24)	13	25)	$2\frac{1}{2}$
26)	20	27)	23	28)	$9\frac{1}{4}$	29)	$2\frac{3}{4}$	30)	$2\frac{1}{2}$
31)	12	32)	6	33)	8	34)	3	35)	$3\frac{1}{2}$
36)	3	37)	5	38)	3	39)	$2\frac{1}{2}$	40)	6

Lesson 8 Worksheet 2 A

1)	16	2)	1	3)	$2\frac{1}{2}$	4)	1	5)	$1\frac{3}{4}$
6)	6	7)	2	8)	2	9)	6	10)	6

Lesson 8 Worksheet 2 B

11)	2	12)	6	13)	$\frac{1}{2}$	14)	30	15)	24
16)	3	17)	9	18)	15	19)	12	20)	6

Basic Answers of Chapter 4

Lesson 9 Review 4 A

1)	$\frac{3}{5}$	2)	$4\frac{1}{4}$	3)	$1\frac{1}{4}$	4)	14	5)	$\frac{3}{4}$
6)	$\frac{4}{5}$	7)	$2\frac{1}{4}$	8)	$\frac{1}{4}$	9)	3	10)	2

Lesson 9 Review 4 B

1)	$\frac{2}{5}$	2)	$3\frac{1}{4}$	3)	$2\frac{4}{5}$	4)	4	5)	$\frac{3}{5}$
6)	$\frac{1}{5}$	7)	$1\frac{4}{5}$	8)	$\frac{1}{2}$	9)	2	10)	2

Lesson 10 Test 4 A

1)	$\frac{1}{2}$	2)	$3\frac{3}{5}$	3)	$3\frac{4}{5}$	4)	20	5)	$\frac{3}{4}$
6)	1	7)	$1\frac{1}{2}$	8)	$\frac{1}{2}$	9)	2	10)	20

Lesson 10 Test 4 B

1)	$\frac{1}{5}$	2)	$2\frac{4}{5}$	3)	$1\frac{3}{4}$	4)	6	5)	$\frac{3}{5}$
6)	$\frac{2}{5}$	7)	$\frac{4}{5}$	8)	$\frac{1}{10}$	9)	$1\frac{1}{5}$	10)	6

Basic Chapter 5

Converting Fractions and Decimals

Students!

Students, you are smarter than a calculator!
Students, you are smarter than a computer!
Students, you need to believe in yourself.
Students, be patient and do your best.

Do not be afraid to make mistakes.
Tell yourself positive words.
I am young! I am smart!
I am strong! I can do anything!

Smart students use their brain!
Your brain is always with you!
A calculator is not always with you.
Mathematics is work for the brain.

Be open to new ideas in math!
Work hard and try your best!
Check answers and ask questions!
Practice! Practice! Practice!

Basic Chapter 5 Lessons
Converting Fractions and Decimals
Lessons 1 – 10
Basic Problems

Lesson 1 Table of Multiplication. Pretest

Lesson 2 Converting Fractions to Repeating Decimals

Lesson 3 Converting Fractions to Terminating Decimals
Converting Decimals to Fractions

Lesson 4 Operations with Fractions and Decimals. Examples

Lesson 5 Order of Operations 1

Lesson 6 Order of operations 2

Lesson 7 Operations with numbers. Laws for numbers

Lesson 8 Worksheet 3A for Final Exam
Worksheet 3 B for Final Exam

Lesson 9 Review 5 A, Review 5

Lesson 10 Test 5 A, Test 5 B

Answers Chapter 5

Basic Chapter 5 Contents
Converting Fractions and Decimals

Contents have topics and exercises, which students need to solve before tests.1)Work with lesson 1, read rules and examples, 2) Solve the problems from contents, 3) Check answers in the end of chapter 1, 4) Work with the next lessons 2, 3, 4, 5, …, 5) Solve Test 1 A, 6) Grading Scale without minuses

A=90–100% B+=85–89% B=80–84% C+=75–79%
C=70–74% D+=65–69% D=60–64 % F=0 - 59%

7)If you solve 10 problems correctly, you get an A, 8) If have an A, you can move on to next chapter, 9) If your grade is lower than an A, work with the mistakes, then solve Test 1B, 10) When you finish all of the chapters, solve Final Exam in last chapter.

Basic Chapter 5 Lesson 1
Table 1
Multiplication and Division

A + 0 = A	0 + A = A	A – 0 = A	A – A = 0
A × 1 = A	1 × A = A	A ÷ 1 = A	A ÷ A = 1
A × 0 = 0	0 × A = 0	0 ÷ A = 0	A ÷ 0 = Ø
A + B = B + A		A × B = B × A	Ø no solutions

Number 5 from the left column multiply by number 4 from first row to get answer 20. 5 × 4 = 20 20 ÷ 5 = 4 20 ÷ 4 = 5

Memorize Table 1

×	1	2	3	<u>4</u>	5	6	7	8	9	10
1	1	2	3	4	5	6	7	8	9	10
2	2	4	6	8	10	12	14	16	18	20
3	3	6	9	12	15	18	21	24	27	30
4	4	8	12	16	20	24	28	32	36	40
<u>5</u>	5	10	15	<u>20</u>	25	30	35	40	45	50
6	6	12	18	24	30	36	42	48	54	60
7	7	14	21	28	35	42	49	56	63	70
8	8	16	24	32	40	48	56	64	72	80
9	9	18	27	36	45	54	63	72	81	90
10	10	20	30	40	50	60	70	80	90	100

Basic Chapter 5 Lesson 1

Date **Name**

Pretest Chapter 5

Solve operations in fractions or in decimals but final answers put in decimals.

1) **Divide**

$0.57 \div \frac{3}{4}$

1) ________

2) **Divide**

$6\frac{7}{8} \div 2.75$

2) ________

Evaluate

3) $3^2 \div (1\frac{1}{2})$

3) ________

4) $(3.5 - 3\frac{1}{3}) \div 1\frac{1}{3}$

4) ________

5) $6.5 \div (3\frac{1}{3} - 1\frac{1}{6})$

5) ______

Basic Chapter 5 Lesson 2

Converting Fractions to Repeating Decimals

Repeating Decimal

Decimals called repeating decimals if the decimal digits after the point repeated: 0.1818... $\approx 0.\overline{18}$ 0.1666... $\approx 0.1\overline{6}$

Period

A digit which is repeated is called a period.
The repeated decimal $0.\overline{18}$ has a period 18.
The repeated decimal $0.1\overline{6}$ has a period 6

Simple Repeating Decimal

If repetition digits begin at once after the decimal point called simple repeating decimal.
To write a fraction to simple repeating decimals, always try to reduce the fraction first, after that, divide the numerator by the denominator.

Reduced fractions can convert to repeating decimals when the denominator's prime factorization contains not just digit 2 and/or 5.
Denominator: $6 = 2 \times 3$ Denominator: $33 = 3 \times 11$

$\frac{1}{3} \approx 0.333... \approx 0.\overline{3}$

$1\frac{4}{6} = 1\frac{2}{3} \approx 1.666... \approx 1.\overline{6}$

$2\frac{2}{11} \approx 2.1818... \approx 2.\overline{18}$

$\frac{7}{33} \approx 0.2121... \approx 0.\overline{21}$

$$\begin{array}{r} 0.33 \\ 3\overline{)1.00} \\ \underline{-0} \\ 10 \\ \underline{-9} \\ 10 \\ \underline{-9} \\ 1 \end{array}$$

$$\begin{array}{r} 0.21 \\ 33\overline{)7.00} \\ \underline{-66} \\ 40 \\ \underline{-33} \\ 70 \\ \underline{-66} \\ 4 \end{array}$$

Basic Chapter 5 Lesson 2

Converting Fractions to Composite Repeating Decimals

Repeating Decimal

Decimals called repeating decimals if the decimal digits after the point are repeated: $0.2727... \approx 0.\overline{27}$ $0.8333... \approx 0.8\overline{3}$

Period

A digit which is repeated is called a period.
The repeating decimal $0.\overline{27}$ has a period 27.
The repeating decimal $0.8\overline{3}$ has a period 3.
The number with period is 83; number before the period is 8.

Simple Repeating Decimal

If repetition digits begin at once after the decimal point called simple repeating decimal:

$\frac{1}{3} \approx 0.333... \approx 0.\overline{3}$ $1\frac{4}{6} = 1\frac{2}{3} \approx 1.666... \approx 1.\overline{6}$ $2\frac{3}{11} \approx 2.2727... \approx 2.\overline{27}$

Composite Repeating Decimal

If repetition digits begin not at once after the decimal point, after some digits, called composite repeating decimal:

$3\frac{1}{6} \approx 3.1666... \approx 3.1\overline{6}$ $2\frac{5}{6} \approx 2.8333... \approx 2.8\overline{3}$ $1\frac{14}{30} = 1\frac{7}{15} \approx 1.466... \approx 1.4\overline{6}$

To convert a fraction to composite repeating decimals, always try to reduce the fraction first, after that, divide the numerator by the denominator.

Reduced fractions can convert to repeating decimals when the denominator's prime factorization contains not just digit 2 and/or 5.

Denominator: $14=2\times7$ **Denominator: $30=6\times5=2\times3\times5$**

Basic Chapter 5 Lesson 3

Converting Fractions to Terminating Decimals

Terminating Decimals

To convert a fraction to a terminating decimal, always try to reduce the fraction first then divide the numerator by the denominator. When dividing the numerator by the denominator, if the remainder equals zero, this decimal called terminating.

$\frac{3}{10}=0.3$ $\frac{9}{100}=0.09$ $1\frac{7}{1000}=1.007$

$\frac{1}{2}=0.5$ $\frac{1}{4}=0.25$ $\frac{3}{4}=0.75$

$\frac{12}{24}=\frac{1}{2}=0.5$ $\frac{6}{24}=\frac{1}{4}=0.25$ $\frac{15}{20}=\frac{3}{4}=0.75$

Fractions can change to the terminating decimals if the fraction has a factorization of the denominator containing only digits 2 and/or 5.

Fractions can be converted to terminating decimals if the reduced fraction has denominator like 2, 4, 5, 8, 10, 16, 20, 25, 32, 40, 50, 64, 80, 100 because prime factorization for these numbers content just digits 2 and/or 5:

$4=2^2$ $8=2^3$ $10=2\times 5$ $16=2^4$ $20=2^2\times 5$ $25=5^2$

$32=2^5$ $40=2^3\times 5$ $50=2\times 5^2$ $64=2^6$ $80=2^4\times 5$ $100=2^2\times 5^2$

First way to change fraction to decimal is division, lets learn second way. Denominators in decimals are 10, 100, 1000. Prime factorization these numbers are $10=2\times 5$ $100=4\times 25=2\times 2\times 5\times 5$

$1000=8\times 125=2\times 2\times 2\times 5\times 5\times 5$ $\frac{4}{5}=\frac{4\times 2}{5\times 2}=\frac{8}{10}=0.8$

$\frac{7}{25}=\frac{7\times 4}{5\times 5\times 2\times 2}=\frac{28}{10^2}=\frac{28}{100}=0.28$ $\frac{9}{20}=\frac{9}{2\times 2\times 5}=\frac{9\times 5}{2\times 2\times 5\times 5}=\frac{9\times 5}{100}=\frac{45}{100}=0.45$

$\frac{3}{4}=\frac{3\times 5\times 5}{2\times 2\times 5\times 5}=\frac{75}{100}=0.75$ $\frac{1}{125}=\frac{1\times 8}{5\times 5\times 5\times 2\times 2\times 2}=\frac{8}{1000}=0.008$

Converting Mixed Number to the Decimals

$4\frac{11}{50}=4\frac{11\times 2}{2\times 5\times 5\times 2}=4\frac{22}{100}=4.22$ $50=2\times 5\times 5$

Basic Chapter 5 Lesson 3
Converting Decimals to Fractions

Converting Terminating Decimals to Fraction

Terminating decimals always can converted to fractions: put the digits after the decimal point in the same order in the numerator, and in the denominator, put digit 1 one time and 0 as many times, as there are digits after the decimal point. $3.2 = 3\frac{2}{10} = 3\frac{1}{5}$ $\quad 2.45 = 2\frac{45}{100} = 2\frac{9}{20}$

Converting Simple Repeating Decimals to Fractions

a) $x = 0.3333\ldots\ (\times 10)$ $\quad 10x = 3.333\ldots$ $\quad 10x = 3 + 0.333\ldots$

$10x = 3 + x$ $\quad 10x - x = 3$ $\quad 9x = 3$ $\quad 0.\overline{3} = \frac{3}{9} = \frac{1}{3}$

b) $x = 0.1818\ldots$ $\quad 100x = 18.18\ldots$ $\quad 100x = 18 + 0.1818\ldots$

$100x = 18 + x$ $\quad 100x - x = 18$ $\quad 99x = 18$ $\quad 0.\overline{18} = \frac{18}{99} = \frac{2}{11}$

To convert simple repeating decimals to fraction, put the period in the numerator and in the denominator write the digit 9 as many times as there are digits in the period:

$0.\overline{3} = \frac{3}{9} = \frac{1}{3}$ $\quad 0.\overline{6} = \frac{6}{9} = \frac{2}{3}$ $\quad 2.\overline{18} = 2\frac{18}{99} = 2\frac{2}{11}$ $\quad 1.\overline{27} = 1\frac{27}{99} = 1\frac{3}{11}$

Converting Composite Repeating Decimals in the Fractions

c) $x = 0.8333\ldots\ (\times 10)$ $\quad 10x = 8.333\ldots$ $\quad 10x = 8 + 0.333\ldots$

$10x = 8 + \frac{3}{9}$ $\quad 10x = 8\frac{3}{9}$ $\quad 10x = \frac{75}{9}$

$10x = \frac{83-8}{9}$ $\quad \underline{x = \frac{83-8}{90}}$ $\quad 0.8\overline{3} = \frac{75}{90} = \frac{5}{6}$

To converting composite repeating decimals $0.8\overline{3}$ to fractions, put in the numerator, the difference between the number with the period (83) and the number before the period (8). Put in the denominator, the digit 9 as many times as there are digits in the period and the digit, 0 as many times as there are digits before the period.

$0.1\overline{3} = \frac{13-1}{90} = \frac{12}{90} = \frac{2}{15}$ $\quad 1.1\overline{6} = 1\frac{16-1}{90} = 1\frac{15}{90} = 1\frac{1}{6}$

$1.8\overline{3} = 1\frac{83-8}{90} = 1\frac{75}{90} = 1\frac{5}{6}$ $\quad 1.4\overline{6} = 1\frac{46-4}{90} = 1\frac{42}{90} = 1\frac{7}{15}$

Basic Chapter 5 Lesson 4

Operation with Fractions and Decimals

Decimals always can convert to fraction:

$$1.125 = 1\frac{125}{1000} = 1\frac{125 \div 5}{1000 \div 5} = 1\frac{25}{200} = 1\frac{25 \div 5}{200 \div 5} = 1\frac{5}{40} = 1\frac{1}{8}$$

Fractions can be converted to the terminating decimal when the denominator's prime factorization contains only digits 2 and/or 5.

Fractions $\frac{1}{2} = 0.5$ $\frac{1}{4} = 0.25$ $\frac{3}{4} = 0.75$ $\frac{1}{5} = 0.2$ $\frac{1}{8} = 0.125$ $\frac{1}{10} = 0.1$
can converted to terminating decimals. These are exact numbers.

Memorize this fractions and decimals.
To do operations with fractions and decimals, use only terminating decimals because it's exact numbers.

To do operations, always check at the fraction because the fraction cannot always converted in terminating decimal.
If a fraction can convert to a terminating decimal, work in decimals.

If a fraction cannot convert to a terminating decimal, work in fractions.

Converting the Terminating Decimals to Fractions

a) $1.875 = 1\frac{875}{1000} = 1\frac{875 \div 5}{1000 \div 5} = 1\frac{175}{200} = 1\frac{175 \div 5}{200 \div 5} = 1\frac{35}{40} = 1\frac{7}{8}$

Converting the Repeating Decimals to Fractions

b) $1.\bar{6} + 0.\bar{3} = 1\frac{6}{9} + \frac{3}{9} = 1\frac{2}{3} + \frac{1}{3} = 1\frac{3}{3} = 2$

c) $0.8\bar{3} - 0.4\bar{6} = \frac{83-8}{90} - \frac{46-4}{90} = \frac{75}{90} - \frac{42}{90} = \frac{33}{90} = \frac{11}{30}$

Basic Chapter 5 Lesson 4

Examples:

$$\text{a)}\ \frac{2}{3}\times 0.36 = 0.36\times\frac{2}{3} = 0.36\div 3\times 2 = 0.12\times 2 = 0.24$$

$$\text{b)}\ 3.6\div 4\frac{1}{2} = 3.6\div 4.5 = 36\div 45 = \frac{36}{45} = \frac{4}{5} = 0.8$$

$$\text{c)}\ (0.66+\frac{31}{50})\div(-\frac{2}{10}) = (\frac{66}{100}+\frac{31}{50})\div(-\frac{1}{5}) = (\frac{33}{50}+\frac{31}{50})(-\frac{5}{1}) = \frac{64}{50}(-\frac{5}{1}) = -6.4$$

$$\text{d)}\ -0.25\div(\frac{39}{50}-\frac{37}{50}) = -\frac{1}{4}\div\frac{2}{50} = -\frac{1}{4}\times\frac{25}{1} = -\frac{25}{4} = -6\frac{1}{4} = -6.25$$

$$\text{e)}\ (-\frac{43}{50}+\frac{41}{50})\div 0.16 = (-\frac{2}{50})\div\frac{16}{100} = -\frac{1}{25}\div\frac{4}{25} = -\frac{1}{25}\times\frac{25}{4} = -\frac{1}{4} = -0.25$$

$$\text{f)}\ 9.6\div(-\frac{49}{50}-\frac{47}{50}) = 9.6\div(-\frac{96}{50}) = 9\frac{6}{10}\times(-\frac{50}{96}) = 9\frac{3}{5}(-\frac{25}{48}) = -\frac{48}{5}\times\frac{25}{48} = -5$$

Addition and subtraction is easy work in decimals. Multiplication and division is easy work in fractions. Always try to find the easiest way to calculate, that way, fewer mistakes will be made.

$$\text{g)}\ \frac{1}{2} - 0.25 = 0.5 - 0.25 = 0.25$$

$$\text{h)}\ 0.25\div\frac{1}{8} = \frac{1}{4}\div\frac{1}{8} = \frac{1}{4}\times\frac{8}{1} = 2$$

Basic Chapter 5 Lesson 5

Order of Operations 1

If an expression has five operations,

first work on operations at the third level of operations (a^n),

then work on the operations at second level of operations (×,÷).

last, work on operations at first level of operations (+,-),

in order, as they occur, from left to right.

in short

Grouping Symbols (1) [2] {3}

Exponents or Power – third level of operations a^n

Multiplication or Division from left to right–second level

× **or** ÷ **L → R**

Addition or Subtract from left to right – first level

\+ **or** – **L → R**

Always try to reduce fractions first, convert improper fractions to mixed numbers, and convert fractions to terminating decimals.

Work in fractions or decimals, but if you can the final answers should give in decimals. Try to find easiest way.

Basic Chapter 5 Lesson 5
Review 1 A (# 1 - # 15)
Perform the Operations

1) $(1.25 - \frac{1}{4}) \div \frac{2}{7}$

2) $(1.08 - \frac{2}{25}) \div \frac{4}{7}$

3) $(\frac{1}{6} + \frac{1}{3}) \div 0.25$

4) $(0.3 - \frac{3}{20}) \div \frac{1}{20}$

5) $(2.5 - 2\frac{1}{3}) \div \frac{5}{6}$

6) $(4.5 - 1\frac{1}{6}) \div \frac{5}{9}$

7) $(1.5 - 1\frac{1}{3}) \div \frac{1}{3}$

8) $1.2 \div (\frac{8}{9} - \frac{2}{3})$

9) $1.5 \div (\frac{5}{6} - \frac{2}{3})$

10) $3.6 \div (3\frac{1}{5} + 2\frac{4}{5})$

11) $3.5 \div (3\frac{1}{5} - 2\frac{7}{10})$

12) $2.4 \div (2\frac{3}{4} + 1\frac{1}{4})$

13) $6.75 \div (1\frac{1}{8} + 2\frac{1}{4})$

14) $29 \div (2.5 + 1\frac{1}{8})$

15) $4.2 \div (3\frac{1}{2} + 1\frac{3}{4})$

16) $1.1 \div (3\frac{1}{2} - 1\frac{3}{10})$

Review 1 B (# 16 - # 30)

17) $4.2 \div (3\frac{1}{5} - 1\frac{1}{10})$

18) $\frac{5}{6} \div (1\frac{2}{3} - 1.25)$

19) $6.5 \div (4\frac{1}{3} - 2\frac{1}{6})$

20) $\frac{8}{15} \div (1\frac{1}{6} + 1.5)$

21) $\frac{7}{15} \div (5\frac{5}{6} - 3.5)$

22) $\frac{4}{15} \div (2.8 - 2\frac{2}{3})$

23) $\frac{5}{6} \div (1\frac{8}{12} - 1.25)$

24) $\frac{8}{15} \div (1\frac{1}{6} + 1.5)$

25) $\frac{7}{15} \div (5\frac{5}{6} - 3.5)$

26) $1\frac{2}{3} \div (7.5 - 3\frac{1}{3})$

27) $2.8 \div (2\frac{3}{4} + 1.25)$

28) $2\frac{1}{9} \div (\frac{4}{9} + 0.4)$

29) $0.4 \div (\frac{8}{9} - \frac{2}{3})$

30) $3\frac{4}{5} \div (3\frac{1}{3} - 2.7)$

Basic Chapter 5 Lesson 6

Order of operations and laws for numbers

If you can, always in answer try to reduce fractions to lowest term, convert improper fraction to mixed number, and convert fraction to terminating decimal.

If you can, do operations in fractions or decimals but final answers put in decimals. Easy work: + and – in decimals, × and ÷ in fractions.

$\frac{3}{4}+0.25=0.75+0.25=1$ $\qquad$ $0.5\div(\frac{1}{8})=\frac{1}{2}\div(\frac{1}{8})=\frac{1}{2}\times(\frac{8}{1})=4$

Review 1 A

1) $(1\frac{3}{8}+1.75)\div 6\frac{1}{4}$

2) $(1.75-1\frac{5}{8})\div\frac{5}{8}$

3) $(1\frac{1}{2}+1.8)\div 2.2$

4) $(2\frac{3}{5}+1.6)\div 2.4$

5) $1.65\div(0.6+\frac{1}{2})$

6) $2.2\div(3\frac{2}{5}-1.2)$

7) $1\frac{7}{8}\div(1\frac{1}{5}+1.3)$

8) $2.1\div(4.5-3\frac{1}{3})$

9) $5\frac{1}{4}\div(2\frac{1}{4}+1.25)$

10) $3.8\div(4\frac{2}{3}-3.4)$

Basic Chapter 5 Lesson 6

Laws for numbers
Common Factor

AB + AC = A (B + C) AB - AC = A (B - C)
$16 \cdot 5 + 4 \cdot 5 = 5\,(16 + 4) = 5\,(20) = 100$
$22 \cdot 5 - 2 \cdot 5 = 5\,(22 - 2) = 5\,(20) = 100$

Review 1 B
Evaluate

1) $1\frac{3}{5}(2\frac{1}{8})+1\frac{3}{5}(2\frac{7}{8})$

2) $\frac{3}{4}(4\frac{2}{3})-\frac{3}{4}(3\frac{1}{3})$

3) $\frac{5}{6}(4\frac{3}{5})-\frac{5}{6}\times1\frac{3}{5}$

4) $4\frac{1}{2}\times\frac{1}{6}+\frac{1}{6}\times1\frac{1}{2}$

5) $3\frac{3}{4}\times1\frac{1}{6}-2\frac{1}{4}\times1\frac{1}{6}$

6) $(\frac{1}{3}-\frac{1}{5})\div\frac{1}{9}$

7) $(1\frac{3}{4}-1\frac{1}{2})\div\frac{1}{8}$

8) $(1\frac{1}{2}-1\frac{1}{6})\div\frac{1}{9}$

9) $(2\frac{1}{2}+1\frac{1}{3})\div\frac{1}{6}$

10) $(1\frac{1}{3}+1\frac{3}{4})\div\frac{1}{12}$

Basic Chapter 5 Lesson 7
Laws for operations
Distributive laws of multiplication

(A + B) C = AC + BC **(A - B) C = AC – B C**

Example a) $\frac{5}{6}\times 3=\frac{5}{6}\times\frac{3}{1}=\frac{5}{2}=2\frac{1}{2}=2.5$

Example b) $\frac{5}{3}\times 6=6\times\frac{5}{3}=6\div 3\times 5=2\times 5=10$

Example c) $\left(1\frac{1}{3}-\frac{5}{6}\right)\times 3=\frac{4}{3}\times\frac{3}{1}-\frac{5}{6}\times\frac{3}{1}=4-\frac{5}{2}=4-2.5=1.5$

Review 1 A
Evaluate by using laws

1) $9(1-\frac{4}{9})$

2) $(\frac{2}{3}-\frac{3}{5})\times 15$

3) $9(3-\frac{5}{9})$

4) $(1+\frac{5}{6})\times 6$

5) $7(2+\frac{3}{7})$

6) $(1-\frac{1}{6})\times 6$

7) $6(1\frac{1}{2}-\frac{2}{3})$

8) $(\frac{5}{7}+\frac{4}{7})\times 7$

9) $7(\frac{5}{7}-\frac{4}{7})$

10) $(3+\frac{2}{3})\times 3$

11) $3(2+\frac{1}{6})$

12) $(2\frac{1}{4}-\frac{5}{8})\times 4$

13) $3(1\frac{5}{6}-1\frac{1}{3})$

14) $(3\frac{1}{2}-\frac{1}{4})\times 4$

15) $3(1\frac{2}{3}-\frac{5}{6})$

16) $(1\frac{2}{7}-\frac{1}{3})\times 21$

17) $6(2\frac{1}{2}+1\frac{1}{3})$

18) $(4+\frac{5}{8})\times 2$

19) $3(1\frac{1}{3}-\frac{5}{12})$

20) $(1\frac{1}{4}-\frac{5}{8})\times 4$

Basic Chapter 5 Lesson 7
Solve by finding common factor
Common Factor (Put C before parentheses)

$AC + BC = C\,(A + B)$ $\qquad\qquad$ $AC - B\,C = C\,(A - B)$

Example $\quad 1\frac{1}{4}\times\frac{3}{4}+\frac{3}{4}\times2\frac{3}{4}=\frac{3}{4}(1\frac{1}{4}+2\frac{3}{4})=\frac{3}{4}\times\frac{4}{1}=3$

Evaluate by using laws
Review 1 B

1) $8\times1\frac{3}{4}-8\times\frac{5}{8}$

2) $3\frac{2}{9}\times\frac{3}{5}+\frac{3}{5}\times6\frac{7}{9}$

3) $7\times1\frac{1}{7}-7\times\frac{5}{14}$

4) $4\times2\frac{1}{4}-4\times1\frac{7}{8}$

5) $2\frac{6}{7}\times\frac{7}{10}+2\frac{1}{7}\times\frac{7}{10}$

6) $5\frac{5}{6}\times\frac{3}{4}-1\frac{5}{6}\times\frac{3}{4}$

7) $4\frac{4}{5}\times\frac{5}{8}+\frac{5}{8}\times3\frac{1}{5}$

8) $3\frac{2}{3}\times1\frac{1}{2}-1\frac{2}{3}\times1\frac{1}{2}$

9) $\frac{5}{6}\times5\frac{1}{5}-\frac{5}{6}\times2\frac{1}{5}$

10) $3\frac{1}{3}\times1\frac{1}{5}+1\frac{1}{5}\times1\frac{2}{3}$

11) $1\frac{3}{5}\times2\frac{1}{2}+1\frac{3}{5}\times7\frac{1}{2}$

12) $4\frac{2}{3}\times\frac{3}{4}-3\frac{1}{3}\times\frac{3}{4}$

13) $4\frac{3}{5}\times\frac{5}{6}-\frac{5}{6}\times1\frac{3}{5}$

14) $4\frac{1}{2}\times\frac{1}{6}+\frac{1}{6}\times1\frac{1}{2}$

15) $3\frac{3}{4}\times1\frac{1}{6}-2\frac{1}{4}\times1\frac{1}{6}$

16) $\frac{1}{3}\times15-\frac{1}{5}\times15$

17) $1\frac{3}{4}\times8-1\frac{1}{2}\times8$

18) $1\frac{1}{2}\times6-1\frac{1}{6}\times6$

19) $2\frac{1}{2}\times6-1\frac{1}{2}\times6$

20) $1\frac{3}{4}\times12-1\frac{1}{4}\times12$

21) $1\frac{3}{4}+1\frac{1}{4})\div\frac{1}{4}$

22) $(1\frac{3}{5}-1\frac{2}{5})\div\frac{2}{5}$

23) $(4\frac{3}{5}+1\frac{2}{5})\div\frac{1}{5}$

24) $(1\frac{3}{4}-1\frac{1}{4})\div\frac{1}{8}$

25) $(2\frac{1}{2}+1\frac{1}{2})\div\frac{1}{6}$

26) $(3\frac{2}{3}-1\frac{2}{3})\div\frac{2}{3}$

27) $(1\frac{5}{6}+1\frac{1}{6})\div\frac{1}{6}$

28) $(4\frac{3}{5}-1\frac{3}{5})\div\frac{1}{5}$

29) $(1\frac{2}{3}+1\frac{1}{3})\div\frac{1}{3}$

30) $(4\frac{5}{6}-2\frac{1}{6})\div\frac{1}{6}$

Basic Chapter 5 Lesson 8

Worksheet 3 A for Final Exam

Perform the indicated operations

1) $\left(0.25+\frac{1}{8}\right)\div 1.5$

2) $\left(\frac{1}{16}+\frac{1}{32}\right)\div\frac{3}{16}$

3) $\left(0.1+\frac{1}{25}\right)\div 1.4$

4) $\left(\frac{1}{64}+\frac{1}{32}\right)\div\frac{3}{16}$

5) $\left(0.05+\frac{1}{40}\right)\div\frac{3}{20}$

6) $\left(0.02+\frac{1}{25}\right)\div\frac{6}{40}$

7) $\left(\frac{1}{80}+\frac{1}{40}\right)\div\frac{3}{64}$

8) $\left(0.01+\frac{1}{50}\right)\div\frac{1}{20}$

9) $\left(0.01+\frac{1}{25}\right)\div 0.25$

10) $\left(0.01+\frac{1}{20}\right)\div\frac{12}{20}$

Basic Chapter 5 Lesson 8

Worksheet 3 B for Final Exam

Perform the indicated operations

1) $\frac{3}{8} \div \left(0.25 - \frac{1}{8}\right)$

2) $\frac{5}{32} \div \left(\frac{1}{16} - \frac{1}{32}\right)$

3) $\frac{6}{50} \div \left(0.1 - \frac{1}{25}\right)$

4) $\frac{5}{32} \div \left(\frac{1}{32} - \frac{1}{64}\right)$

5) $\frac{3}{20} \div \left(0.05 - \frac{1}{40}\right)$

6) $\frac{7}{25} \div \left(\frac{1}{25} - 0.02\right)$

7) $\frac{13}{40} \div \left(\frac{1}{40} - \frac{1}{80}\right)$

8) $\frac{9}{50} \div \left(\frac{1}{50} - 0.01\right)$

9) $\frac{9}{25} \div \left(\frac{1}{25} - 0.01\right)$

10) $\frac{8}{25} \div \left(\frac{1}{20} - 0.01\right)$

Basic Chapter 5 Lesson 9
Reviews for Test 5
Review 5 A

Date **Name**
Read directions carefully. **Answers must be in indicated place.**
Show all steps for full credits. **Each problem is worth 5 points.**
Do only 1 operation per 1 step. **Solve 1 problem 5 minutes.**

1) Convert decimal 0.125 to fraction.

1) ________

2) Convert fraction $\frac{3}{25}$ to terminating decimal.

2) ________

3) Add $0.875 + \frac{3}{6}$

3) ________

4) Subtract $1\frac{1}{4} - 0.45$

4) ________

5) Multiply $1.6 \times \frac{3}{8}$

5) ________

Basic Chapter 5 Lesson 9
Review 5 A

Date — **Name**

Read directions carefully.
Show all steps for full credits.
Do only 1 operation per 1 step.

Answers must be in indicated place.
Each problem is worth 5 points.
Solve 1 problem 5 minutes.

Divide

6) $0.15 \div \frac{3}{4}$

6)________

7) $3.75 \div 3\frac{1}{8}$

7)________

Evaluate

8) $3^3 \div \left(1\frac{1}{2}\right)^2$

8)________

9) $(0.3 - \frac{3}{20}) \div \frac{1}{20}$

9) ________

10) $\frac{5}{6} \div (1\frac{2}{3} - 1.25)$

10) ______

Basic Chapter 5 Lesson 9
Review 5 B

Date **Name**
Read directions carefully. **Answers must be in indicated place.**
Show all steps for full credits. **Each problem is worth 5 points.**
Do only 1 operation per 1 step. **Solve 1 problem 5 minutes.**

1) Convert terminating decimal 0.375 to fraction.

1) ________

2) Convert fraction $\frac{7}{25}$ to terminating decimal.

2) ________

3) Add $1\frac{3}{6}+1.125$

3) ________

4) Subtract $2\frac{1}{4}-1.75$

4) ________

5) Multiply $3.2\times\frac{5}{8}$

5) ________

Basic Chapter 5 Lesson 9
Review 5 B

Date **Name**

Read directions carefully.
Show all steps for full credits.
Do only 1 operation per 1 step.
Answers must be in indicated place.
Each problem is worth 5 points.
Solve 1 problem 5 minutes.

Divide

6) $0.45 \div \frac{3}{4}$

6) ________

7) $4\frac{1}{8} \div 2.75$

7) ________

Evaluate

8) $(1\frac{1}{2})^2 \div 3^2$

8) ________

9) $(1.5 - 1\frac{1}{3}) \div \frac{1}{3}$

9) ________

10) $6.5 \div (4\frac{1}{3} - 2\frac{1}{6})$

10) ______

Basic Chapter 5 Lesson 10
Tests for Chapter 5
Test 5 A

Date | **Name**
Read directions carefully. | **Answers must be in indicated place.**
Show all steps for full credits. | **Each problem is worth 5 points.**
Do only 1 operation per 1 step. | **Solve 1 problem 5 minutes.**

1) Convert decimal 0.625 to fraction

1)________

2) Convert fraction $\frac{9}{25}$ to terminating decimal

2) ________

3) Add $1.25 + \frac{4}{8}$

3) ________

4) Subtract $1\frac{1}{4} - 0.75$

4) ________

5) Multiply $2.4 \times \frac{3}{8}$

5) ________

Basic Chapter 5 Lesson 10

Test 5 A

Date | **Name**

Read directions carefully.
Show all steps for full credits.
Do only 1 operation per 1 step.

Answers must be in indicated place.
Each problem is worth 5 points.
Solve 1 problem 5 minutes.

Divide

6) $0.36 \div \frac{3}{4}$

6)________

7) $4.25 \div 2\frac{1}{8}$

7) ________

Evaluate

8) $3^2 \div \left(1\frac{1}{2}\right)^2$

8) ________

9) $(0.6 - \frac{3}{20}) \div \frac{3}{20}$

9) ________

10) $\frac{5}{6} \div (2\frac{2}{3} - 2.25)$

10) __

Basic Chapter 5 Lesson 10
Test 5 B

Date **Name**

Read directions carefully. **Answers must be in indicated place.**

Show all steps for full credits. **Each problem is worth 5 points.**

Do only 1 operation per 1 step. **Solve 1 problem 5 minutes.**

1) **Convert terminating decimal 0.875 to fraction**

1) ________

2) **Convert fraction** $\frac{11}{25}$ **to terminating decimal**

2) ________

3) **Add** $1\frac{4}{8}+1.25$

3) ________

4) **Subtract** $2\frac{1}{4}-1.75$

4) ________

5) **Multiply** $4.8\times\frac{5}{8}$

5) ________

Basic Chapter 5 Lesson 10
Test 5 B

Date

Read directions carefully.

Show all steps for full credits.

Do only 1 operation per 1 step.

Name

Answers must be in indicated place.

Each problem is worth 5 points.

Solve 1 problem 5 minutes.

Divide

6) $0.57 \div \frac{3}{4}$

6) ________

7) $6\frac{7}{8} \div 2.75$

7) ________

Evaluate

8) $3^2 \div (1\frac{1}{2})$

8) ________

9) $(3.5 - 3\frac{1}{3}) \div 1\frac{1}{3}$

9) ________

10) $6.5 \div (3\frac{1}{3} - 1\frac{1}{6})$

10) ______

Basic Answers Chapter 5

Lesson 1

1) 0.76 2) 2.5 3) 6 4) 0.125 5) 3

Lesson 5 Review 1 A

1)	3.5	2)	1.75	3)	2	4)	3	5)	0.2
6)	6	7)	0.5	8)	5.4	9)	9	10)	0.6
11)	7	12)	0.6	13)	2	14)	8	15)	0.8

Lesson 5 Review 1 B

16)	0.5	17)	2	18)	2	19)	3	20)	0.2
21)	0.2	22)	2	23)	2	24)	0.2	25)	0.2
26)	0.4	27)	0.7	28)	2.5	29)	1.8	30)	6

Lesson 6 Review 1 A

1) 0.5	2) 0.2	3) 1.5	4) 1.75	5) 1.5
6) 1	7) 0.75	8) 1.8	9) 1.5	10) 3

Lesson 6 Review 1 B

1)	8	2)	1	3)	2.5	4)	1	5)	1.75
6)	1.2	7)	2	8)	3	9)	23	10)	37

Basic Answers Chapter 5

Lesson 7 Review 1 A

1)	5	2)	1	3)	22	4)	11	5)	17
6)	5	7)	5	8)	9	9)	1	10)	11
11)	6.5	12)	6.5	13)	1.5	14)	13	15)	2.5
16)	20	17)	23	18)	9.25	19)	2.75	20)	2.5

Lesson 7 Review 1 B

1)	9	2)	6	3)	5.5	4)	1.5	5)	3.5
6)	3	7)	5	8)	3	9)	2.5	10)	6
11)	16	12)	1	13)	2.5	14)	1	15)	1.75
16)	2	17)	2	18)	2	19)	6	20)	6
21)	12	22)	0.5	23)	30	24)	4	25)	24
26)	3	27)	18	28)	15	29)	9	30)	16

Lesson 8 Worksheet 3 A for Final Exam

1)	0.25	2)	0.5	3)	0.1	4)	0.25	5)	0.5
6)	0.4	7)	0.8	8)	0.6	9)	0.2	10)	0.1

Lesson 8 Worksheet 3 B for Final Exam

1)	3	2)	5	3)	2	4)	10	5)	6
6)	14	7)	26	8)	18	9)	12	10)	8

Lesson 9 Review 5 A

1)	$\frac{1}{8}$	2)	0.12	3)	1.375	4)	0.8	5)	0.6
6)	0.2	7)	1.2	8)	12	9)	3	10)	2

Lesson 9 Review 5 B

1)	$\frac{3}{8}$	2)	0.28	3)	2.625	4)	0.5	5)	2
6)	0.6	7)	1.5	8)	0.25	9)	0.5	10)	3

Lesson 10 Test 5 A

1)	$\frac{5}{8}$	2)	0.36	3)	1.75	4)	0.5	5)	0.9
6)	0.48	7)	2	8)	4	9)	3	10)	2

Lesson 10 Test 5 B

1)	$\frac{7}{8}$	2)	0.44	3)	2.75	4)	0.5	5)	3
6)	0.76	7)	2.5	8)	6	9)	0.125	10)	3

Basic

Chapter 6

Rational

Numbers

Students!

Students, you are smarter than a calculator!
Students, you are smarter than a computer!
Students, you need to believe in yourself.
Students, be patient and do your best.

Do not be afraid to make mistakes.
Tell yourself positive words.
I am young! I am smart!
I am strong! I can do anything!

Smart students use their brain!
Your brain is always with you!
A calculator is not always with you.
Mathematics is work for the brain.

Be open to new ideas in math!
Work hard and try your best!
Check answers and ask questions!
Practice! Practice! Practice!

Basic Chapter 6 Lessons
Rational Numbers
Lessons 1 – 10
Basic Problems

Lesson 1 Table Perfect Square Numbers. Pretest

Lesson 2 Addition and Subtraction of Rational Numbers

Lesson 3 Multiplication, Division, Powers of Rational Numbers

Lesson 4 Order of Five Operations

Lesson 5 Laws for Rational Numbers

Lesson 6 Tables 7, 8. Mental Math 9 – 12

Lesson 7 Square Roots. Mental Math 13, 14

Lesson 8 Worksheet 4A, Worksheet 4 B for Final Exam

Lesson 9 Review 6 A, Review 6 B

Lesson 10 Test 6 A, Test 6 B

Answers Chapter 6

Basic Chapter 6 Contents
Rational Numbers

Contents have topics and exercises, which students need to solve before tests.1)Work with lesson 1, read rules and examples, 2) Solve the problems from contents, 3) Check answers in the end of chapter 1, 4) Work with the next lessons 2, 3, 4, 5, …, 5) Solve Test 1 A, 6) Grading Scale without minuses

A=90–100% B+=85–89% B=80–84% C+=75–79%
C=70–74% D+=65–69% D=60–64 % F=0 - 59%

7)If you solve 10 problems correctly, you get an A, 8) If have an A, you can move on to next chapter, 9) If your grade is lower than an A, work with the mistakes, then solve Test 1B, 10) When you finish all of the chapters, solve Final Exam in last chapter.

#1 - #20 odd or # 1, 3, 5, 7, 9, 11, 13, 15, 17, 19

Lesson 1 Memorize table perfect square numbers.
Solve the Pretest 1#1–#5
Lesson 2 Addition and Subtraction of Rational Numbers

Lesson 3 Multiplication, Division, Powers of Rational Numbers.
Lesson 4 Order of Five Operations
Solve Review 1 A # 1 – # 15 odd. Solve Review 1 B # 16 – # 30 odd

Lesson 5 Laws for Rational Numbers
Solve Review 1 A # 1 - #30 odd. Solve Review 1 B # 1 – # 30 odd
Lesson 6 Work with Table 7, Table 8.
Solve Mental Math 9, 10, 11, 12 odd

Lesson 7 Square Roots
Solve Review 1 A # 1 – # 20 odd. Solve Review 1 B # 21 – # 40 odd
Review 1 C # 1 - # 20 odd, Review 1 D # 21 - # 40 odd
Lesson 8 Solve Mental Math 13, 14 odd
Solve Worksheet 4 A for Final Exam # 1 – # 10 all
Solve Worksheet 4 B for Final Exam # 1 – # 10 all

Lesson 9
Solve Review 6 A # 1 - # 10 all. Solve Review 6 B # 1 - # 10 all
Lesson 10
Solve Test 6 A # 1 – # 10 all. Solve Test 6 B # 1 – # 10 all
Answer Chapter 6

Basic Chapter 6 Lesson 1

Table 2

Perfect Square Numbers

Take number 20 from the left column add number 5 from the first row to get number 25

Answer is in intersection row of 20 and column of 5

$25^2 = 25 \cdot 25 = 625$

Memorize Table of Perfect Square Numbers first 2 rows and first column.

a^2	0	1	2	3	4	5	6	7	8	9
	0	1	4	9	16	25	36	49	64	81
10	100	121	144	169	196	225	256	289	324	361
20	400	441	484	529	576	625	676	729	784	841
30	900	961	1024	1089	1156	1225	1296	1369	1444	1521
40	1600	1681	1764	1849	1936	2025	2116	2209	2304	2401
50	2500	2601	2704	2809	2916	3025	3136	3249	3364	3481
60	3600	3721	3844	3969	4096	4225	4356	4489	4624	4761
70	4900	5041	5184	5329	5476	5625	5776	5929	6084	6241
80	6400	6561	6724	6889	7056	7225	7396	7569	7744	7921
90	8100	8281	8464	8649	8836	9025	9216	9409	9604	9801

Basic Chapter 6 Lesson 1

Date **Name**

Pretest

Evaluate

1) $-3\frac{1}{3} \times 1\frac{1}{5} - 1\frac{1}{5} \times 1\frac{2}{3}$

1) ________

2) $1\frac{1}{4} \div (-4\frac{5}{8} + 3\frac{1}{16})$

2)________

3) $(-2\frac{1}{5} + 0.6) \div 3.2$

3)________

4) $-\frac{4}{25} \div (-0.6 + \frac{4}{5})$

4)________

5) $-\frac{9}{20} \div (-0.3 + \frac{7}{20})$

5)______

Basic Chapter 6 Lesson 2
Addition and Subtraction Rational Numbers

A line with a chosen origin, unit and direction called a number line. Origin is point O represents the number zero. Numbers standing to the right of zero called positive numbers. Positive numbers written with a plus sign in front of the numbers or written without a sign at all.

**Numbers standing to the left of zero called negative numbers.
Negative numbers must always have a minus sign in front of them.
The number zero is not a positive or a negative number.
A number line is a line where any points on it represent numbers.**

A number that corresponds to a given point on the number line called a coordinate of point. A (-2) O (0) B (4) _____- 2__ 0_____4__

An example of a number line is a thermometer for measuring temperature.

Opposite Numbers

**Two numbers, differing from each other only in signs, called opposite numbers. Opposite numbers found at the same distance from the origin but in the opposite directions. For every number there is only one opposite number. The number zero is opposite into itself.
The opposite number of 2 is – 2 and the opposite number of - 2 is 2.
Minus is the word used for "opposite":**

$-(-2) = 2$ $-(+2) = -2$ $-(+a) = -a$ $-(-a) = a$

Natural numbers, their opposites and zero called integers numbers
… -2, -1, 0, 1, 2 …
Each integer can show on a number line:

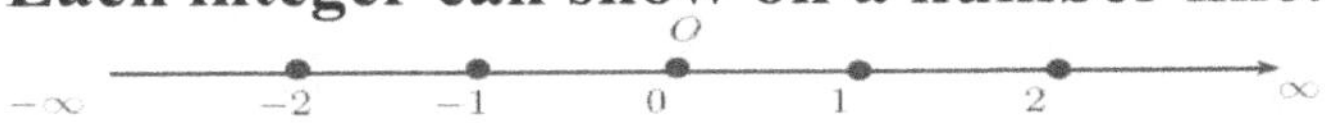

Integer's numbers, decimals, fractions together represent rational numbers. Rational numbers consist of all numbers that can expressed as a terminating decimal or as a repeating decimal. The irrational number can be in the form of an infinite non-periodic decimals. The real numbers include all the rational numbers, such as the integer −6 and the fraction $\frac{1}{2}$,and all the irrational numbers, such as $\sqrt{2} \approx 1.414..$ $\sqrt{3} \approx 1.73...$, $\pi \approx 3.14...$

Basic Chapter 6 Lesson 2

Absolute Value

The distance from the origin to a point that represents a number called the absolute value of the number. The absolute value of a number is always a positive number.

The absolute value of a positive number and zero is itself.
The absolute value of a negative number is opposite its number.
Opposite numbers have equal absolute value:

$|+4| = 4$ $\quad$ $|0| = 0$ $\quad$ $|-4| = 4$

Comparison of Numbers

In a number line, a point with a greater coordinate lies further to the right than a point with a small coordinate. A (-2) O (0) B (4)
Any positive number greater than zero: $4 > 0$ $\quad$ $a > 0$
Any negative number less the zero: $-2 < 0$ $\quad$ $a < 0$
Any positive number greater than any negative number: $4 > -2$
Any negative number less than any positive number: $-2 < 4$
To compare two negative numbers, use this rule: the smallest number has the greater absolute value: $-2 > -4$ **because** $2 < 4$
The order of numbers means putting numbers from smaller to larger number: Example: - 4, - 2, 0, 2, 4

Addition and Subtraction

Two add two numbers with the same signs, put in the result, the sign of the addends and add the absolute value of the numbers:
$(+0.8) + (+1.25) = +2.05$ $\quad$ $(-1.25) + (-0.8) = -1.25 - 0.8 = -2.05$
To add two numbers with different signs, put in the result, the sign of the addend with the greater absolute value and subtract its absolute value: $0.8 - 1.25 = -0.45$ $\quad$ $-1.25 + 0.8 = -0.45$
The sum of two opposite numbers equals zero: $-4 + 4 = 0$ $\quad$ $-A + A = 0$
Subtracting numbers: $A - B = A + (-B)$, $-7-(-4) = x$ $\quad$ $x+(-4)=-7$ $\quad$ $x=-3$
To subtract two numbers, add to the first number the opposite of the second number: $-1.25 - (-0.8) = -1.25 + 0.8 = -0.45$

Basic Chapter 6 Lesson 3
Multiplication and Division Rational Numbers

Multiplication of Rational Numbers

Problem 1

The thermometer shows 0 F, the temperature of air decreases every hour by 2F. What was temperature of the air 3 hours ago? (-2)(-3)=6F

Problem 2

The thermometer shows 0 F, the temperature of air increases every hour by 2F. What was temperature of the air 3 hours ago? (+2)(-3)=-6F

To multiply two numbers with the same signs, the product is positive and multiplies the absolute values of the numbers:

	(+ 6)(+ 2) = + 12	(– 6) (- 2) = + 12
in short	(+) (+) = +	(-) (-) = +

To multiply two numbers with different signs, the product is negative and multiplies the absolute values of the numbers:

	(+ 6)(- 2) = - 12	(- 6) (+ 2) = - 12
In short	(+) (-) = -	(-) (+) = -

If the product of two numbers equals one, these numbers called reciprocals: $(-2)(-\frac{1}{2})=1$ $\quad$ $(a)(\frac{1}{a})=1$

The product of any number and one is the same number:

A · 1 = A $\quad$ **(- 5) (1) = - 5**

The product of any number and zero is zero:

A · 0 = 0 $\quad$ **(- 5)(0) = 0**

The product of zero and any number is zero:

0 · A = 0 $\quad$ **0 (- 6) = 0**

Multiply decimals by 10, 100, 1000 …

(- 0.25) (10) = - 2.5 $\quad$ (0.25) (- 100) = - 25 $\quad$ (- 0.25) (- 1000) = 250

Multiply decimals:

Put in first place number with more digits for easy multiplication.

(- 4) (2.5) = - 10 $\quad$ 8 (- 12.5) = - 100 $\quad$ (1.25) (- 0.8) = - 1

Basic Chapter 6 Lesson 3

Division of Rational Numbers

To divide two numbers (-15) ÷ (-5) = x we need x (-5) = -15.
This means x = + 3 or (-15) ÷ (-5) =+3
To divide two numbers with the same signs, the quotient is positive and divides the absolute values of the numbers:

(+ 6) ÷ (+ 2) = + 3 (– 6) ÷ (- 2) = + 3

in short (+) ÷ (+) = + (-) ÷ (-) = +

To divide two numbers with different signs, the quotient is negative and divides the absolute value of the numbers:

(- 6) ÷ (+ 2) = - 3 (+ 6) ÷ (- 2) = - 3

in short (-) ÷ (+) = - (+) ÷ (-) = -

Any number divided by 1 is the same number:
A ÷ 1 = A (- 5) ÷ 1 = - 5
Any number other than zero divided by itself is 1:
A ÷ A = 1 (- 5) ÷ (- 5) = 1
Zero divided by any number other than zero is zero
0 ÷ A = 0 0 ÷ (- 5) = 0
Division by zero not allowed:
A ÷ 0 no solutions - 5 ÷ 0 none numbers 0 ÷ 0 undefined

Divide by 10, 100, 1000…
(- 5) ÷10 = - 0.5 5 ÷ (- 100) = - 0.05 (- 5) ÷ (- 1000) = 0.005
Divide the numbers:
(- 2) ÷ 5 = - 0.4 0.2 ÷ (- 5) = - 0.04 (- 2) ÷ (- 0.5) = 20÷5 = 4
0.2 ÷ (- 0.5) = (- 2) ÷ 5 = - 0.4 (- 1.2) ÷ 0.8 = (- 12)÷8 = - 1.5

Power of Numbers

Positive numbers in any exponent are always positive numbers:
$3^2 = 3 \cdot 3 = 9$ $2^3 = 2 \cdot 2 \cdot 2 = 8$
Negative numbers in an even exponent are always positive numbers:
$(-2)^4 = (-2)(-2)(-2)(-2) = 16$ $(-0.1)^2 = (-0.1)(-0.1) = 0.01$
Negative numbers in an odd exponent are always negative numbers:
$(-2)^3 = (-2)(-2)(-2) = -8$ $(-0.1)^3 = (-0.1)(-0.1)(-0.1) = -0.001$
An expression without parentheses: $-0.3^2 = -0.3 \cdot 0.3 = -0.09$
$-0.2^3 = -0.2(0.2)(0.2) = -0.008$ $-0.2^4 = -0.2(0.2)(0.2)(0.2) = -0.0016$

Basic Chapter 6 Lesson 4

Order of Five Operations

Operations have three levels: (+, - , ×, ÷, a^n)

Operations of the first level: addition (+) and subtraction (-)

Operations of the second level: multiplication (×) and division (÷)

Operations of the third level: power (a^n), root, and logarithm

Arithmetic expressions contain numbers, sign of operations, and sometimes parentheses.

Order of Operations with the Same Level of Operations

Expressions without parentheses that have operations of only addition or of only multiplication can done in any order:
-80 - 60 - 20 = - 80 – 20 - 60= - 160   -25(- 6) (- 4) = - 25×4×6= - 600

Expression without parentheses that have operations at only the same level of operations can done in order from left to right:
-80 - 40 + 20 = -120 + 20 = -100   -160 ÷ (- 4) × (- 2) =40× (- 2) = -80

Order of Operations of Two Levels of Operations (+, - , ×, ÷)

Expressions without parentheses that have operations at two levels: first work on operations at the second level (×, ÷) from left to right, then work on operations at the first level (+, -) in order from left to right:
(-20) (4) + 20 = -80 + 20 = - 60   -180 ÷ 9 - 40 = -20 - 40 = -60
- 60 – 40(-2) = -60 + 80 = 20   - 80 + 150 ÷ 3 = - 80 + 50 = - 30

Basic Chapter 6 Lesson 4

Order of Operations in Expression with Parentheses

When expressions have parentheses: at first work on operations in the parentheses, then work on operations before or behind the parentheses:

$3 \times (-90 + 70) = 3(-20) = -60$ $\quad$ $(-50 - 40) \times 2 = -90 \times 2 = -180$

$200 \div (-40 + 20) = 200 \div (-20) = -10$ $\quad$ $(-180 + 40) \div 7 = -140 \div 7 = -20$

Grouping Symbols

Expressions have three kinds of grouping symbols: first work on operations in the parentheses, and then work on operations in the brackets, and last, work on operations in the braces.

Parentheses ( ) Brackets [ ] Braces { }

Order of five operations ($+, -, \times, \div, a^n$)

When an expression has five operations: first work on operations at the third level of operations (a^n), then work on operations at second level of operations ($\times, \div$), last, work on operations at first level of operations ($+, -$), in the order, they occur from left to right:

in short

Grouping Symbols	**(1) [2]**	**{3}**
Exponents (Power)	**a^n**	
Multiplication or Division from left to right	**$\times$ or $\div$**	**L → R**
Addition or Subtract from left to right	**+ or –**	**L → R**

Example 1

$5(-3^2 - 2^2 \cdot 2^3) = 5(-9 - 4 \cdot 8) = 5(-9 - 32) = 5(-41) = -205$

Example 2

$20 - \{-200 \div [-60 - (-30 - 20)]\} = 20 - \{-200 \div [-60 - (-50)]\} =$

$20 - \{-200 \div [-60 + 50]\} = 20 - \{-200 \div (-10)\} = 20 - \{20\} = 20 - 20 = 0$

Basic Chapter 6 Lesson 4
Operations with Rational Numbers

If you can, always in answer try to reduce fractions to lowest term, convert improper fraction to mixed number, and convert fraction to terminating decimal. If you can, do operations in fractions or decimals but final answers put in decimals. Easy work: + and – in decimals, × and ÷ in fractions.

$-\frac{3}{4}+0.25=-0.75+0.25=-0.5 \qquad 0.5\div(-\frac{1}{8})=\frac{1}{2}\div(-\frac{1}{8})=\frac{1}{2}\times(-\frac{8}{1})=-4$

Evaluate

1) $(-1\frac{3}{8}-1.75)\div 6\frac{1}{4}=(-1\frac{3}{8}-1\frac{3}{4})\div\frac{25}{4}=-2\frac{9}{8}\times\frac{4}{25}=-3\frac{1}{8}\times\frac{4}{25}=-0.5$

2) $(-1.75+1\frac{5}{8})\div\frac{5}{8}=(-1\frac{3}{4}+1\frac{5}{8})\div\frac{5}{8}=(-1\frac{6}{8}+1\frac{5}{8})\div\frac{5}{8}=-\frac{1}{8}\times\frac{8}{5}=-0.2$

3) $(-1\frac{1}{2}-1.8)\div 2.2=(-1\frac{5}{10}-1\frac{8}{10})\div 2\frac{1}{5}=-3\frac{3}{10}\times\frac{5}{11}=-\frac{33}{10}\times\frac{5}{11}=-1.5$

4) $(-2\frac{3}{5}-1.6)\div 2.4=(-2\frac{6}{10}-1\frac{6}{10})\div 2\frac{2}{5}=-3\frac{12}{10}\div\frac{12}{5}=-4\frac{1}{5}\times\frac{5}{12}=-1.75$

5) $1.65\div(-0.6-\frac{1}{2})=1\frac{13}{20}\div(-\frac{3}{5}-\frac{1}{2})=\frac{33}{20}\div(-\frac{6}{10}-\frac{5}{10})=\frac{33}{20}\div(-\frac{11}{10})=-1.5$

6) $2.2\div(-3\frac{2}{5}+1.2)=2\frac{1}{5}\div(-3\frac{4}{10}+1\frac{2}{10})=2\frac{1}{5}\div(-2\frac{2}{10})=2\frac{1}{5}\div(-2\frac{1}{5})=-1$

7) $1\frac{7}{8}\div(-1\frac{1}{5}-1.3)=1\frac{7}{8}\div(-1.2-1.3)=\frac{15}{8}\div(-2.5)=-\frac{15}{8}\times\frac{2}{5}=-0.75$

8) $2.1\div(-4.5+3\frac{1}{3})=2.1\div(-4\frac{3}{6}+3\frac{2}{6})=\frac{21}{10}\div(-1\frac{1}{6})=\frac{21}{10}(-\frac{6}{7})=-1.8$

9) $5\frac{1}{4}\div(-2\frac{1}{4}-1.25)=5\frac{1}{4}\div(-2.25-1.25)=\frac{21}{4}\div(-3.5)=-\frac{21}{4}\times\frac{2}{7}=-1.5$

10) $3.8\div(-4\frac{2}{3}+3.4)=3\frac{4}{5}\div(-4\frac{10}{15}+3\frac{6}{15})=\frac{19}{5}\div(-1\frac{4}{15})=\frac{19}{5}(-\frac{15}{19})=-3$

Basic Chapter 6 Lesson 4
Review 1 A (# 1 - # 15)
Perform the indicated operations

1) $(-1.08+\frac{2}{25})\div\frac{4}{7}$

2) $(-2.5+2\frac{1}{4})\div 1\frac{1}{4}$

3) $(-3\frac{3}{5}+2.4)\div 1\frac{3}{5}$

4) $(-0.5+\frac{3}{4})\div\frac{5}{8}$

5) $(-3\frac{1}{2}+2.4)\div\frac{11}{15}$

6) $(-4\frac{1}{4}+2.125)\div 1\frac{5}{12}$

7) $(-2\frac{1}{2}-3.75)\div 2\frac{1}{2}$

8) $(-3\frac{4}{5}+1.4)\div 2\frac{2}{5}$

9) $(-3\frac{1}{2}+1.7)\div 0.9$

10) $(-3\frac{1}{5}+1.1)\div 4.2$

11) $(-3\frac{1}{2}+2.3)\div 4.8$

12) $(\frac{1}{5}-0.3)\div 0.2$

13) $(-1\frac{1}{2}-2\frac{3}{4})\div 8.5$

14) $(-2\frac{1}{2}+1.4)\div 2.2$

15) $(-3.5+1\frac{1}{6})\div 1\frac{1}{3}$

16) $(-1\frac{1}{2}-1.25)\div 2.2$

Review 1 B (# 16 - # 30)

17) $2.2\div(-2\frac{1}{2}+1.4)$

18) $\frac{8}{25}\div(-0.8+\frac{2}{5})$

19) $5.25\div(-4\frac{7}{10}+2.6)$

20) $0.15\div(-\frac{7}{10}+0.4)$

21) $1.2\div(-2\frac{1}{3}+2\frac{1}{6})$

22) $3.6\div(-2\frac{7}{8}+1\frac{3}{4})$

23) $0.5\div(-\frac{7}{8}+0.75)$

24) $6.75\div(-3\frac{1}{10}-1.4)$

25) $2.2\div(-\frac{1}{2}+0.6)$

26) $2.1\div(-2\frac{2}{5}-1.8)$

27) $\frac{5}{8}\div(-2.25+2\frac{1}{2})$

28) $\frac{7}{20}\div(-0.3+\frac{1}{20})$

29) $3.3\div(-\frac{2}{3}+\frac{1}{6})$

30) $\frac{3}{20}\div(-2.8+2\frac{1}{2})$

Basic Chapter 6 Lesson 5

Laws for Rational Numbers

First Law of Addition (commutative)

Changing the order of addends does not change the sum.
Add in any order: $A + B = B + A$ $\quad -2 - 3 = -5$
$-3 - 2 = -5 \quad -2 - 3 = -3 - 2$

Second Law of Addition (associative)

Changing the grouping of addends does not change the sum:
$A + (B + C) = (A + B) + C \quad -8 + (-2 - 5) = (-8 - 2) - 5 = -10 - 5 = -15$

First Law of Multiplication (commutative)

Changing the order of factors does not change the product.
Multiply in any order: $A\,B = B\,A \quad (-2)(-3) = 6$
$(-3)(-2) = 6 \quad (-2)(-3) = (-3)(-2)$

Second Law of Multiplication (associative)

Changing the grouping of factors does not change the product:
$A\,(B\,C) = (AB)\,C \quad -5(-4)(-2) = (-5)(-2)(-4) = 10 \times (-4) = -40$

Distributive Laws of Multiplication

To multiply numbers by the sum or difference, multiply that number by each of the numbers in parentheses, then add or subtract the products:
$A\,(B + C) = AB + AC \qquad A\,(B - C) = AB - AC$
$(B+C)A = AB + AC \qquad (B - C)\,A = AB - AC$
$-14 \cdot 5 = (-10 - 4) \cdot 5 = -10 \cdot 5 - 4 \cdot 5 = -50 - 20 = -70$
$-18 \cdot 5 = (-20 + 2) \cdot 5 = -20 \cdot 5 + 2 \cdot 5 = -100 + 10 = -90$

Common Factor (Put A before parentheses)

$AB + AC = A\,(B + C) \qquad AB - AC = A\,(B - C)$
If two products have a common factor, put the common factor before parentheses: $AB + AC$ **A – common factor**
In parentheses, the remaining numbers is as result of dividing the product by the common factor: $AB \div A = B \quad AC \div A = C$
$-16 \cdot 5 - 4 \cdot 5 = 5\,(-16 - 4) = 5\,(-20) = -100$
$-22 \cdot 5 + 2 \cdot 5 = 5\,(-22 + 2) = 5\,(-20) = -100$

Basic Chapter 6 Lesson 5
Common Factor

If first term negative, put (- A) before parentheses and change the sign before each term: -AB – AC = - A (B + C) - AB + AC = - A (B - C)

- 16 · 5 - 4 · 5 = - 5 (16 + 4) = - 5 (20) = - 100

- 22 · 5 + 2 · 5 = - 5 (22 - 2) = - 5 (20) = - 100

Examples: Evaluate

1) $-1\frac{3}{5}(2\frac{1}{8})-1\frac{3}{5}(2\frac{7}{8})=-1\frac{3}{5}(2\frac{1}{8}+2\frac{7}{8})=-\frac{8}{5}(5)=-8$

2) $-\frac{3}{4}(4\frac{2}{3})+\frac{3}{4}(3\frac{1}{3})=-\frac{3}{4}(4\frac{2}{3}-3\frac{1}{3})=-\frac{3}{4}(1\frac{1}{3})=-\frac{3}{4}\times\frac{4}{3}=-1$

3) $\frac{5}{6}(-4\frac{3}{5})+\frac{5}{6}\times1\frac{3}{5}=-\frac{5}{6}(4\frac{3}{5}-1\frac{3}{5})=-\frac{5}{6}(3)=-\frac{5}{2}=-2\frac{1}{2}=-2.5$

4) $-4\frac{1}{2}\times\frac{1}{6}-\frac{1}{6}\times1\frac{1}{2}=-\frac{1}{6}(4\frac{1}{2}+1\frac{1}{2})=-\frac{1}{6}(6)=-1$

5) $-3\frac{3}{4}\times1\frac{1}{6}+2\frac{1}{4}\times1\frac{1}{6}=-1\frac{1}{6}(3\frac{3}{4}-2\frac{1}{4})=-1\frac{1}{6}(1\frac{2}{4})=-\frac{7}{6}\times\frac{3}{2}=-\frac{7}{4}=-1.75$

6) $(-\frac{1}{3}+\frac{1}{5})\div\frac{1}{15}=(-\frac{1}{3}+\frac{1}{5})\times15=-\frac{1}{3}\times\frac{15}{1}+\frac{1}{5}\times\frac{15}{1}=-5+3=-2$

7) $(-1\frac{3}{4}+1\frac{1}{2})\div\frac{1}{8}=(-\frac{7}{4}+\frac{3}{2})\times8=-\frac{7}{4}\times\frac{8}{1}+\frac{3}{2}\times\frac{8}{1}=-14+12=-2$

8) $(-1\frac{1}{2}+1\frac{1}{6})\div\frac{1}{6}=(-\frac{3}{2}+\frac{7}{6})\times6=-\frac{3}{2}\times\frac{6}{1}+\frac{7}{6}\times\frac{6}{1}=-9+7=-2$

9) $(-2\frac{1}{2}-1\frac{1}{3})\div\frac{1}{6}=(-\frac{5}{2}-\frac{4}{3})\times6=-\frac{5}{2}\times\frac{6}{1}-\frac{4}{3}\times\frac{6}{1}=-15-8=-23$

10) $(-1\frac{1}{3}-1\frac{3}{4})\div\frac{1}{12}=(-\frac{4}{3}-\frac{7}{4})\times12=-\frac{4}{3}\times\frac{12}{1}-\frac{7}{4}\times\frac{12}{1}=-16-21=-37$

Basic Chapter 6 Lesson 5
Review 1 A
Evaluate by using laws

1) $-\frac{7}{18}-\frac{1}{3}-\frac{5}{18}$

2) $-\frac{3}{14}-\frac{1}{2}-\frac{4}{14}$

3) $-\frac{3}{16}-\frac{3}{4}-\frac{1}{16}$

4) $-\frac{1}{15}-\frac{4}{5}-\frac{2}{15}$

5) $(-\frac{1}{4}-\frac{9}{16})-\frac{7}{16}$

6) $-\frac{7}{8}-\frac{3}{4}-\frac{1}{8}$

7) $-\frac{3}{7}-\frac{5}{9}-\frac{4}{7}-\frac{4}{9}$

8) $(-\frac{1}{7}-\frac{11}{18})+(-\frac{6}{7}-\frac{7}{18})$

9) $-\frac{5}{8}+(-\frac{7}{24}-\frac{3}{8})-\frac{5}{24}$

10) $(-\frac{8}{15}-\frac{1}{5})+(-\frac{4}{15}-\frac{4}{5})$

11) $9(-1+\frac{4}{9})$

12) $(-\frac{2}{3}+\frac{3}{5})\times 15$

13) $9(-3+\frac{5}{9})$

14) $(-1-\frac{5}{6})\times 6$

15) $7(-2-\frac{3}{7})$

16) $(-1+\frac{1}{6})\times 6$

17) $6(-1\frac{1}{2}+\frac{2}{3})$

18) $(-\frac{5}{7}-\frac{4}{7})\times 7$

19) $7(-\frac{5}{7}+\frac{4}{7})$

20) $(-3-\frac{2}{3})\times 3$

21) $3(-2-\frac{1}{6})$

22) $(-2\frac{1}{4}+\frac{5}{8})\times 4$

23) $3(-1\frac{5}{6}+1\frac{1}{3})$

24) $(-3\frac{1}{2}+\frac{1}{4})\times 4$

25) $3(-1\frac{2}{3}+\frac{5}{6})$

26) $(-1\frac{2}{7}+\frac{1}{3})\times 21$

27) $6(-2\frac{1}{2}-1\frac{1}{3})$

28) $(-4-\frac{5}{8})\times 2$

29) $3(-1\frac{1}{3}+\frac{5}{12})$

30) $(-1\frac{1}{4}+\frac{5}{8})\times 4$

Basic Chapter 6 Lesson 5
Review 1 B
Evaluate by using laws

1) $-8\times1\frac{3}{4}+8\times\frac{5}{8}$

2) $-3\frac{2}{9}\times\frac{3}{5}-\frac{3}{5}\times6\frac{7}{9}$

3) $-7\times1\frac{1}{7}+7\times\frac{5}{14}$

4) $-4\times2\frac{1}{4}+4\times1\frac{7}{8}$

5) $-2\frac{6}{7}\times\frac{7}{10}-2\frac{1}{7}\times\frac{7}{10}$

6) $-5\frac{5}{6}\times\frac{3}{4}+1\frac{5}{6}\times\frac{3}{4}$

7) $-4\frac{4}{5}\times\frac{5}{8}-\frac{5}{8}\times3\frac{1}{5}$

8) $-3\frac{2}{3}\times1\frac{1}{2}+1\frac{2}{3}\times1\frac{1}{2}$

9) $-\frac{5}{6}\times5\frac{1}{5}+\frac{5}{6}\times2\frac{1}{5}$

10) $-3\frac{1}{3}\times1\frac{1}{5}-1\frac{1}{5}\times1\frac{2}{3}$

11) $-1\frac{3}{5}\times2\frac{1}{2}-1\frac{3}{5}\times7\frac{1}{2}$

12) $-4\frac{2}{3}\times\frac{3}{4}+3\frac{1}{3}\times\frac{3}{4}$

13) $-4\frac{3}{5}\times\frac{5}{6}+\frac{5}{6}\times1\frac{3}{5}$

14) $-4\frac{1}{2}\times\frac{1}{6}-\frac{1}{6}\times1\frac{1}{2}$

15) $-3\frac{3}{4}\times1\frac{1}{6}+2\frac{1}{4}\times1\frac{1}{6}$

16) $-\frac{1}{3}\times15+\frac{1}{5}\times15$

17) $-1\frac{3}{4}\times8+1\frac{1}{2}\times8$

18) $-1\frac{1}{2}\times6+1\frac{1}{6}\times6$

19) $-2\frac{1}{2}\times6+1\frac{1}{2}\times6$

20) $-1\frac{3}{4}\times12+1\frac{1}{4}\times12$

21) $(-1\frac{3}{5}-1\frac{2}{5})\div\frac{1}{5}$

22) $(-1\frac{3}{5}+1\frac{2}{5})\div\frac{2}{5}$

23) $(-4\frac{3}{5}-1\frac{2}{5})\div\frac{1}{5}$

24) $(-3\frac{2}{3}+1\frac{2}{3})\div\frac{2}{3}$

25) $(-1\frac{3}{4}+1\frac{1}{2})\div\frac{1}{8}$

26) $(-2\frac{1}{2}-1\frac{1}{2})\div\frac{1}{6}$

27) $(-1\frac{4}{5}-1\frac{1}{5})\div\frac{1}{5}$

28) $(-4\frac{3}{5}+1\frac{3}{5})\div\frac{1}{5}$

29) $(-1\frac{1}{3}-1\frac{2}{3})\div\frac{1}{3}$

30) $(-4\frac{1}{5}+2\frac{1}{5})\div\frac{1}{5}$

Basic Chapter 6 Lesson 6

Table 7 Fractions and Decimals

Converting Fractions to Terminating Decimals

$10=2\times5$ $\qquad$ $100=4\times25$ $\qquad$ $1000 = 8\times125$

$\frac{1}{10}=0.1$ $\qquad$ $\frac{1}{100}=0.01$ $\qquad$ $\frac{1}{1000}=0.001$

$\frac{1}{5}=\frac{1\times2}{5\times2}=\frac{2}{10}=0.2$ $\qquad$ $\frac{15}{20}=\frac{3}{4}=\frac{3\times25}{4\times25}=\frac{75}{100}=0.75$ $\qquad$ $\frac{1}{2}=\frac{1\times5}{2\times5}=\frac{5}{10}=0.5$

$\frac{1}{4}=\frac{1\times25}{4\times25}=\frac{25}{100}=0.25$ $\qquad$ $2\frac{1}{8}=2\frac{1\times125}{8\times125}=2\frac{125}{1000}=2.125$

Memorize

$\frac{1}{2}=0.5$ $\qquad$ $\frac{1}{4}=0.25$ $\qquad$ $\frac{1}{5}=0.2$ $\qquad$ $\frac{1}{8}=0.125$ $\qquad$ $\frac{3}{4}=0.75$

Converting Terminating Decimals to Mixed Numbers

a) $1.125=1\frac{125}{1000}=1\frac{125\div5}{1000\div5}=1\frac{25}{200}=1\frac{25\div5}{200\div5}=1\frac{5}{40}=1\frac{1}{8}$

Combine Operations with Fractions and Decimals

b) $\frac{2}{3}\times0.36=0.36\times\frac{2}{3}=0.36\div3\times2=0.12\times2=0.24$

c) $3.6\div4\frac{1}{2}=3.6\div4.5=36\div45=\frac{36}{45}=\frac{4}{5}=0.8$

d) $(0.66+\frac{31}{50})\div(-\frac{2}{10})=(\frac{66}{100}+\frac{31}{50})\div(-\frac{1}{5})=(\frac{33}{50}+\frac{31}{50})(-\frac{5}{1})=\frac{64}{50}(-\frac{5}{1})=-6.4$

e) $-0.25\div(\frac{39}{50}-\frac{37}{50})=-\frac{1}{4}\div\frac{2}{50}=-\frac{1}{4}\times\frac{25}{1}=-\frac{25}{4}=-6\frac{1}{4}=-6.25$

f) $(-\frac{43}{50}+\frac{41}{50})\div0.16=(-\frac{2}{50})\div\frac{16}{100}=-\frac{1}{25}\div\frac{4}{25}=-\frac{1}{25}\times\frac{25}{4}=-\frac{1}{4}=-0.25$

g) $9.6\div(-\frac{49}{50}-\frac{47}{50})=9.6\div(-\frac{96}{50})=9\frac{6}{10}\times(-\frac{50}{96})=9\frac{3}{5}(-\frac{25}{48})=-\frac{48}{5}\times\frac{25}{48}=-5$

Easy: + and - in decimals, $\times$ and $\div$ in fractions

$\frac{1}{2}-0.25=0.5-0.25=0.25$ $\qquad$ $0.25\div\frac{1}{8}=\frac{1}{4}\div\frac{1}{8}=\frac{1}{4}\times\frac{8}{1}=2$

Basic Chapter 6 Lesson 6
Table 8 Rational Numbers

Addition and Subtraction of Rational Numbers

$(-1.25) + (-0.8) = -1.25 - 0.8 = -2.05$ $\quad -1.25 + 0.8 = -0.45$

$-1.25 - (-0.8) = -1.25 + 0.8 = -0.45$ $\quad 0.8 - 1.25 = -0.45$

Multiplication of Rational Numbers

$(+6)(+2) = +12$ $\quad (-6)(-2) = +12$ $\quad (+)(+) = +$ $\quad (-)(-) = +$

$(+6)(-2) = -12$ $\quad (-6)(+2) = -12$ $\quad (+)(-) = -$ $\quad (-)(+) = -$

$(-0.25)(10) = -2.5$ $\quad (0.25)(-100) = -25$ $\quad (-0.25)(-1000) = 250$

$(-4)(2.5) = -10$ $\quad 8(-12.5) = -100$ $\quad (1.25)(-0.8) = -1$

Division of Rational Numbers

$(+6) \div (+2) = +3$ $\quad (-6) \div (-2) = +3$ $\quad (+) \div (+) = +$ $\quad (-) \div (-) = +$

$(-6) \div (+2) = -3$ $\quad (+6) \div (-2) = -3$ $\quad (+) \div (-) = -$ $\quad (-) \div (+) = -$

$(-5) \div 10 = -0.5$ $\quad 5 \div (-100) = -0.05$ $\quad (-5) \div (-1000) = 0.005$

$(-2) \div 5 = -0.4$ $\quad 0.2 \div (-5) = -0.04$ $\quad (-2) \div (-0.5) = 20 \div 5 = 4$

$0.2 \div (-0.5) = (-2) \div 5 = -0.4$ $\quad (-1.2) \div 0.8 = (-12) \div 8 = -1.5$

Powers of Rational Numbers

$2^3 = 2 \times 2 \times 2 = 8$ $\quad (-3)^2 = (-3)(-3) = 9$

$(-2)^3 = -2(-2)(-2) = -8$ $\quad (-2)^4 = (-2)(-2)(-2)(-2) = 16$

$-2^3 = -2 \times 2 \times 2 = -8$ $\quad -3^2 = -3 \times 3 = -9$

$(-0.1)^2 = (-0.1)(-0.1) = 0.01$ $\quad -0.3^2 = -0.09$

Order of Operations

$-25(-6)(-4) = -25 \times 4 \times 6 = -600$ $\quad 20(-4) + 20 = -80 + 20 = -60$

$-60 - 40(-2) = -60 + 80 = 20$ $\quad -16 \div (-4)(-20) = 4(-20) = -80$

$(-50 - 40) \times 2 = -90 \times 2 = -180$ $\quad 3(-90 + 70) = 3(-20) = -60$

$-180 \div 9 - 40 = -20 - 40 = -60$ $\quad -80 + 150 \div 3 = -80 + 50 = -30$

$(-180 + 40) \div 7 = -140 \div 7 = -20$ $\quad 200 \div (-40 + 20) = 200 \div (-20) = -10$

Examples

a) $5(-3^2 - 2^2 \times 2^3) = 5(-9 - 4 \times 8) = 5(-9 - 32) = 5(-41) = -205$

b) $\{-200 \div [-60 - (-30 - 20)]\} = \{-200 \div [-60 + 50]\} = -200 \div (-10) = 20$

Basic Chapter 6 Lesson 6

Mental Math 9

Law for Decimals

1) $8\times7.5+8\times2.5$ 2) $3.7\times4+4\times1.3$

3) $2.5\times4-1.5\times4$ 4) $5.4\times5-3.4\times5$

5) $7\times3.7+6.3\times7$ 6) $4.9\times2+2\times5.1$

7) $7.5\times6-3.5\times6$ 8) $3.9\times8-8\times1.9$

9) $3\times2.7+3\times2.3$ 10) $2.4\times3+1.6\times3$

11) $9.3 \times 2 - 2 \times 5.3$

12) $5.8 \times 3 + 4.2 \times 3$

13) $4.3 \times 4 + 4 \times 1.7$

14) $9.6 \times 2 - 2.6 \times 2$

15) $5 \times 2.8 - 5 \times 1.8$

16) $3.8 \times 4 - 1.8 \times 4$

17) $2.9 \times 3 + 3.1 \times 3$

18) $0.3 \times 6 + 0.7 \times 6$

19) $6.7 \times 4 - 4 \times 4.7$

20) $8 \times 5.7 - 8 \times 4.7$

Basic Chapter 6 Lesson 6

Mental Math 10

Write the Fractions as Decimals

1) $\frac{3}{10}$

2) $\frac{9}{100}$

3) $\frac{7}{1000}$

4) $\frac{1}{2}$

5) $\frac{12}{24}$

6) $\frac{1}{5}$

7) $\frac{2}{5}$

8) $\frac{3}{5}$

9) $\frac{4}{5}$

10) $\frac{12}{15}$

11) $\frac{1}{4}$

12) $\frac{2}{4}$

13) $\frac{3}{4}$

14) $\frac{3}{12}$

15) $\frac{15}{20}$

16) $\frac{7}{25}$

17) $\frac{7}{50}$

18) $\frac{11}{50}$

19) $\frac{3}{20}$

20) $\frac{9}{20}$

Basic Chapter 6 Lesson 6

Mental Math 11
Negative Fractions

1) $12(-\frac{5}{6})$

2) $-\frac{5}{9}\times 18$

3) $24(-\frac{3}{4})$

4) $-\frac{5}{8}\times 32$

5) $\frac{1}{2}\div(-\frac{1}{2})$

6) $\frac{1}{3}\div(-\frac{1}{6})$

7) $-\frac{1}{4}\div\frac{1}{12}$

8) $-2\div\frac{1}{2}$

9) $(-\frac{1}{2})^2$

10) $(-\frac{1}{5})^2$

11) $(-\frac{2}{5})^2$

12) $-(-\frac{3}{5})^2$

13) $-\frac{1}{4}-\frac{3}{4}$

14) $-\frac{1}{2}-\frac{3}{2}$

15) $-\frac{7}{8}+\frac{5}{8}$

16) $-\frac{7}{8}+\frac{3}{8}$

17) $-\frac{1}{2}-\frac{1}{4}$

18) $-\frac{3}{4}-\frac{1}{2}$

19) $-\frac{1}{2}+\frac{1}{4}$

20) $-\frac{3}{4}+\frac{1}{2}$

Mental Math 12
Negative Mixed Numbers

1) $-1-\frac{1}{2}$

2) $-\frac{1}{4}-2$

3) $-1+\frac{1}{2}$

4) $-2\frac{1}{2}+1$

5) $-2\frac{3}{8}-1\frac{1}{8}$

6) $-3\frac{3}{4}-2\frac{1}{4}$

7) $-4\frac{3}{4}+2\frac{1}{4}$

8) $-3\frac{5}{8}+2\frac{1}{8}$

9) $-3\frac{1}{2}-2\frac{1}{2}$

10) $-4\frac{1}{2}-2\frac{1}{2}$

11) $-2\frac{3}{4}+1\frac{1}{4}$

12) $-4\frac{5}{6}+1\frac{2}{6}$

13) $-2\frac{3}{8}+1\frac{1}{8}$

14) $-3\frac{3}{8}+2\frac{1}{8}$

15) $-4\frac{7}{8}+3\frac{3}{8}$

16) $-5\frac{5}{8}+4\frac{1}{8}$

17) $-2+1\frac{1}{2}$

18) $-2+1\frac{3}{4}$

19) $-3+1\frac{1}{4}$

20) $-3+1\frac{1}{5}$

Basic Chapter 6 Lesson 7

Square Roots

The first operation with numbers is addition.
Addition combines numbers together.
Addend + Addend = Sum $3 + 2 = 5$

The second operation with numbers is subtraction.
Subtraction means to take something away.
Minuend – Subtrahend = Difference $5 - 2 = 3$

The third operation with numbers is multiplication.
Multiplication is the addition of the same numbers.
Factor × Factor = Product $3 \times 2 = 3 + 3 = 6$

The fourth operation with numbers is division.
Division means to separate number by some of equal groups.
Dividend ÷ Divisor = Quotient $6 \div 2 = 3$

The fifth operation with numbers is power.
Power is the multiplication of the same numbers.
Base ^{Exponent} = Power $3^2 = 3 \cdot 3 = 9$

The sixth operation with numbers is root.
The reverse process of squaring is finding a square root.
A square root of number is a number whose square is a number.

$4^2 = 16$ $\sqrt[2]{16} = 4$ **or** $\sqrt{16} = 4$

Perfect square numbers are 4, 9, 16, 25, 36, 49, 64, 81, 100.

$0^2 = 0$ $1^2 = 1$ $2^2 = 4$ $3^2 = 9$ $4^2 = 16$ $5^2 = 25$

$6^2 = 36$ $7^2 = 49$ $8^2 = 64$ $9^2 = 81$ $10^2 = 100$

A square root of perfect square numbers is whole numbers.

$\sqrt{0} = 0$ $\sqrt{1} = 1$ $\sqrt{4} = 2$ $\sqrt{9} = 3$ $\sqrt{16} = 4$ $\sqrt{25} = 5$

$\sqrt{36} = 6$ $\sqrt{49} = 7$ $\sqrt{64} = 8$ $\sqrt{81} = 9$ $\sqrt{100} = 10$

Basic Chapter 6 Lesson 7

Rule for Finding Square Roots from Perfect Square Numbers

The square root of three and four-digit perfect square numbers is always a two-digit number. Divide the number under the root by groups of two digits, from right to left. The last group can be one or two digits.

To find the first digit, find the number whose square is smaller than or equal to the last group.

To find the second digit, use the next rule:

Products $1 \times 1 = 1$ **or** $9 \times 9 = 81$ **have last digit 1.**

Products $2 \times 2 = 4$ **or** $8 \times 8 = 64$ **have last digit 4.**

Products $3 \times 3 = 9$ **or** $7 \times 7 = 49$ **have last digit 9.**

Products $4 \times 4 = 16$ **or** $6 \times 6 = 36$ **have last digit 6.**

Product $5 \times 5 = 25$ **has a last digit 5.**

Last digit of given number	1	4	5	6	9
Last digit of answer	1 or 9	2 or 8	5	4 or 6	3 or 7

Logarithm

The seventh operation with numbers is logarithms

Logarithm = Exponent **If** $b^x = b^y$ **than** $x = y$

$5^x = 125 \quad 5^x = 5^3 \quad x = 3$ $\quad 10^x = 100 \quad 10^x = 10^2 \quad x = 2$

Find the exponents

a)	$2^x = 2$	b)	$3^x = 3$	c)	$4^x = 4$	d)	$5^x = 5$
e)	$6^x = 6$	f)	$2^x = 4$	g)	$2^x = 8$	h)	$2^x = 16$
i)	$2^x = 32$	j)	$2^x = 64$	k)	$3^x = 9$	l)	$3^x = 27$
m)	$3^x = 81$	n)	$4^x = 16$	o)	$4^x = 64$	p)	$5^x = 25$
q)	$6^x = 36$	r)	$7^x = 49$	s)	$8^x = 64$	t)	$9^x = 81$

Answers

a)	1	b)	1	c)	1	d)	1
e)	1	f)	2	g)	3	h)	4
i)	5	j)	6	k)	2	l)	3
m)	4	n)	2	o)	3	p)	2
q)	2	r)	2	s)	2	t)	2

Basic Chapter 6 Lesson 7
Find the Square Roots

Examples 1 Find the square root of: $\sqrt{8'41}$
First digit is 2 because: $2^2 = 4$ $3^2 = 9$ $4 < 8$ $9 > 8$
Second digit can be 1 or 9 since the number 841 has a last digit 1, but $21^2 = 441$ and $29^2 = 841$ or $\sqrt{8'41} = 29$

Review 1 A

1) $\sqrt{121}$	2) $\sqrt{289}$	3) $\sqrt{169}$	4) $\sqrt{361}$
5) $\sqrt{256}$	6) $\sqrt{144}$	7) $\sqrt{324}$	8) $\sqrt{196}$
9) $\sqrt{441}$	10) $\sqrt{729}$	11) $\sqrt{529}$	12) $\sqrt{841}$
13) $\sqrt{676}$	14) $\sqrt{484}$	15) $\sqrt{784}$	16) $\sqrt{576}$
17) $\sqrt{961}$	18) $\sqrt{1369}$	19) $\sqrt{1089}$	20) $\sqrt{1521}$

Example 2 Find the square root of: $\sqrt{73'96} = 86$
First digit is 8 because $8^2 = 64$ and $64 < 73$
Second digit can be 4 or 6, check $84^2 = 7056$ and $86^2=7396$

Review 1 B

21) $\sqrt{1296}$	22) $\sqrt{1024}$	23) $\sqrt{1444}$	24) $\sqrt{1156}$
25) $\sqrt{1681}$	26) $\sqrt{2209}$	27) $\sqrt{1849}$	28) $\sqrt{2401}$
29) $\sqrt{2116}$	30) $\sqrt{1764}$	31) $\sqrt{2304}$	32) $\sqrt{1936}$
33) $\sqrt{2601}$	34) $\sqrt{3249}$	35) $\sqrt{2809}$	36) $\sqrt{3481}$
37) $\sqrt{3136}$	38) $\sqrt{2704}$	39) $\sqrt{3364}$	40) $\sqrt{2916}$

Basic Chapter 6 Lesson 7
Find the Square Roots

Review 1 C

Example 3 Find $\sqrt{14'44} = 38$ First digit is 3 because $3^2 = 9$ and $9 < 14$
Second digit can be 2 or 8, check $32^2 = 1024$ and $38^2 = 1444$

1) $\sqrt{3721}$	2) $\sqrt{4489}$	3) $\sqrt{3969}$	4) $\sqrt{4761}$
5) $\sqrt{4356}$	6) $\sqrt{3844}$	7) $\sqrt{4624}$	8) $\sqrt{4096}$
9) $\sqrt{5041}$	10) $\sqrt{5929}$	11) $\sqrt{5329}$	12) $\sqrt{6241}$
13) $\sqrt{5776}$	14) $\sqrt{5184}$	15) $\sqrt{6084}$	16) $\sqrt{5476}$
17) $\sqrt{6561}$	18) $\sqrt{7569}$	19) $\sqrt{6889}$	20) $\sqrt{7921}$

Review 1 D

Example 4 Find $\sqrt{22'09} = 47$ First digit is 4 because $4^2 = 16$ and $16 < 22$
Second digit can be 3 or 7, check $43^2 = 1849$ and $47^2 = 2209$

21) $\sqrt{7396}$	22) $\sqrt{6724}$	23) $\sqrt{7744}$	24) $\sqrt{7056}$
25) $\sqrt{8281}$	26) $\sqrt{9409}$	27) $\sqrt{8649}$	28) $\sqrt{9801}$
29) $\sqrt{9216}$	30) $\sqrt{8464}$	31) $\sqrt{9604}$	32) $\sqrt{8836}$
33) $\sqrt{625}$	34) $\sqrt{5625}$	35) $\sqrt{2025}$	36) $\sqrt{9025}$
37) $\sqrt{4225}$	38) $\sqrt{1225}$	39) $\sqrt{7225}$	40) $\sqrt{3025}$

Basic Chapter 6 Lesson 8

Mental Math 13

Exponents

1) 10^2
2) 11^2
3) 12^2
4) 13^2
5) 14^2
6) 15^2
7) 16^2
8) 17^2
9) 18^2
10) 19^2
11) 20^2
12) 21^2
13) 22^2
14) 23^2
15) 24^2
16) 25^2
17) 26^2
18) 27^2
19) 28^2
20) 29^2
21) 30^2
22) 31^2
23) 40^2
24) 50^2
25) 60^2
26) 70^2
27) 80^2
28) 90^2

Mental Math 14
Square Roots

1)	$\sqrt{64}$	16)	$\sqrt{529}$
2)	$\sqrt{81}$	17)	$\sqrt{576}$
3)	$\sqrt{100}$	18)	$\sqrt{625}$
4)	$\sqrt{121}$	19)	$\sqrt{676}$
5)	$\sqrt{144}$	20)	$\sqrt{729}$
6)	$\sqrt{169}$	21)	$\sqrt{784}$
7)	$\sqrt{196}$	22)	$\sqrt{841}$
8)	$\sqrt{225}$	23)	$\sqrt{900}$
9)	$\sqrt{256}$	24)	$\sqrt{961}$
10)	$\sqrt{289}$	25)	$\sqrt{1600}$
11)	$\sqrt{324}$	26)	$\sqrt{2500}$
12)	$\sqrt{361}$	27)	$\sqrt{3600}$
13)	$\sqrt{400}$	28)	$\sqrt{4900}$
14)	$\sqrt{441}$	29)	$\sqrt{6400}$
15)	$\sqrt{484}$	30)	$\sqrt{8100}$

Basic Chapter 6 Lesson 8

Worksheet 4 A Rational Numbers for Final Exam

Perform the indicated operations

1) $(-0.25-\frac{1}{8})\div 1.5$

2) $\left(\frac{1}{16}+\frac{1}{32}\right)\div\left(-\frac{3}{16}\right)$

3) $\left(-0.1-\frac{1}{25}\right)\div 1.4$

4) $\left(\frac{1}{64}+\frac{1}{32}\right)\div\left(-\frac{3}{16}\right)$

5) $\left(-0.05-\frac{1}{40}\right)\div\frac{3}{20}$

6) $\left(0.02+\frac{1}{25}\right)\div\left(-\frac{3}{20}\right)$

7) $\left(-\frac{1}{80}-\frac{1}{40}\right)\div\frac{3}{64}$

8) $\left(0.01+\frac{1}{50}\right)\div\left(-\frac{1}{20}\right)$

9) $\left(-0.01-\frac{1}{25}\right)\div 0.25$

10) $\left(0.01+\frac{1}{20}\right)\div\left(-\frac{12}{20}\right)$

Basic Chapter 6 Lesson 8

Worksheet 4 B Rational Numbers for Final Exam

Perform the indicated operations

1) $\frac{3}{8} \div \left(-0.25 + \frac{1}{8} \right)$

2) $-\frac{5}{32} \div \left(\frac{1}{16} - \frac{1}{32} \right)$

3) $\frac{6}{50} \div \left(-0.1 + \frac{1}{25} \right)$

4) $-\frac{5}{32} \div \left(\frac{1}{32} - \frac{1}{64} \right)$

5) $\frac{3}{20} \div \left(-0.05 + \frac{1}{40} \right)$

6) $-\frac{7}{25} \div \left(\frac{1}{25} - 0.02 \right)$

7) $\frac{13}{40} \div \left(-\frac{1}{40} + \frac{1}{80} \right)$

8) $-\frac{9}{50} \div \left(\frac{1}{50} - 0.01 \right)$

9) $\frac{9}{25} \div \left(-\frac{1}{25} + 0.01 \right)$

10) $-\frac{8}{25} \div \left(\frac{1}{20} - 0.01 \right)$

Basic Chapter 6 Lesson 9
Reviews for Chapter 6
Review 6 A

Date **Name**
Read directions carefully. **Answers must be in indicated place.**
Show all steps for full credits. **Each problem is worth 5 points.**
Do only 1 operation per 1 step. **Solve 1 problem for 5 minutes.**

Evaluate

1) $\frac{1}{5}+(-2)$

1) ________

2) $-(+1\frac{3}{10})-(-2\frac{7}{10})-(-\frac{1}{5})$

2) ________

3) $(-1\frac{2}{5})(0.35)$

3) ________

4) $(-4.5)\div(+1\frac{1}{2})$

4) ________

5) $(-\frac{2}{5})^3+(-\frac{3}{5})^3$

5) ________

Basic Chapter 6 Lesson 9
Review 6 A

Date | **Name**

Read directions carefully. **Answers must be in indicated place.**
Show all steps for full credits. **Each problem is worth 5 points.**
Do only 1 operation per 1 step. **Solve 1 problem for 5 minutes.**

Evaluate

6) $-4\frac{4}{5}\times\frac{5}{8}-\frac{5}{8}\times3\frac{1}{5}$

6) ________

7) $3\frac{1}{3}\div(-4\frac{1}{6}+3\frac{1}{2})$

7) ________

8) $(\frac{1}{5}-0.3)\div0.2$

8) ________

9) $0.15\div(-\frac{7}{10}+0.4)$

9) ________

10) $\frac{3}{40}\div(-2.8+2\frac{1}{2})$

10) ______

Basic Chapter 6 Lesson 9
Review 6 B

Date **Name**

Read directions carefully. **Answers must be in indicated place.**
Show all steps for full credits. **Each problem is worth 5 points.**
Do only 1 operation per 1 step. **Solve 1 problem for 5 minutes.**

Evaluate

1) $-4-(-3\frac{2}{5})$

1)________

2) $-\frac{1}{5}-(-\frac{3}{10})-(-2\frac{1}{10})$

2)________

3) $-1\frac{4}{5}(0.45)$

3)________

4) $1.2\div(-1\frac{1}{2})$

4) ________

5) $(\frac{3}{5})^3-(-\frac{2}{5})^3$

5) ________

Basic Chapter 6 Lesson 9
Review 6 B

Date **Name**

Read directions carefully. **Answers must be in indicated place.**

Show all steps for full credits. **Each problem is worth 5 points.**

Do only 1 operation per 1 step. **Solve 1 problem for 5 minutes.**

Evaluate

6) $-3\frac{2}{3} \times 1\frac{1}{2} + 1\frac{2}{3} \times 1\frac{1}{2}$

6) _______

7) $1\frac{11}{16} \div (-4\frac{1}{4} + 3\frac{1}{8})$

7)_______

8) $(-3\frac{1}{5} + 1.1) \div 4.2$

8)_______

9) $\frac{8}{25} \div (-0.8 + \frac{2}{5})$

9) _______

10) $\frac{9}{20} \div (-0.3 + \frac{1}{20})$

10) ______

Basic Chapter 6 Lesson 10
Tests for Chapter 6
Test 6 A

Date **Name**

Read directions carefully. **Answers must be in indicated place.**

Show all steps for full credits. **Each problem is worth 5 points.**

Do only 1 operation per 1 step. **Solve 1 problem for 5 minutes.**

Evaluate

1) $\frac{2}{5}+(-3)$

1) ________

2) $-(+2\frac{7}{10})-(-1\frac{3}{10})-(-\frac{3}{5})$

2)________

3) $(-1\frac{3}{5})(0.75)$

3)________

4) $(4.5)\div(-1\frac{1}{4})$

4) ________

5) $-(-\frac{1}{2})^2-(-\frac{1}{5})^2$

5)________

Basic Chapter 6 Lesson 10
Test 6 A

Date **Name**

Read directions carefully. **Answers must be in indicated place.**

Show all steps for full credits. **Each problem is worth 5 points.**

Do only 1 operation per 1 step. **Solve 1 problem for 5 minutes.**

Evaluate

6) $-\frac{5}{6}\times 5\frac{1}{5}+\frac{5}{6}\times 2\frac{1}{5}$

6)_______

7) $4\frac{1}{2}\div(-4\frac{5}{6}+3\frac{1}{3})$

7)_______

8) $(\frac{1}{5}-0.4)\div 0.5$

8)_______

9) $0.25\div(-\frac{8}{10}+0.3)$

9) _______

10) $5\frac{1}{5}\div(-2.6+2\frac{1}{2})$

10) ______

Basic Chapter 6 Lesson 10
Test 6 B

Date **Name**

Read directions carefully. **Answers must be in indicated place.**

Show all steps for full credits. **Each problem is worth 5 points.**

Do only 1 operation per 1 step. **Solve 1 problem for 5 minutes.**

Evaluate

1) $-5-(-4\frac{2}{5})$

1)________

2) $-\frac{2}{5}-(-\frac{7}{10})-(-2\frac{3}{10})$

2)________

3) $-1\frac{1}{5}(0.25)$

3) ________

4) $1.5\div(-1\frac{1}{4})$

4) ________

5) $(-\frac{1}{5})^2-(-\frac{1}{2})^2$

5)________

Basic Chapter 6 Lesson 10

Test 6 B

Date **Name**

Read directions carefully. **Answers must be in indicated place.**

Show all steps for full credits. **Each problem is worth 5 points.**

Do only 1 operation per 1 step. **Solve 1 problem for 5 minutes.**

Evaluate

6) $-3\frac{1}{3} \times 1\frac{1}{5} - 1\frac{1}{5} \times 1\frac{2}{3}$

6) ________

7) $1\frac{1}{4} \div (-4\frac{5}{8} + 3\frac{1}{16})$

7)________

8) $(-2\frac{1}{5} + 0.6) \div 3.2$

8)________

9) $-\frac{4}{25} \div (-0.6 + \frac{4}{5})$

9)________

10) $-\frac{9}{20} \div (-0.3 + \frac{7}{20})$

10)______

Basic Answers of Chapter 6

Lesson 1 Pretest 1) - 6 2) - 0.8 3) - 0.5 4) - 0.8 5) - 9

Lesson 4 Operations with Rational numbers. Review 1 A

1)	- 1.75	2)	- 0.2	3)	- 0.75	4)	0.4	5)	- 1.5
6)	- 1.5	7)	- 2.5	8)	- 1	9)	- 2	10)	- 0.5
11)	- 0.25	12)	- 0.5	13)	- 0.5	14)	- 0.5	15)	- 1.75

Review 1 B

16)	- 1.25	17)	- 2	18)	- 0.8	19)	- 2.5	20)	- 0.5
21)	- 7.2	22)	- 3.2	23)	- 4	24)	- 1.5	25)	22
26)	- 0.5	27)	2.5	28)	- 1.4	29)	- 6.6	30)	- 0.5

Lesson 5 Laws for Rational Numbers. Review 1 A

1)	-1	2)	-1	3)	-1	4)	-1	5)	-1.25
6)	-1.75	7)	-2	8)	-2	9)	-1.5	10)	-1.8
11)	-5	12)	-1	13)	-22	14)	-11	15)	-17
16)	-5	17)	-5	18)	-9	19)	-1	20)	-11
21)	-6.5	22)	-6.5	23)	-1.5	24)	- 13	25)	-2.5
26)	-20	27)	-23	28)	-9.25	29)	-2.75	30)	-2.5

Review 1 B

1)	-9	2)	-6	3)	-5.5	4)	-1.5	5)	-3.5
6)	- 3	7)	-5	8)	-3	9)	-2.5	10)	-6
11)	- 16	12)	- 1	13)	- 2.5	14)	- 1	15)	- 1.75
16)	- 2	17)	- 2	18)	- 2	19)	- 6	20)	- 6
21)	- 15	22)	- 0.5	23)	- 30	24)	- 3	25)	- 2
26)	- 24	27)	- 15	28)	- 15	29)	- 9	30)	- 10

Lesson 6 Mental Math 10 1) 0.3, 2) 0.09, 3)0.007, 4)0.5, 5)0.5, 6)0.2, 7)0.4, 8)0.6, 9)0.8, 10)0.8, 11)0.25, 12)0.5, 13)0.75, 14)0.25, 15)0.75, 16)0.28, 17)0.14, 18)0.22, 19)0.15, 20) 0.45

Lesson 7 Square Roots. Review 1 A

1)	11	2)	17	3)	13	4)	19
5)	16	6)	12	7)	18	8)	14
9)	21	10)	27	11)	23	12)	29
13)	26	14)	22	15)	28	16)	24
17)	31	18)	37	19)	33	20)	39

Review 1 B

21)	36	22)	32	23)	38	24)	34
25)	41	26)	47	27)	43	28)	49
29)	46	30)	42	31)	48	32)	44
33)	51	34)	57	35)	53	36)	59
37)	56	38)	52	39)	58	40)	54

Basic Answers of Chapter 6

Lesson 7 Review 1 C

1)	61	2)	67	3)	63	4)	69
5)	66	6)	62	7)	68	8)	64
9)	71	10)	77	11)	73	12)	79
13)	76	14)	72	15)	78	16)	74
17)	81	18)	87	19)	83	20)	89

Review 1 D

21)	86	22)	82	23)	88	24)	84
25)	91	26)	97	27)	93	28)	99
29)	96	30)	92	31)	98	32)	94
33)	25	34)	75	35)	45	36)	95
37)	65	38)	35	39)	85	40)	55

Lesson 8 Worksheet 4 A

1)	- 0.25	2)	- 0.5	3)	- 0.1	4)	- 0.25	5)	- 0.5
6)	- 0.4	7)	- 0.8	8)	- 0.6	9)	- 0.2	10)	- 0.1

Worksheet 4 B

1)	- 3	2)	- 5	3)	- 2	4)	- 10	5)	- 6
6)	- 14	7)	- 26	8)	- 18	9)	- 12	10)	- 8

Lesson 9

Review 6 A

1)	- 1.8	2)	1.6	3)	- 0.49	4)	-3	5)	- 0.28
6)	- 5	7)	- 5	8)	- 0.5	9)	- 0.5	10)	- 0.25

Review 6 B

1)	- 0.6	2)	2.2	3)	- 0.81	4)	- 0.8	5)	0.28
6)	- 3	7)	- 1.5	8)	- 0.5	9)	- 0.8	10)	- 1.8

Lesson 10

Test 6 A

1)	– 2.6	2)	- 0.8	3)	-1.2	4)	- 3.6	5)	- 0.29
6)	- 2.5	7)	- 3	8)	- 0.4	9)	- 0.5	10)	- 52

Test 6 B

1)	– 0.6	2)	2.6	3)	- 0.3	4)	- 1.2	5)	- 0.21
6)	- 6	7)	- 0.8	8)	- 0.5	9)	- 0.8	10)	- 9

Algebra

Chapter 7

Equations

Students!

Students, you are smarter than a calculator!
Students, you are smarter than a computer!
Students, you need to believe in yourself.
Students, be patient and do your best.

Do not be afraid to make mistakes.
Tell yourself positive words.
I am young! I am smart!
I am strong! I can do anything!

Smart students use their brain!
Your brain is always with you!
A calculator is not always with you.
Mathematics is work for the brain.

Be open to new ideas in math!
Work hard and try your best!
Check answers and ask questions!
Practice! Practice! Practice!

Algebra Chapter 7 Lessons
Equations
Lessons 1 – 10
Advanced Problems

Algebra Chapter 7 Contents
Equations

Contents have topics and exercises, which students need to solve before tests.1)Work with lesson 1, read rules and examples, 2) Solve the problems from contents, 3) Check answers in the end of chapter 1, 4) Work with the next lessons 2, 3, 4, 5, ..., 5) Solve Test 1 A, 6) Grading Scale without minuses

A=90–100% B+=85–89% B=80–84% C+=75–79%
C=70–74% D+=65–69% D=60–64 % F=0 - 59%

7)If you solve 10 problems correctly, you get an A, 8) If have an A, you can move on to next chapter, 9) If your grade is lower than an A, work with the mistakes, then solve Test 1B, 10) When you finish all of the chapters, solve Final Exam in last chapter.

Algebra Chapter 7 Lesson 1

Solve the Equations

Mental Math Equations

1) $x + 6 = 10$
2) $7 + x = 10$

3) $x - 9 = 5$
4) $x - 8 = 9$

5) $x - 7 = 8$
6) $12 - y = 3$

7) $13 - y = 8$
8) $14 - y = 5$

9) $15 - y = 7$
10) $16 - y = 9$

11) $8 \cdot x = 72$
12) $x \cdot 9 = 63$

13) $x \div 8 = 7$
14) $x \div 7 = 6$

15) $x \div 6 = 9$
16) $72 \div z = 9$

17) $72 \div z = 8$
18) $63 \div z = 9$

19) $56 \div z = 7$
20) $54 \div z = 9$

Algebra Chapter 7 Lesson 1

Pretest

Date **Name**

Solve the equations

1) $0.6p - 0.3p - 0.2 = 2.5$

1)________

2) $\frac{1}{10}g + \frac{2}{5}g + \frac{1}{2} = 1\frac{1}{2}$

2)________

3) $-8(\frac{1}{2} - \frac{1}{4}h) - 3(\frac{2}{3} - \frac{1}{3}h) = 0$

3)________

4) $\frac{x}{4.7} = \frac{2.7}{0.9}$

4) ________

5) $2 \div y = 1\frac{4}{7} \div 3\frac{1}{7}$

5)______

Algebra Chapter 7 Lesson 2

Algebraic expressions

If Arithmetic is the first part of Mathematics, Algebra is the second part. Algebra includes operations on variables and solving algebraic equations.Arithmetic expressions can contain numbers, sign of operations, and parentheses. $2 + 3 = 5$

The value of expression is the result after all operations done. The value of the expression $2 + 3 = 5$ is 5. Algebraic expressions can contain numbers, variables, signs of operations, and parentheses.

Variables mean letters that represent numbers. Any numbers can represent letter: $x = 3$ $2x = 2(3) = 6$ Addition represented by the words: plus, sum, total, increased by, more than. Subtraction represented by the words: minus, difference, decreased by and less than, subtract from.

Multiplication represented by the words: times, of, product of, twice. Division represented by the words: divided by, quotient, ratio, and rates. Some power represented by the words: squared, cubed. Expressions: Sum $a + b$, Difference $a - b$, Product ab, Quotient $a \div b$

Results of comparison can write in equality ($5 = 5$) or inequality:
$2 < 5$ $6 > 4$ $3 \leq 7$ $8 \geq 1$
Compare two expressions $a + b$ and $a - b$ if $a = -6$ $b = 2$
$a + b = -6 + 2 = -4$ $a - b = -6 - 2 = -8$ $-4 > -8$ $a + b > a - b$

Algebra Chapter 7 Lesson 2

Special Algebraic Expressions

The sum of the squares of two terms a and b is the expression $a^2 + b^2$
The difference of the square of two terms a and b is the expression $a^2 - b^2$

The square of the sum of two terms a and b is $(a + b)^2$
The square of the difference of two terms a and b is $(a - b)^2$

The perfect square of the sum of two terms a and b is the expression $a^2 + 2ab + b^2$
The perfect square of the difference of two terms a and b is the expression $a^2-2ab+b^2$

The sum of cubes of two terms a and b is the expression $a^3 + b^3$
The difference of cubes of two terms a and b is the expression $a^3 - b^3$

If $a = -6$ and $b = -2$ find value of expressions $a + b$, $a - b$, ab, $a \div b$

$a + b = -6 - 2 = -8$ **$a - b = -6 + 2 = -4$**

$ab = (-6)(-2) = 12$ **$a \div b = (-6) \div (-2) = 3$**

Evaluating Algebraic Expressions

$x = -5$	$y = -4$	$x^2 - y^2 = (-5)^2 - (-4)^2 = 25 - 16 = 9$
$a = -3$	$b = -2$	$a^3 - b^3 = (-3)^3 - (-2)^3 = -27-(-8) = -27 + 8 = -19$
$c = -2$	$d = -1$	$c^2 - 2cd + d^2 = (-2)^2 - 2(-2)(-1)+(-1)^2 = 4-4+1=1$

To compare two arithmetic expressions means to know which value of expressions is biggest or smallest, or if they are equal

Algebra Chapter 7 Lesson 2

Identities

Two expressions with the same corresponding value that is equal for any of the variables called identically equal: $a + b = b + a$
If one expression is changed $2(x - 3)$ for another identically equal expression it is called identical transformation of expression: $2x - 6$

Equality that is true for any value of variables called identity.
Equality $a + b = b + a$ or $ab = ba$ is identity

Equality can be identity and can be equation
Identity: $ab = ba$ Equation: $x + 2 = 6$

Identically transformation of expressions uses laws and properties of numbers: $(-a)(-b) = ab$ $a(-b) = -ab$

Commutative and associative laws of addition and multiplication help simplify expressions
$2 + x + 5 = 2 + 5 + x = 7 + x = x + 7$ $(25)(x)(4) = (25)(4)(x) = 100x$

The distributive law of multiplication used if combining like terms to needed to simplify expressions or to remove parentheses:

$a(b + c) = ab + ac$ $a(b - c) = ab - ac$
$4(x + y) = 4x + 4y$ $-1(t - z) = -t + z$
$(b + c)a = ab + ac$ $(b - c)a = ab - ac$
$(a + b)(-5) = -5a - 5b$ $(m - n)(3) = 3m - 3n$
$-5(x + y) - 3(x - y) = -5x - 5y - 3x + 3y = -8x - 2y$

Remove parentheses

If before parentheses there is a plus sign or no sign, remove the plus sign and remove the parentheses and rewrite each term with the same sign:
$a + (b - c) = a + b - c$ $(a - b) + (c - d) = a - b + c - d$
If before parentheses there is a minus sign, remove the minus sign, remove the parentheses, and rewrite each term with the opposite sign:
$a - (b - c) = a - b + c$ $(a - b) - (c - d) = a - b - c + d$

Algebra Chapter 7 Lesson 2
Review 1 A
Simplify the expressions

1) $3(2x + 8) - (5x + 2)$
2) $5(7 - 3x) + 7(2 + 2x)$

3) $8(3 - 2y) + 5(3y + 5)$
4) $3(x - 5) - 2(x - 4)$

5) $5(7y - 2) - 6(5y + 2)$
6) $4(8x + 3) - 6(4x - 3)$

7) $8(2x - 6) - 2(4x + 3)$
8) $3(3y - 4) - 4(2y - 1)$

9) $5(y + 3) - 8(10 - y)$
10) $8(9 + 2x) - 5(2 - 3x)$

11) $-5(0.3y + 1) - 11(y - 1)$
12) $-2(3.1x - 1) - 3(1.2x + 1)$

13) $-5(4.2y + 1) - 5(1.4y - 2)$
14) $-1.6(z - 6) + 1.2(z - 8)$

15) $-0.9(5z - 3) + 0.3(7z + 1)$
16) $0.15(x + 4) - 9.9 - 0.3(x - 1)$

17) $1.6(y - 4) - 0.6 - 3(0.4y - 7)$
18) $3z + 2(2z - 3) - 8 - 7(z - 2)$

19) $10(0.1x - 4) - 100(0.01x - 5)$
20) $100(0.01y - 0.03) - 1000(0.001y - 0.05)$

Algebra Chapter 7 Lesson 2

Review 1 B

Simplify the expressions

21) $5(\frac{4}{5}x - 2) - 7(\frac{3}{7}x - 3)$

22) $7(\frac{6}{7}y - 3) - 5(\frac{3}{5}y - 2)$

23) $4(\frac{3}{4} - \frac{1}{4}y) - 3(\frac{1}{3}y + \frac{2}{3})$

24) $5(\frac{2}{5}x - \frac{4}{5}) - 4(\frac{1}{4}x - \frac{3}{4})$

25) $3(\frac{2}{3}x - 2) - 4(\frac{1}{2}x + \frac{3}{4})$

26) $10(\frac{1}{2}y - \frac{3}{5}) - 4(\frac{1}{4}y - \frac{1}{2})$

27) $5(\frac{2}{5}z - 1) - 6(\frac{2}{3}z + \frac{1}{2})$

28) $4(\frac{1}{2}x - \frac{1}{4}y) - 6(\frac{1}{6}x + \frac{1}{3}y)$

29) $5(\frac{2}{5}x + \frac{4}{5}) - 6(\frac{5}{6}x - \frac{2}{3})$

30) $14(\frac{3}{7}y - \frac{5}{7}) - 16(\frac{3}{8}y - \frac{5}{8})$

Algebra Chapter 7 Lesson 2

Review 1 B

Simplify the expressions

31) $-6(\frac{1}{6} - \frac{1}{3}x) - 2(1 - \frac{1}{2}x)$

32) $-8(\frac{1}{8} - \frac{1}{4}y) - 3(1 - \frac{1}{3}y)$

33) $-5(\frac{2}{5}x + \frac{4}{5}) + 4(\frac{1}{4}x - \frac{3}{4})$

34) $-\frac{1}{2}(6x - 2) - \frac{1}{5}(15x + 5)$

35) $-\frac{1}{9}(18y - 63) - 9(\frac{1}{3}y - \frac{2}{9})$

36) $12(\frac{1}{2}x - \frac{2}{3}y) - \frac{1}{2}(4x - 2y)$

37) $16(\frac{3}{4}x + \frac{1}{4}y) - \frac{2}{3}(9x + 27y)$

38) $\frac{3}{7}(14x - 21y) - 15(\frac{2}{3}x - \frac{1}{3}y)$

39) $4(\frac{1}{2}x - \frac{1}{4}y) - \frac{2}{3}(6x + 3y)$

40) $\frac{4}{9}(18x - 9y) - 12(\frac{5}{6}x - \frac{1}{6}y)$

Algebra Chapter 7 Lesson 3
Equations

Equality means two expressions joined with an equality sign: $3 + 2 = 5$
Identity is equality that always true. $A + B = B + A$ $AB = BA$

A formula is an equation that expresses known relationships between two or more variables. Formula: $A + B = B + A$ or $AB = BA$
An equality that contains one letter called an equation with one variable: $x + 2 = 5$

An expression that stands to the left of the equal sign is called the left side of the equation: $x + 2$

An expression that stands to the right of the equal sign is called the right side of the equation: 5

The value of a variable that makes an equation true equality,is called the solution of equation.

The solution of an equation is the value of the letter that makes the equation true.
$x + 2 = 5$ $x = 3$ solution $3 + 2 = 5$ True

If the product of factors equals zero, one of the factors equals zero:
$AB = 0$ than $A = 0$ or $B = 0$
$2 \cdot 0 = 0$ $0 \cdot 2 = 0$

Equations having the same solution called equivalent equations.
First equation: $x - 6 = 0$ $x = 6$
Second equation: $2(x - 6) = 0$ $x = 6$
Equations $x - 6 = 0$ and $2(x - 6) = 0$ are equivalent

Algebra Chapter 7 Lesson 3

Properties of Equation

1. First Property

To solve an equation, add the same number on both sides of the equation, to obtain equivalent equations

$6 = 6$	$6 + 2 = 6 + 2$	$8 = 8$
$a = b$	$a + c = b + c$	
$x - 2 = 10$	$x - 2 + 2 = 10 + 2$	$x = 12$

2. Second Property

To solve an equation, subtract the same number on both sides of the equation, to obtain equivalent equations

$6 = 6$	$6 - 2 = 6 - 2$	$4 = 4$
$a = b$	$a - c = b - c$	
$x + 2 = 10$	$x + 2 - 2 = 10 - 2$	$x = 8$

3. Third Property

To solve an equation, multiply by the same nonzero number on both sides of the equation, to obtain equivalent equations

Do not multiply by zero.

$6 = 6$	$6 \times 2 = 6 \times 2$	$12 = 12$
$a = b$	$ac = bc$	$c \neq 0$
$\frac{1}{2}x = 10$	$2 \times \frac{1}{2}x = 10 \times 2$	$x = 20$

4. Fourth Property

To solve an equation, divide by the same nonzero number on both sides of the equation, to obtain equivalent equations

Do not divide by zero

$6 = 6$	$6 \div 2 = 6 \div 2$	$3 = 3$
$a = b$	$a \div c = b \div c$	$c \neq 0$
$2x = 10$	$2x \div 2 = 10 \div 2$	$x = 5$

Algebra Chapeer 7 Lesson 3

Equations with Like Terms

A term is an expression with numbers, letters, product, and power:$2a^3 b^4$
A number standing before letters in a term is called a coefficient: 2
Terms with the same letters with the same exponent called like terms.

Like terms $3a^2$ and $2a^2$ and unlike terms $2x^2$ and $3y^2$ or $2x^2$ and $3x^3$
Like terms can have different coefficients: $2b^3$ and $4b^3$
To combine or simplify like terms, add their coefficients and multiply by their common letters and exponents:

$3a^2 + 3a^2 = 6a^2$	$2b^3 + 4b^3 = 6b^3$
$3x^2y - 2x^2y = x^2y$	$x + x + x = 3x$
$x + 3x = 4x$	$3x - x = 2x$
$x + 3 + x = x + x + 3 = 2x + 3$	$2 + x + 3 = 2 + 3 + x = 5 + x$
$- 5m - 8n - 4m + 3n =- 9m - 5n$	$2x - 3x = - 1\ x = - x$

Fifth Property of Equations
If both sides of an equation have the same terms with the same sign, drop this term from both sides: $2x + 6 = 6$ $2x = 0$ $x = 0$
If $a + c = b + c$ than $a = b$ $2x + 3 = 4 + 3$ $2x = 4$ $x = 2$
Sixth Property of Equations
If any term is moved from one side of an equation to another side, change sign to opposite sign: $2x - 6 = 0$ $2x = 6$ $x = 3$
If $a + b = c$ than $a = c - b$ $2x + 3 = 7$ $2x = 7 - 3$ $2x = 4$ $x = 2$

To solve equations with like terms, simplify or combine like terms.

Example a)	$7x + 4x = - 66$	$11x = - 66$	$x = - 6$
Example b)	$x + 3x = 60$	$4x = 60$	$x = 15$
Example c)	$3x + 8 - x = 36$	$3x - x + 8 = 36$	$2x + 8 = 36$
	$2x = 36 - 8$	$2x = 28$	$x = 14$

Algebra Chapter 7 Lesson 3
Review 1 A

Solve the Equations with Like Terms

1) $2x + 4x = 72$
2) $8y + 4y = 240$
3) $z + 30z = 341$
4) $x + 19x = 2100$

5) $13y - 2y = 44$
6) $9z - 5z = 44$
7) $7x - x = 84$
8) $8y - y = 707$

9) $30z - z = 319$
10) $40x - x = 7956$
11) $y + 2y + 3y = 108$
12) $z + 4z + 5z = 150$

13) $5 + 2 - y = 3$
14) $50 + 20 - z = 30$
15) $13x - 2x + 3 = 14$
16) $8y - 3y - 30 = 15$

17) $25 + 4 + 2z = 29$
18) $75 + 9 + 5x = 1614$
19) $124 - 49 - 2x = 889$
20) $548 - 679 - 3y = 1393$

Review 1 B

21) $z + 5 + z = 15$
22) $x + 4 + x = 104$
23) $y + 50 + y = 100$
24) $z + 125 + z = 133$

25) $x + 1234 + x = 568$
26) $2y + 6678 + y = 789$
27) $2 + z + 5 = 17$
28) $1 + x + 30 = 31$

29) $12 + y + 50 = 100$
30) $123 + 2z + 456 = -987$
31) $4x + 28 - x = 136$
32) $15y + 123 - y = 5807$

33) $-234 + 23z = 11427$
34) $-65x - 345 + 9x = 33759$
35) $-6y - 456 - y = 258$
36) $-34z - 567 + 8z = 4711$

37) $-78x - 678 + 9x = 20298$
38) $9y + 789 - 28y = 8484$
39) $28z + 987 - 45z = 9589$
40) $125x - 876 - 301x = 105956$

Algebra Chapter 7 Lesson 4

Finding Unknowns
Advanced Problems

Symbol → means follows.

Addition

The first number in addition is called the addend: 2
The second number in addition is called the addend: 3
The result of addition is called the sum: 5
Addend + Addend = Sum $2 + 3 = 5$

$x + 2 = 8 \rightarrow x + 2 - 2 = 8 - 2 \rightarrow \underline{x = 8 - 2} \rightarrow x = 6$

$2 + x = 8 \rightarrow 2 + x - 2 = 8 - 2 \rightarrow \underline{x = 8 - 2} \rightarrow x = 6$

To find one addend, subtract the other addend from the sum.
First Addend = Sum – Second Addend

$x + 3 = 5 \rightarrow x = 5 - 3 \rightarrow x = 2$

Second Addend = Sum – First Addend

$2 + x = 5 \rightarrow x = 5 - 2 \rightarrow x = 3$

Subtraction

The first number in subtraction is called the minuend: 5
The second number in subtraction is called the subtrahend: 2
The result of subtraction is called the difference: 3
Minuend – Subtrahend = Difference $5 - 2 = 3$

Finding the minuend:

$x - 2 = 8 \rightarrow x - 2 + 2 = 8 + 2 \rightarrow \underline{x = 8 + 2} \rightarrow x = 10$

To find minuend, add the difference to the subtrahend
Minuend = Difference + Subtrahend

$x - 2 = 3 \rightarrow x = 3 + 2 \rightarrow x = 5$

Finding the subtrahend

$8 - x = 2 \rightarrow 8 - x - 8 = 2 - 8 \rightarrow -x = -6 \rightarrow$

$(-1)(-x) = (-1)(-6) \rightarrow x = 6 \rightarrow \underline{x = 8 - 2}$

To find the subtrahend, subtract the difference from the minuend.
Subtrahend = Minuend – Difference

$5 - x = 3 \rightarrow x = 5 - 3 \rightarrow x = 2$

Algebra Chapter 7 Lesson 4

Multiplication

The first number in multiplication is called the factor: 2
The second number in multiplication is called the factor: 3
The result of multiplication is called the product: 6
Factor × Factor = Product $2 \times 3 = 6$

Finding the first factor:

$x \cdot 2 = 8 \quad \rightarrow \quad x \cdot 2 \div 2 = 8 \div 2 \rightarrow \quad \underline{x = 8 \div 2} \quad \rightarrow \quad x = 4$

Finding the second factor:

$2 \cdot x = 8 \quad \rightarrow \quad 2 \cdot x \div 2 = 8 \div 2 \rightarrow \quad \underline{x = 8 \div 2} \quad \rightarrow \quad x = 4$

To find one of the factors, divide the product by the other factor

First Factor = Product ÷ Second Factor

$x \cdot 3 = 6 \quad \rightarrow \quad x = 6 \div 3 \quad \rightarrow \quad x = 2$

Second Factor = Product ÷ First Factor

$2 \cdot x = 6 \quad \rightarrow \quad x = 6 \div 2 \quad \rightarrow \quad x = 3$

Division

The first number in division is called the dividend: 8
The second number in division is called the divisor: 4
The result of division is called the quotient: 2
Dividend ÷ Divisor = Quotient $8 \div 4 = 2$

Finding the dividend:

$x \div 2 = 8 \quad \rightarrow \quad x \div 2 \cdot 2 = 8 \cdot 2 \quad \rightarrow \quad x = \underline{8 \cdot 2} \quad \rightarrow \quad x = 16$

To find the dividend, multiply the quotient by the divisor.

Dividend = Quotient × Divisor

$x \div 2 = 3 \quad \rightarrow \quad x = 3 \cdot 2 \quad \rightarrow \quad x = 6$

Finding the divisor:

$8 \div x = 2 \qquad 8 \div x \cdot x = 2 \cdot x \qquad 8 = 2x$

$8 \div 2 = 2 \cdot x \div 2 \qquad 4 = x \qquad \underline{8 \div 2 = x}$

To find the divisor, divide the dividend by the quotient.

Divisor = Dividend ÷ Quotient

$6 \div x = 3 \quad \rightarrow \quad x = 6 \div 3 \quad \rightarrow \quad x = 2$

Algebra Chapter 7 Lesson 4
Find unknowns
Solve equations by definition
Review 1 A

1) $X + 9 = 12$

2) $99 + X = 100$

3) $X + 199 = 888$

4) $345 + X = 789$

5) $X + 989 = 1478$

6) $X - 8 = 43$

7) $X - 87 = 542$

8) $X - 876 = 1894$

9) $X - 267 = 378$

10) $X - 865 = 678$

11) $80 - Y = 9$

12) $87 - Y = 48$

13) $547 - Y = 89$

14) $946 - Y = 787$

15) $794 - Y = 485$

16) $264 - Y = 178$

17) $2857 - Y = 998$

18) $567 - Y = 123$

19) $654 - Y = 246$

20) $1234 - Y = 876$

Algebra Chapter 7 Lesson 4
Find unknowns
Solve equations by definition.
Review 1 B

21) $X \cdot 9 = 72$

22) $87 \cdot X = 261$

23) $X \cdot 408 = 22848$

24) $609 \cdot X = 47502$

25) $X \cdot 307 = 19648$

26) $X \div 9 = 81$

27) $X \div 78 = 26$

28) $X \div 24 = 76$

29) $X \div 305 = 248$

30) $X \div 256 = 308$

31) $81 \div Z = 9$

32) $90 \div Z = 18$

33) $162 \div Z = 27$

34) $1152 \div Z = 144$

35) $16940 \div Z = 605$

36) $15873 \div Z = 407$

37) $117936 \div Z = 567$

38) $119615 \div Z = 235$

39) $104442 \div Z = 206$

40) $368742 \div Z = 906$

Algebra Chapter 7 Lesson 5
Equations with Decimals
Examples:

a)	$x + 0.2 = 0.6$	$x = 0.6 - 0.2$	$x = 0.4$
b)	$x - 0.2 = 0.6$	$x = 0.6 + 0.2$	$x = 0.8$
c)	$0.6 - x = 0.2$	$x = 0.6 - 0.2$	$x = 0.4$
d)	$0.2\,x = 0.6$	$x = 0.6 \div 0.2$	$x = 3$
e)	$x \div 0.2 = 0.6$	$x = 0.6\,(0.2)$	$x = 0.12$
f)	$0.6 \div x = 0.2$	$x = 0.6 \div 0.2$	$x = 3$

Review 1 A
Solve the equations

1) $0.8z + 0.4z = 2.4$

2) $0.1y + 3y = 3.41$

3) $0.9z - 0.5z = 4.4$

4) $3y - 0.1y = 3.19$

5) $1.7z - z = 8.4$

6) $0.8y - 0.1y = 7.07$

7) $2.5 + 0.4 + y = 2.9$

8) $1.2 + z + 5 = 10$

9) $y + 1.25 + 0.08 = 1.33$

10) $0.1 + y + 3 = 3.1$

11) $0.2 + y + 0.5 = 1.7$

12) $0.5 + 0.2 - z = 0.3$

13) $y + 0.5 + y = 1.5$

14) $0.1y + 5 + 0.1y = 10$

15) $z + 0.4 + z = 10.4$

16) $0.3y - 0.2y - 0.3 = 1.2$

17) $0.8z - 0.3z - 3 = 1.5$

18) $0.6x - 0.3x - 0.2 = 2.5$

19) $0.1y + 0.2\,y + 0.3y = 1.08$

20) $0.1z + 0.4z + 0.5z = 1.5$

Algebra Chapter 7 Lesson 5
Equations with Fractions
Examples:

a) $x+\frac{1}{5}=\frac{3}{5}$ $\quad x=\frac{3}{5}-\frac{1}{5}$ $\quad x=\frac{2}{5}$ $\quad x=0.4$

b) $x-\frac{1}{5}=\frac{3}{5}$ $\quad x=\frac{3}{5}+\frac{1}{5}$ $\quad x=\frac{4}{5}$ $\quad x=0.8$

c) $\frac{3}{5}-x=\frac{1}{5}$ $\quad x=\frac{3}{5}-\frac{1}{5}$ $\quad x=\frac{2}{5}$ $\quad x=0.4$

d) $\frac{1}{5}x=\frac{3}{5}$ $\quad x=\frac{3}{5}\div\frac{1}{5}$ $\quad x=\frac{3}{5}\cdot\frac{5}{1}$ $\quad x=3$

e) $x\div\frac{1}{5}=\frac{3}{5}$ $\quad x=\frac{3}{5}\cdot\frac{1}{5}$ $\quad x=\frac{3}{25}$ $\quad x=0.12$

f) $\frac{3}{5}\div x=\frac{1}{5}$ $\quad x=\frac{3}{5}\div\frac{1}{5}$ $\quad x=\frac{3}{5}\cdot\frac{5}{1}$ $\quad x=3$

Review 1 B
Solve the equations

1) $x+\frac{1}{5} = \frac{4}{5}$

2) $\frac{3}{5}+x = \frac{4}{5}$

3) $x-\frac{1}{5} = \frac{3}{5}$

4) $\frac{4}{5}-x = \frac{1}{5}$

5) $x \cdot 2 = \frac{4}{5}$

6) $\frac{2}{5}x = \frac{4}{5}$

7) $x\div\frac{1}{5} = \frac{2}{5}$

8) $\frac{4}{5}\div x = \frac{2}{5}$

9) $1\frac{1}{5}x = 6$

10) $1\frac{2}{5}x = 5\frac{3}{5}$

Algebra Chapter 7 Lesson 5
Equations with Fractions
Review 1 B

11) $2\frac{2}{5}\ x = 6$

12) $12\frac{1}{2}\ x = 10$

13) $2\frac{1}{2}z = 5$

14) $z \cdot \frac{4}{5} = 100$

15) $z \div \frac{3}{10} = \frac{2}{5}$

16) $\frac{4}{5} \div z = 0.02$

17) $z - \frac{3}{10} = 0.9$

18) $10 - z = 1\frac{23}{100}$

19) $\frac{1}{5}y + \frac{2}{5}y = 7\frac{1}{5}$

20) $\frac{1}{10}y + 3y = 3.41$

21) $\frac{4}{5}z + \frac{2}{5}z = 2.4$

22) $0.9z - \frac{1}{2}z = 4.4$

23) $3y - \frac{1}{10}y = 3.19$

24) $\frac{3}{10}y - \frac{1}{5}y = 4\ \frac{2}{5}$

25) $1\frac{7}{10}z - z = 8.4$

26) $\frac{4}{5}y - \frac{1}{10}y = 7\frac{7}{10}$

Algebra Chapter 7 Lesson 5
Equations with Fractions
Review 1 B

27) $1\frac{1}{5} + z + 5 = 10$

28) $\frac{1}{10} + y + 3 = 3.1$

29) $2.5 + \frac{2}{5} + y = 2.9$

30) $0.1z + \frac{2}{5}z + \frac{1}{2}z = 1.5$

31) $\frac{1}{2} + \frac{1}{5} - z = 0.3$

32) $\frac{1}{2} + \frac{3}{5} - y = \frac{7}{10}$

33) $\frac{1}{5} + y + \frac{1}{2} = 1\frac{7}{10}$

34) $\frac{1}{10}y + \frac{1}{5}y + \frac{3}{10}y = 1\frac{2}{25}$

35) $y + \frac{1}{2} + y = 1\frac{1}{2}$

36) $\frac{2}{5}z + z = 14$

37) $\frac{1}{10}y + 5 + \frac{1}{10}y = 10$

38) $y + 1\frac{1}{4} + 0.08 = 1.33$

39) $\frac{3}{10}y - \frac{1}{5}y - \frac{3}{10} = 1\frac{1}{5}$

40) $\frac{4}{5}z - 0.3z - 3 = 1.5$

Algebra Chapter 7 Lesson 6

Conditional equations

An equation in the form $ax = b$, where x is the variable, a and b are any numbers, is called a linear equation with one variables.
All equations have a left side and a right side.
A conditional equation has one solution: $2x = 8$ $\quad x = 4$
A solution of conditional equation: $ax = b$ if $a \neq 0$ then $x = b \div a$

The value of the variable that makes an equation a true equality called the solution of equation:
$x + 2 = 8$ $\quad x = 6$ Solution $\quad 6 + 2 = 8$ True
Equations having the same solutions called equivalent equations.
To solve an equation means to find all solutions to the equation.

Six properties to solve the equation
if $a = b$ then

1)	$a + c = b + c$	2) $a - c = b - c$		
3)	$ac = b\ c$	4) $a \div c = b \div c$		$c \neq 0$
5)	$a + c = b + c$	$a = b$		
	$2x + 3 = 4 + 3$	$2x = 4$	$x = 2$	
6)	$a - b = c$	$a = c + b$		
	$2x + 3 = 7$	$2x = 7 - 3$	$2x = 4$	$x = 2$

Solve an equation in 5 steps

$2(x - 1) = 3(x + 2) - 4$	1) Open parentheses $a\,(b + c) = ab + ac$
$2x - 2 = 3x + 6 - 4$	2) Combine like terms
$2x - 2 = 3x + 2$	3) Eliminate numbers or letter (Addition)
$2x - 2 + 2 = 3x + 2 + 2$	
$2x = 3x + 4$	
$2x - 3x = 4$	4) Move 3x from right to left and put minus 3x
$- x = 4$	
$(- 1)(- x) = (-1)(4)$	5) Multiplication property
$\underline{x = - 4}$	

Algebra Chapter 7 Lesson 6
Review 1 A
Solve the equations

1) $3(2x + 8) - (5x + 2) = 0$
2) $-3(3y + 4) + 4(2y - 1) = 0$

3) $5(7 - 3x) + 7(2 + 2x) = 0$
4) $8(3 - 2y) + 5(3y + 5) = 9$

5) $3(x - 5) - 2(x - 4) = 8$
6) $5(7y - 2) - 6(5y + 2) = -24$

7) $4(8x + 3) - 6(4x - 3) = 36$
8) $-5(3y + 1) - 11(y - 1) = -20$

9) $5(x - 7) = 3(x - 4) - 27$
10) $4(y - 3) - 16 = 5(y - 5)$

11) $0.8(2x - 6) = 0.2(4x + 3)$
12) $-1.6(z - 6) = -1.2(z - 8)$

13) $0.8(9 + 2x) = 0.5(2 - 3x)$
14) $0.5(y + 3) = 0.8(10 - y)$

15) $-0.9(5z - 3) = -0.3(7z + 1)$
16) $0.15(x - 4) = 9.9 - 0.3(x - 1)$

17) $1.6(y - 4) - 0.6 = 3(0.4y - 7)$
18) $-2(3.1x - 1) + 3(1.2x + 1) = -14.5$

19) $-5(4.2y + 1) + 4(1.4y - 2) = -20.7$
20) $0.3z + 0.2(2z - 3) = 0.8 - 0.7(z - 2)$

Algebra Chapter 7 Lesson 6
Review 1 B

Solve the equations

21) $5(\frac{4}{5}x - 2) - 7(\frac{3}{7}x - 3) = 0$

22) $7(\frac{6}{7}y - 3) - 5(\frac{3}{5}y - 3) = 0$

23) $3(\frac{1}{3}x - 2) - 4(\frac{1}{2}x + \frac{1}{4}) = 0$

24) $10(\frac{1}{2}y - \frac{3}{5}) - 4(\frac{1}{4}y - \frac{1}{2}) = 0$

25) $1.8(\frac{8}{9}z + \frac{1}{6}) - 1.5(\frac{2}{3}z - \frac{1}{5}) = 0$

26) $1.6(\frac{3}{4} - \frac{3}{8}y) - \frac{2}{3}(0.9 - 2.7y) = 0$

27) $\frac{2}{9}(1.8y - 2.7) - 0.9(\frac{1}{3}y - \frac{2}{9}) = 0$

28) $\frac{3}{7}(1.4x - 2.1) - 0.6(\frac{1}{3}x - \frac{1}{6}) = 0$

29) $1.5(\frac{2}{5}x - \frac{4}{5}) - 1.2(\frac{1}{6}x - \frac{2}{3}) = 0$

30) $1.6(\frac{3}{4}y - \frac{5}{8}) - 1.8(\frac{2}{9}y - \frac{7}{9}) = 0$

Algebra Chapter 7 Lesson 6
Review 1 B

Solve the equations

31) $\frac{1}{2}(1.2x - 4) - \frac{1}{5}(1.5x + 5) = 0$

32) $\frac{3}{7}(2.8x + 5.6) - 1.5(\frac{2}{3}x - \frac{2}{3}) = 0$

33) $0.4(\frac{1}{2}x - \frac{1}{2}) - \frac{2}{3}(0.6x - 0.6) = 0$

34) $\frac{1}{3}(1.8x - 3.6) - 4.2(\frac{6}{7}x - \frac{1}{7}) = 0$

35) $1.2(\frac{1}{2}x + \frac{2}{3}) - \frac{1}{2}(2x - 6) = 0$

36) $\frac{1}{7}(1.4 - 2.1y) - 1.6(\frac{1}{4} - \frac{1}{8}y) = 0$

37) $-6(\frac{1}{6} - \frac{1}{2}x) - 2(1 - \frac{1}{2}x) = 0$

38) $-8(\frac{1}{2} - \frac{1}{4}y) - 3(\frac{2}{3} - \frac{1}{3}y) = 0$

39) $-8(\frac{1}{4} - \frac{3}{8}y) + 6(\frac{1}{3}y + \frac{1}{2}) = 0$

40) $-5(\frac{2}{5}x + \frac{4}{5}) - 4(\frac{3}{4}x - \frac{1}{4}) = 0$

Algebra Chapter 7 Lesson 7
Proportions

The quotient of two numbers called the ratio or rate of these numbers. A ratio shows us how many times a first number is greater than second number, or what portion first number from second number.

Find the ratio of numbers 40 to 20. $40 \div 20 = 2$

Find the ratio of numbers 20 to 40. $20 \div 40 = \frac{1}{2}$

Find the Ratio of Numbers – divide (the same units)

$$\frac{10\text{ft}}{10\text{yd}} = \frac{10\text{ft}}{30\text{ft}} = \frac{1}{3} = 1 \div 3$$

Find the Rates of Numbers – divide (different units)

$$\frac{250\text{mi}}{10\text{hr}} = 25\frac{\text{mi}}{\text{hr}} = 25\text{mph}$$

A special type of equation called a proportion.
Equality of two ratios or rates called proportion

$8 \div 2 = 16 \div 4$ $\quad \frac{8}{2} = \frac{16}{4}$ $\quad 4 = 4$ **or** $\quad a \div b = c \div d$ $\quad \frac{a}{b} = \frac{c}{d}$

$a \neq 0 \quad b \neq 0 \quad c \neq 0 \quad d \neq 0$

Numbers a and d called exterior numbers: 8 and 4 or 'a' and 'd'
Numbers b and c called interior numbers: 16 and 2 or 'b' and 'c'

Properties of proportions

If the proportion is true, the product of exterior numbers equals the product of the interior numbers. $a \div b = c \div d$ $\quad a\,d = b\,c$

$8 \div 2 = 16 \div 4$ $\quad 8 \cdot 4 = 32$ $\quad 16 \cdot 2 = 32$ $\quad 8 \cdot 4 = 16 \cdot 2$

If the product of the exterior numbers equals the product of the interior numbers of the proportion, then the proportion is true.

$8 \cdot 4 = 32$ $\quad 16 \cdot 2 = 32$ $\quad 8 \div 2 = 16 \div 4$ $\quad 4 = 4$

If the proportion is true interchange exterior or interior numbers, then new proportions is true too.

$4 \div 2 = 16 \div 8$ $\quad 8 \div 16 = 2 \div 4$ $\quad 4 \div 16 = 2 \div 8$ **or**

$\frac{4}{2} = \frac{16}{8}$ $\quad 2 = 2$ $\quad \frac{8}{16} = \frac{2}{4}$ $\quad \frac{1}{2} = \frac{1}{2}$ $\quad \frac{4}{16} = \frac{2}{8}$ $\quad \frac{1}{4} = \frac{1}{4}$

Algebra Chapter 7 Lesson 7

To check proportions means that proportions is correct or not, check cross products is equal or not.

$\frac{0.8}{8}=\frac{4}{40}$ 0.8(40)=32 8·4=32 yes, proportion is correct

$\frac{0.2}{4}=\frac{2}{80}$ 80(0.2)=16 4·2=8 no, proportion is not correct

Equations 1

To find one of the exterior numbers of a proportion, divide the product of the two interior numbers by another exterior number:

$\frac{x}{b}=\frac{c}{d}$ or $x \div b = c \div d$ $dx = b\,c$ $x = \frac{bc}{d}$

$\frac{1.2}{3}=\frac{2.4}{x}$ or $1.2 \div 3 = 2.4 \div x$ $1.2x = 2.4 \cdot 3$ $1.2x = 7.2$ $x = 6$

Equations 2

To find one of the interior numbers of the proportion, divide the product of the two the exterior numbers by another interior number.

$\frac{a}{b}=\frac{x}{d}$ or $a \div b = x \div d$ $bx = ad$ $x = \frac{ad}{b}$

$\frac{1.2}{x}=\frac{2.4}{6}$ or $1.2 \div x = 2.4 \div 6$ $2.4\,x = 1.2 \cdot 6$ $2.4\,x = 7.2$ $x = 4$

Application of proportion: to compare the fractions check cross product:

$\frac{1}{2}=\frac{3}{6}$ $1 \cdot 6 = 2 \cdot 3$ $6 = 6$

$\frac{5}{6}<\frac{7}{8}$ $5 \cdot 8 < 6 \cdot 7$ $40 < 42$

Solve the equation

$3\frac{1}{3} \div 2\frac{1}{2} = 2\frac{2}{3} \div x$ $3\frac{1}{3}x = 2\frac{1}{2} \cdot 2\frac{2}{3}$ $x = 2\frac{1}{2} \cdot 2\frac{2}{3} \div 3\frac{1}{3}$

$x = \frac{5}{2} \cdot \frac{8}{3} \div \frac{10}{3}$ $x = \frac{5}{2} \cdot \frac{8}{3} \cdot \frac{3}{10}$ $x = \frac{1 \cdot 4}{1 \cdot 2}$ $x = 2$

Algebra Chapter 7 Lesson 7
Review 1 A

Solve the proportions

1) $\frac{24}{72} = \frac{x}{9}$

2) $\frac{60}{45} = \frac{24}{x}$

3) $\frac{42}{x} = \frac{51}{34}$

4) $\frac{18}{12} = \frac{72}{x}$

5) $\frac{69}{46} = \frac{x}{54}$

6) $\frac{4.2}{x} = \frac{5.7}{3.8}$

7) $\frac{x}{5.1} = \frac{2.7}{0.9}$

8) $\frac{x}{4.7} = \frac{27.3}{9.1}$

9) $\frac{7.8}{2.6} = \frac{4.5}{x}$

10) $\frac{4.4}{x} = \frac{8.8}{2.2}$

11) $\frac{y}{5\frac{2}{3}} = \frac{1\frac{7}{9}}{1\frac{7}{27}}$

12) $\frac{z}{\frac{4}{5}} = \frac{3\frac{1}{8}}{1\frac{1}{4}}$

13) $y \div \frac{3}{5} = 3\frac{1}{8} \div 3\frac{3}{4}$

14) $1\frac{7}{8} \div 8\frac{1}{3} = z \div \frac{4}{9}$

15) $3\frac{3}{4} \div 1\frac{1}{8} = 2\frac{2}{3} \div y$

16) $\frac{3}{8} \div \frac{3}{11} = z \div \frac{4}{11}$

17) $y \div \frac{1}{3} = \frac{3}{4} \div \frac{1}{2}$

18) $z \div \frac{3}{14} = 3\frac{1}{9} \div \frac{4}{9}$

19) $y \div 4\frac{1}{6} = \frac{4}{7} \div \frac{20}{21}$

20) $z \div 3\frac{1}{5} = 4\frac{1}{2} \div 2\frac{1}{4}$

Algebra Chapter 7 Lesson 7
Review 1 B
Solve the proportions

$3\frac{1}{3} \div 2\frac{1}{2} = 2\frac{2}{3} \div x$ $\quad$ $3\frac{1}{3}x = 2\frac{1}{2} \cdot 2\frac{2}{3}$ $\quad$ $x = 2\frac{1}{2} \cdot 2\frac{2}{3} \div 3\frac{1}{3}$

$x = \frac{5}{2} \cdot \frac{8}{3} \div \frac{10}{3}$ $\quad$ $x = \frac{5}{2} \cdot \frac{8}{3} \cdot \frac{3}{10}$ $\quad$ $x = \frac{1 \cdot 4}{1 \cdot 2}$ $\quad$ $x = 2$

21) $\frac{3}{14} \div 3\frac{1}{7} = y \div 3\frac{2}{3}$

22) $1\frac{1}{2} \div y = 4\frac{3}{4} \div 2\frac{3}{8}$

23) $y \div \frac{4}{25} = \frac{3}{4} \div \frac{4}{5}$

24) $y \div 2\frac{1}{3} = \frac{4}{7} \div 4\frac{1}{6}$

25) $7\frac{1}{3} \div 2\frac{1}{2} = 3\frac{2}{3} \div y$

26) $6\frac{1}{2} \div 3\frac{1}{4} = 3\frac{1}{2} \div k$

27) $m \div 2\frac{1}{2} = 3\frac{1}{4} \div 6\frac{1}{2}$

28) $n \div 2.7 = 8\frac{1}{3} \div 0.5$

29) $2.8 \div p = 1\frac{2}{3} \div 2\frac{6}{7}$

30) $d \div 6.4 = 1\frac{7}{8} \div 7\frac{1}{2}$

31) $c \div 3\frac{1}{3} = 0.9 \div 0.4$

32) $3\frac{1}{3} \div d = 3\frac{3}{5} \div 2.7$

33) $1\frac{7}{9} \div g = 9\frac{1}{3} \div 2\frac{5}{8}$

34) $h \div \frac{7}{12} = \frac{2}{3} \div \frac{7}{9}$

35) $3\frac{3}{5} \div k = 1\frac{4}{5} \div 1\frac{1}{4}$

36) $2\frac{1}{2} \div 1\frac{1}{2} = 6\frac{2}{3} \div m$

37) $3\frac{3}{5} \div \frac{4}{15} = n \div \frac{5}{9}$

38) $2 \div p = 1\frac{4}{7} \div 3\frac{1}{7}$

39) $5\frac{1}{4} \div 3\frac{1}{2} = 4\frac{1}{2} \div u$

40) $3\frac{1}{3} \div 2\frac{1}{2} = 2\frac{2}{3} \div w$

Algebra Chapter 7 Lesson 8

Worksheets 5 for Final Exam

Worksheet 5 A

1) $-2^4 \div 2^2 - 2$

2) $-3^4 \div 3^2 - 3$

3) $-4^3 \div 2^2 - 2^3$

4) $-6^3 \div 2^2 + 3^3$

5) $-8^3 \div 2^2 - 4^3$

6) $-9^3 \div 3^2 - 3^3$

7) $-3^4 - 3^3 \div 3^2$

8) $-4^3 - 4^2 \div 2^2$

9) $-6^3 - 6^2 \div 3^2$

10) $-8^2 - 6^2 \div 2^2$

11) $-2^2 - 8^2 \div 4^2$

12) $-4^2 + 4^2 \div 2^2$

13) $-5^2 - 6^2 \div 2^2$

14) $-6^2 - 8^2 \div 4^2$

15) $-7^2 - 9^2 \div 3^2$

16) $-8^2 - 6^2 \div 3^2$

17) $-4^3 \div (-4 + 2)^2$

18) $-9^2 \div (-6 + 3)^2$

19) $-6^3 \div (-8 + 2)^2$

20) $-5^3 \div (-8 + 3)^2$

21) $-4^3 \div (-6 + 4)^2$

22) $-3^3 \div (-6 + 3)^2$

23) $-2^5 + 2^4 \div 2^2$

24) $-2^5 + 2^4 \div 2^3$

25) $-2^6 + 2^5 \div 2^4$

26) $-2^6 + 2^5 \div 2^3$

27) $-2^8 \div 2^7 - 2^6$

28) $-2^9 \div 2^8 - 2^7$

29) $-2^9 \div 2^7 - 2^8$

30) $-4^3 \div 2^3 - 1^3$

Algebra Chapter 7 Lesson 8

Worksheet 5 B for Final Exam

1) $-3^4 - 3^3 \div 3^2$

2) $-3^5 - 3^4 \div 3^3$

3) $-3^6 - 3^5 \div 3^4$

4) $-3^3 - 4^3 \div 2^3$

5) $-4^2 + 5^2 - 6^2$

6) $-5^4 + 4^4 - 3^4$

7) $-6^3 + 5^3 - 3^3$

8) $-7^2 + 8^2 - 9^2$

9) $-9^3 - 7^3 + 8^3$

10) $-8^3 - 6^3 + 4^3$

11) $[(-13)^2 + (-11)^2] \div (-10)$

12) $[(-15)^2 - (-13)^2] \div (-7)$

13) $[(-17)^2 - (-15)^2] \div (-8)$

14) $[(-19)^2 - (-17)^2] \div (-8)$

15) $[(-21)^2 - (-19)^2] \div (-8)$

16) $[(-23)^2 - (-21)^2] \div (-2)$

17) $[(-25)^2 - (-23)^2] \div (-6)$

18) $-1040 \div [(-27)^2 - (-25)^2)$

19) $[(-6)^2 - (-4)^2][(-4)^2 - (-3)^2]$

20) $[(-8)^2 - (-7)^2] \div [(-3)^2 - (-2)^2]$

21) $-8^2 \div 4^2 - 6^2 \div 3^2$

22) $-9^2 \div 3^2 - 7^2 \div 7$

23) $-10^2 \div 5^2 - 2^3 \div 2^2$

24) $-8^2 \div 2^2 - 9^2 \div 3^2$

25) $-12^2 \div 4^2 - 8^2 \div 4^2$

26) $(-13^2 + 12^2) \div (-3^2 + 2^2)$

27) $(-17^2 + 15^2) \div (-5^2 + 3^2)$

28) $(-15^2 + 12^2) \div (-5^2 + 4^2)$

29) $(-13^2 + 5^2) \div (-5^2 + 3^2)$

30) $(-29^2 + 20^2) \div (-12^2 + 9^2)$

Algebra Chapter 7 Lesson 9
Reviews for Chapter 7
Review 7 A

Date **Name**

Read directions carefully. **Answers must be in the indicated place.**
Show all steps for full credits. **Each problem is worth 5 points.**
Do only 1 operation per 1 step. **Solve 1 problem 5 minutes.**

Simplify the expressions:

1) $100(0.01y - 0.03) - 1000(0.001y - 0.05)$

1)_______

2) $4(\frac{1}{2}x - \frac{1}{4}y) - \frac{2}{3}(6x + 3y)$

2) _______

Solve the equations:

3) $\frac{1}{10}k + 10 = 50$

3) _______

4) $81072 \div m = 9$

4) _____

5) $28n - 987 - 45n = 9247$

5) _______

Algebra Chapter 7 Lesson 9

Review 7 A

Date **Name**

Read directions carefully. **Answers must be in the indicated place.**

Show all steps for full credits. **Each problem is worth 5 points.**

Do only 1 operation per 1 step. **Solve 1 problem 5 minutes.**

Solve the equations

6) $0.1p + 0.2p + 0.3p = 1.08$

6) ______

7) $\frac{3}{10}g - \frac{1}{5}g - \frac{3}{10} = 1\frac{1}{5}$

7) ________

8) $-5(\frac{2}{5}h + \frac{4}{5}) - 4(\frac{3}{4}h - \frac{1}{4}) = 0$

8) ______

9) Solve the proportion $\frac{7.8}{3.9} = \frac{4.4}{x}$

9)______

10) Solve the proportion $5\frac{1}{4} \div 3\frac{1}{2} = 4\frac{1}{2} \div y$

10) ______

Algebra Chapter 7 Lesson 9
Review 7 B

Date **Name**

Read directions carefully. Answers must be in the indicated place.
Show all steps for full credits. Each problem is worth 5 points.
Do only 1 operation per 1 step. Solve 1 problem 5 minutes.

Simplify the expressions

1) $10(0.1x - 4) - 100(0.01x - 5)$

1) ______

2) Simplify the expression

$$\frac{4}{9}(18c - 9d) - 12(\frac{5}{6}c - \frac{1}{6}d)$$

2) ______

Solve the equations

3) $\frac{1}{11}k - 5 = 55$

3) ______

4) $72056 \div m = 8$

4) ______

5) $125n - 876 - 201n = 37580$

5) ______

Algebra Chapter 7 Lesson 9
Review 7 B

Date **Name**

Read directions carefully. **Answers must be in the indicated place.**
Show all steps for full credits. **Each problem is worth 5 points.**
Do only 1 operation per 1 step. **Solve 1 problem 5 minutes.**

Solve the equations

6) $0.1p + 0.4p + 0.5p = 1.5$

6) ______

7) $\frac{4}{5}g - 0.3g - 3 = 1.5$

7) ________

8) $1.6(\frac{3}{4}h - \frac{5}{8}) - 1.8(\frac{2}{9}h - \frac{7}{9}) = 0$

8)______

9) Solve the proportion $\frac{4.4}{z} = \frac{8.8}{2.2}$

9)______

10) Solve the proportion $3\frac{1}{3} \div 2\frac{1}{2} = 2\frac{2}{3} \div y$

10) ______

Algebra Chapter 7 Lesson 10
Tests for Chapter 7
Test 7 A

Date **Name**

Read directions carefully. Answers must be in the indicated place.
Show all steps for full credits. Each problem is worth 5 points.
Do only 1 operation per 1 step. Solve 1 problem 5 minutes.

Simplify the expressions

1) $0.3z + 0.2(2z - 3) - 0.8 - 0.7(z - 2)$

1) ______

2) $16(\frac{3}{4}c + \frac{1}{4}d) - \frac{2}{3}(9c + 27d)$

2)_______

Solve the equations

3) $\frac{1}{8}k - 10 = 40$

3) ________

4) $49056 \div m = 7$

4) ________

5) $-78n - 678 + 9n = 20298$

5)_______

Algebra Chapter 7 Lesson 10
Test 7 A

Date **Name**

Read directions carefully. **Answers must be in the indicated place.**
Show all steps for full credits. **Each problem is worth 5 points.**
Do only 1 operation per 1 step. **Solve 1 problem 5 minutes.**

Solve the equations

6) $0.8p - 0.3p - 3 = 1.5$

6) ________

7) $\frac{1}{10}g + 5 + \frac{1}{10}g = 10$

7) ________

8) $-6(\frac{1}{6} - \frac{1}{2}h) - 2(1 - \frac{1}{2}h) = 0$

8) ________

9) $\frac{x}{4.1} = \frac{2.7}{0.9}$

9) ________

10) $y \div \frac{7}{12} = \frac{2}{3} \div \frac{7}{9}$

10) ______

Algebra Chapter 7 Lesson 10
Test 7 B

Date **Name**

Read directions carefully. **Answers must be in the indicated place.**
Show all steps for full credits. **Each problem is worth 5 points.**
Do only 1 operation per 1 step. **Solve 1 problem 5 minutes.**

Simplify the expressions

1) $1.6(y - 4) - 0.6 - 0.8(2y - 8)$

1)______

2) $\frac{3}{7}(14c - 21d) - 15(\frac{2}{3}c - \frac{1}{3}d)$

2) ______

Solve equations

3) $\frac{1}{9}k + 15 = 45$

3) ________

4) $64072 \div m = 8$

4)________

5) $9n + 789 - 28n = 8541$

5)_______

Algebra Chapter 7 Lesson 10
Test 7 B

Date **Name**

Read directions carefully. Answers must be in the indicated place.
Show all steps for full credits. Each problem is worth 5 points.
Do only 1 operation per 1 step. Solve 1 problem 5 minutes.

Solve the equations

6) $0.6p - 0.3p - 0.2 = 2.5$

6)________

7) $\frac{1}{10}g + \frac{2}{5}g + \frac{1}{2} = 1\frac{1}{2}$

7)________

8) $-8(\frac{1}{2} - \frac{1}{4}h) - 3(\frac{2}{3} - \frac{1}{3}h) = 0$

8)________

9) $\frac{x}{4.7} = \frac{2.7}{0.9}$

9) ________

10) $2 \div y = 1\frac{4}{7} \div 3\frac{1}{7}$

10) ______

Algebra Answers Chapter 7

Lesson 1 Pretest

1) p =9 2) g = 2 3) h = 2 4) x = 14.1 5) y=4

Lesson 2 Identities. Review 1 A

1)	x + 22	2)	- x + 49	3)	- y + 49	4)	x - 7
5)	5y – 22	6)	8x + 30	7)	8x – 54	8)	y - 8
9)	13y – 65	10)	31x + 62	11)	- 12.5y + 6	12)	- 9.8x - 1
13)	- 28y + 5	14)	- 0.4z	15)	- 2.4z + 3	16)	- 0.15x - 9
17)	0.4y+14	18)	0	19)	460	20)	47

Lesson 2 Review 1 B

21)	x + 11	22)	3y - 11	23)	- 2y + 1	24)	x - 1
25)	- 9	26)	4y - 4	27)	- 2z – 8	28)	x - 3y
29)	- 3x + 8	30)	0	31)	3x – 3	32)	3y - 4
33)	- x - 7	34)	- 6x	35)	- 5y + 9	36)	4x – 7y
37)	6x – 14y	38)	- 4x – 4y	39)	- 2x – 3y	40)	- 2x – 2y

Lesson 3 Equations with Like Terms. Review 1 A

1)	12	2)	20	3)	11	4)	105
5)	4	6)	11	7)	14	8)	101
9)	11	10)	204	11)	18	12)	15
13)	4	14)	40	15)	1	16)	9
17)	0	18)	306	19)	- 407	20)	- 508

Review 1 B

21)	5	22)	50	23)	25	24)	4
25)	- 333	26)	- 1963	27)	10	28)	0
29)	38	30)	- 783	31)	36	32)	406
33)	507	34)	- 609	35)	- 102	36)	- 203
37)	- 304	38)	- 405	39)	- 506	40)	- 607

Lesson 4 Find unknowns. Review 1 A

1)	3	2)	1	3)	689	4)	444
5)	489	6)	51	7)	629	8)	2770
9)	645	10)	1543	11)	71	12)	39
13)	458	14)	159	15)	309	16)	86
17)	1859	18)	444	19)	408	20)	358

Review 1 B

21)	8	22)	3	23)	56	24)	78
25)	64	26)	729	27)	2028	28)	1824
29)	75640	30)	78848	31)	9	32)	5
33)	6	34)	8	35)	28	36)	39
37)	208	38)	509	39)	507	40)	407

Algebra Answers Chapter 7

Lesson 5 Equations with Decimals. Review 1 A

1) 2	2) 1.1	3) 11	4) 1.1	5) 12
6) 10.1	7) 0	8) 3.8	9) 0	10) 0
11) 1	12) 0.4	13) 0.5	14) 25	15) 5
16) 15	17) 9	18) 9	19) 1.8	20) 1.5

Lesson 5 Equations with Fractions. Review 1 B

1) 0.6	2) 0.2	3) 0.8	4) 0.6	5) 0.4
6) 2	7) 0.08	8) 2	9) 5	10) 4
11) 2.5	12) 0.8	13) 2	14) 125	15) 0.12
16) 40	17) 1.2	18) 8.77	19) 12	20) 1.1

21) 2	22) 11	23) 1.1	24) 44	25) 12
26) 11	27) 3.8	28) 0	29) 0	30) 1.5
31) 0.4	32) 0.4	33) 1	34) 1.8	35) 0.5
36) 10	37) 25	38) 0	39) 15	40) 9

Lesson 6 Conditional equations. Review 1 A

1) - 22	2) - 16	3) 49	4) 40	5) 15
6) - 0.4	7) 0.75	8) 1	9) - 2	10) - 3
11) 6.75	12) 0	13) - 2	14) 5	15) 1.25
16) 24	17) - 35	18) 7.5	19) 0.5	20) 2

Lesson 6 Review 1 B

21) - 11	22) 2	23) - 7	24) 1	25) - 1
26) - 0.5	27) 4	28) 2	29) 1	30) - 0.5
31) 10	32) - 17	33) 1	34) -0.2	35) 9.5
36) -2	37) 0.75	38) 2	39) - 0.2	40) - 0.6

Algebra Answers Chapter 7

Lesson 7 Proportions.
Review 1 A

1)	3	2)	18	3)	28	4)	48	5)	81
6)	2.8	7)	15.3	8)	14.1	9)	1.5	10)	1.1
11)	8	12)	2	13)	0.5	14)	0.1	15)	0.8
16)	0.5	17)	0.5	18)	1.5	19)	2.5	20)	6.4

Lesson 7
Review 1 B

21)	0.25	22)	0.75	23)	0.15	24)	0.32	25)	1.25
26)	1.75	27)	1.25	28)	45	29)	4.8	30)	1.6
31)	7.5	32)	2.5	33)	0.5	34)	0.5	35)	2.5
36)	4	37)	7.5	38)	4	39)	3	40)	2

Lesson 8 Worksheets 5 A for Final Exam

1)	- 6	2)	- 12	3)	- 24	4)	- 27	5)	- 192
6)	- 108	7)	- 84	8)	- 68	9)	- 220	10)	- 73
11)	- 8	12)	- 12	13)	- 34	14)	- 40	15)	- 58
16)	- 68	17)	- 16	18)	- 9	19)	- 6	20)	- 5
21)	- 16	22)	- 3	23)	- 28	24)	- 30	25)	- 62
26)	- 60	27)	- 66	28)	- 130	29)	- 260	30)	- 9

Lesson 8 Worksheet 5 B for Final Exam

1)	- 84	2)	- 246	3)	- 732	4)	- 35	5)	- 27
6)	- 450	7)	- 118	8)	- 66	9)	- 560	10)	- 664
11)	- 29	12)	- 8	13)	- 8	14)	- 9	15)	- 10
16)	- 44	17)	- 16	18)	- 10	19)	140	20)	3
21)	- 8	22)	- 16	23)	- 6	24)	- 25	25)	- 13
26)	5	27)	4	28)	9	29)	9	30)	7

Algebra Answers Chapter 7

Lesson 9 Reviews 7

1) 47	2) - 2c – 3d	3) 400	4) 9008	5) - 602
6) p = 1.8	7) g = 15	8) h = - 0.6	9) x=2.2	10) x = 3

Lesson 9 Review 7 B

1) 460	2) - 2c – 2d	3) 660	4) 9007	5) - 506
6) p = 1.5	7) g = 9	8) y= - 0.5	9) z = 1.1	10) y = 2

Lesson 10 Tests 7 A

1) 0	2) 6c – 14d	3) 400	4) 7008	5) - 304
6) p = 9	7) g=25	8) h = 0.75	9) x=12.3	10) y=0.5

Lesson 10 Test 7 B

1) - 0.6	2) - 4c – 4d	3) 270	4) 8009	5) - 408
6) p =9	7) g = 2	8) h = 2	9) x = 14.1	10) y=4

Algebra

Chapter 8

Word

Problems

Students!

Students, you are smarter than a calculator!
Students, you are smarter than a computer!
Students, you need to believe in yourself.
Students, be patient and do your best.

Do not be afraid to make mistakes.
Tell yourself positive words.
I am young! I am smart!
I am strong! I can do anything!

Smart students use their brain!
Your brain is always with you!
A calculator is not always with you.
Mathematics is work for the brain.

Be open to new ideas in math!
Work hard and try your best!
Check answers and ask questions!
Practice! Practice! Practice!

Algebra Chapter 8 Lessons
Word Problems
Lessons 1 – 12
Advanced Problems

Lesson 1 Pretest

Lesson 2 Operations

Lesson 3 Worth

Lesson 4 Decimals

Lesson 5 Fractions

Lesson 6 Direct Proportionality

Lesson 7 Uniform Motion

Lesson 8 Percent

Lesson 9 Tables 11, 12, 13, 14, Solve Mental Math 17, 18 odd

Lesson 10 Worksheet 6 A, Worksheet 6 B

Lesson 11 Review 8 A, Review 8 B

Lesson 12 Tests 8 A, Test 8 B

Answers for Chapter 8

Algebra Chapter 8 Contents
Word Problems

Contents have topics and exercises, which students need to solve before tests.1)Work with lesson 1, read rules and examples, 2) Solve the problems from contents, 3) Check answers in the end of chapter 1, 4) Work with the next lessons 2, 3, 4, 5, …, 5) Solve Test 1 A, 6) Grading Scale without minuses

A=90–100%   B+=85–89%   B=80–84%   C+=75–79%
C=70–74%   D+=65–69%   D=60–64 %   F=0 - 59%

7)If you solve 10 problems correctly, you get an A, 8) If have an A, you can move on to next chapter, 9) If your grade is lower than an A, work with the mistakes, then solve Test 1B, 10) When you finish all of the chapters, solve Final Exam in last chapter.

Lesson 1 Work with Table Word Problems. Solve the Pretest #1–#5
Lesson 2 Operations. Solve Review 1 A # 1 - # 5
Solve Review 1 B # 1 - # 5. Solve Review 1 C # 1 - # 5
Lesson 3 Worth. Solve Review 1 A # 1 - # 5
Solve Review 1 B # 1 - # 5. Solve Review 1 C # 1 - # 5
Lesson 4 Decimals
Solve Review 1 A # 1 – # 5. Solve Review 1 B # 1 – # 5
Lesson 5 Fractions
Solve Review 1 A # 1 – # 5. Solve Review 1 B # 6 – # 10
Lesson 6 Direct Proportionality
Solve Review 1 A # 1 – # 5. Solve Review 1 B # 1 – # 5
Lesson 7 Union Motion
Solve Review 1 A # 1 - # 10. Solve Review 1 B # 1 - # 10
Lesson 8 Precent
Solve Review 1 A # 1 - # 20 odd. Solve Review 1 B # 1 - # 20 odd
Lesson 9 Tables 11, 12, 13, 14. Solve Mental Math 17, 18 odd
Lesson 10 Worksheet 6 A for Final Exam # 1 – # 20 odd
Worksheet 6 B for Final Exam # 1 – # 30 odd
Lesson 11 Reviews
Solve Review 8 A # 1 – # 10 all. Solve Review 8 B # 1 – # 10 all
Lesson 12 Tests
Solve Test 8 A # 1 – # 10 all. Solve Test 8 B # 1 – # 10 all

Algebra Chapter 8 Lesson 1

Table 11 Word Problems

1) Decimals

a) To find the amount by known base, multiply the base by a decimal:

Find the amount $\quad$ **0.6 of 40** $\quad$ $40 \times 0.6 = 24$

b) To find the base by known amount, divide the amount by a decimal:

Finding the base: $\quad$ **0.6 is equivalent \$ 24** $\quad$ $24 \div 0.6 = 40$

c) To find the ratio between two numbers, divide the numbers.

Find the ratio $\quad$ **24 from 40** $\quad$ $24 \div 40 = 0.6$

2) Fractions

a) To find the amount by known base, multiply the base by a fraction.

Finding the amount: $\quad$ $\frac{3}{5}$ **of 40** $\quad$ $40 \times \frac{3}{5} = \$24$

b) To find the base by known amount, divide amount by a fraction.

Finding the base: $\quad$ $\frac{3}{5}$ **equivalent 24** $\quad$ $24 \div \frac{3}{5} = \$40$

c) To find the ratio between two numbers, divide the numbers.

Find the ratio: $\quad$ **24 from \$40** $\quad$ $24 \div 40 = \frac{24}{40} = \frac{3}{5}$

3) Uniform Motion: Distance - d $\quad$ Time - t $\quad$ Rate – r (Speed)

a) To find distance, multiply rate by time: $d = rt$

What distance did a man go for 2 hours if his rate was 5 mi/hr?

$d = 5 \cdot 2 = 10$ mi

b) To find time, divide distance by speed. $t = d \div r$

How much time does a man needs if he runs 8 mi with a rate 4 mi/hr?

$t = 8 \div 4 = 2$ hr

c) To find speed, divide distance by time. $r = d \div t$

What is the rate a girl has, if she runs 6 mi for 2 hours

$r = 6 \div 2 = 3$ mi / hr

4) Directly proportional quantities

If 4 apples sell for \$1.32. How much do 11 apples cost?

$132 \div 4 = 33$ c $\qquad$ $33 \times 11 = 363$ c = \$3.63

Algebra Chapter 8 Lesson 1

Date **Name**

Pretest

1) **A bus drove 46 mi at a speed of 20 mi/h. How much time it spent?**

1) ______

2) **What speed does a boy have, if he drove 65 mi in 4 hours?**

2) ______

3) **Of 24 children, 62.5% were boys. How many were girls?**

3) ______

4) **A factory has 6 new machines and 60% were old machine. How many machines are in the factory?**

4) ______

5) **In 75 minutes of lesson, students solve an equation in 30 minutes. What percent of lesson time did students use to solve the equation?**

5) _____

Algebra Chapter 8 Lesson 2

Word Problems on Operations

a) Straight problem
First girl has 3 apples
Second girl has 7apples
Sum of apples is x apples
$x = 3 + 7 = 10$ apples

b) Opposite problem
1 girl has 3 apples
2 girl has y apples
Sum of apples is 10 apples
$y = 10 - 3 = 7$ apples

c) Straight problem
I have $10
I spend $ 4
Left $ x
$x = 10 - 4 = \$\,6$

d) Opposite problem
I have $ y
I spend $ 4
Left $ 6
$y = 4 + 6 = \$\,10$

Review 1 A

Solve the Word Problems

1) There are 35 flowers. 24 more are planted.
   How many are there total?
2) There are 30 red roses and 59 white roses.
   How many flowers are there in total?

3) Amy bought a candle for 25 cents and napkins for 44 cents.
   How much did she spend?
4) There are 28 birds. 30 more come. How many are there now?

5) Olivia bought a pencil for 12 cents, a pen for 34 cents,
   and notebooks for 42 cents. How much did she spend?
6) Ann has 16 pencils. Bob has 14 pencils.
   How many more does Ann have than Bob?

7) There are 18 balloons. Two of the balloons broke.
   How many are left?
8) Bob have 64 roses. 53 are red. The rest are white.
   How many white ones are there?

9) A store has 88 hammers. The store sold 64.
   How many hammers does the store have now?
10) Lisa has 74 cents. She spent 34 cents. How much does she have left?

Algebra Chapter 8 Lesson 2
Review 1 B
Solve the Word Problems on Operations

1) Father works 40 hours each week.
How many hours does he work in 2 weeks?
2) There are 8 rows of seats in the balcony.
There are 35 seats in each row.
How many seats are in the balcony?

3) A student-spelling book has 25 new words on each page.
There are 9 pages in the first section.
How many new words are there in the first section?
4) There are 4 classrooms on the first floor.
Each classroom has 30 seats.
How many seats are on the first floor?

5) Mother works 8 hours each day.
How many hours will she work in 5 days?
6) Michele has 63 peaches. She puts 7 peaches in a package.
How many packages will she have?

7) David worked the same number of hours each day.
He worked 21 hours in 7 days.
How many hours did he work each day?
8) One side of a building has 24 windows.
Each floor has 6 windows on that side.
How many floors does the building have?

9) Mary has 28 balloons for a party.
She will give each person 4 balloons.
How many people will receive balloons?
10) The 45 students in a class separated into 5 groups.
Each group has the same number of students.
How many are students in each group?

Algebra Chapter 8 Lesson 2
Review 1 C
Solve the Word Problems

1) Al had 32 markers. He bought 14 more. How many are there now?

2)Rachel had 26 nickels. She got 43 more. How many does she have now?

3)There are 9 black pens and 4 blue pens. How many more black pens there than blue pens?

4) 24 cups are on the table but 2 are broken. How many are not broken?

5) The teacher gave each student 2 sheets of paper. There were 25 students. How many sheets of paper did he use?

6) There are 25 baseball players on each team. How many players are there total on 2 teams?

7) Michael bought 2 packages of paper. Each package had 30 sheets. How many sheets of paper did he buy?

8) A farmer planted 54 trees in 9 rows. Each row had the same numbers of trees. How many trees were in each row?

9) Yuri worked 20 hours in 5 days. He worked the same numbers of hours each day. How many hours did he work each day?

10) Olivia read 35 pages of a book in 7 days. She read the same number of pages each day. How many pages did she read each day?

Algebra Chapter 8 Lesson 3
Word Problems on Worth
Measurement of worth - $ 1, 1c

To solve word problems at worth, reduce problem to 1 unit.

How many does cost 15m silk if 8 m of silk cost $ 32?

$\$32 \div 8 = \$4$ $\$4 \times 15 = \$60$

Review 1 A
Solve the Word Problems

1) **Amanda bought 5 bags for 2 c for each. How much money did she pay?**

2) **Monica has 4 coins of 5 c each. How much money does she have?**

3) **1 lb of cabbage cost 10 c. How much money does 3 lbs of cabbage cost?**

4) **Nicole bought a journal for 30 c and a book, which cost 3 times more than the journal. How much did the book cost?**

5) **Sophia bought 24 postcards for 3 c each. How much did she spend?**

6) **A cap cost $6 and a coat cost 10 times more than the cap. How much did the coast coat?**

7) **Amy bought 3 meters of fabric for $4 for a meter. She has $38 remaining. How much money did she have?**

8) **In the park are planted 4 rows of roses with 5 plants in each row. How many roses planted?**

9) **Mother spilled 2 liters of milk in each of 4 cans. How many liters of milk does mother have in all?**

10) **Three boxes of oranges, 4 lb each, brought into the restaurant. How many pounds of oranges brought into the restaurant?**

Algebra Chapter 8 Lesson 3
Review 1 B
Solve the Word Problems on Worth

1) Rachel needs to sew 4 buttons onto one coat.
How many buttons does the woman need to sew onto 2 coats?

2) A house has 5 floors and each floor has a height of 4 meters.
How many meters high is this house?

3) A chair cost $40 and table cost 6 times more than the chairs.
How much did the table cost?

4) One fork costs 30 c. How much a set of 3 would forks cost?

5) One balls costs $2. How much do 6 balls cost?

6) Jenna bought 2 ft of silk for $8. How much money does cost 1 ft of silk cost?

7) Linda paid 14 c for 2 packets of salt.
How much money does 1 packet of salt cost?

8) April bought on $1.20 envelopes by 6 c for each.
How many envelopes could she buy?

9) Tiffany bought a dress for $27 and shoes, which cost 3 times less than the dress. How much did the shoes cost?

10) Alana paid for $24 for 1 meter of wool and $6 for 1 meter of silk.
How many times costlier is the wool than the silk?

Algebra Chapter 8 Lesson 3
Review 1 C
Solve the Word Problems

1) A book cost \$20. How many books can Jackie buy for \$140?

2) Brett has 60c. How many tickets to the movie can he buy for this money, if 1 ticket costs 10 c?

3) Mark divided 6 balls by 2 balls for each boy. How many boys were there?

4) Amy laid out 18 pencils from 3 boxes equally. How many pencils were in each box?

5) Claire divided 12 tulips into bouquets with 4 tulips in each. How many bouquets are there?

6) Carl loaded 16 logs on 2 trucks equally. How many logs did they load in each truck?

7) Andrew tied 10 carrots in bunches with 5 carrots in each. How many bunches are there?

8) On the first day, 8 bags were sold for \$32. On the second day 6 of the same bags were sold. How much money paid on the second day?

9) A set of towels cost \$1 and 3 packs of napkins cost 75 c. How many packs of napkin can bought instead of a set of towels?

10) A table cost \$240 and a chair cost 6 times less than the table. How much did the table and chair cost together?

Algebra Chapter 8 Lesson 4

Word Problems on Decimals

In word problems in decimal each word has special meanings
"Base" (B) represents "Whole Quantity",
"Amount" (A) usually represents "Portion of Whole Quantity"

Finding the Amount by Known Base

Amy had whole money \$40. She spent 0.6 of all her whole money. How much amount of money did she spend? To find how much money did she spent, multiply the base, \$40, by the decimal 0.6.

0.6 of \$40 $\quad$ $\$40 \times 0.6 = \$24$

To find the amount of money by known base, multiply the base by a decimal.

Finding the Base by Known Amount

Amy spent 0.6 of all her whole money; this is equals \$24. How much of whole money did she have before? To find how much whole money she had, divide amount of the money, \$24, by the decimal 0.6.

0.6 equivalent \$24 $\quad$ $\$24 \div 0.6 = 240 \div 6 = \$40$

To find the base by known amount, divide the amount by the decimal.

Find the Ratio

Amy has \$40. She spent \$24. What part of the money did she spent? To find the ratio between the money she spent to the whole of her money, divide the money she spent, \$24, by her whole money, \$40.
Write ratio as a decimal.

\$24 from \$40 $\quad$ $\$24 \div \$40 = 0.6$

To find ratio between two numbers, divide the numbers.

Average

The quotient of division of the sum of numbers by the number of addends called an average.
Example: Find the average of numbers 7.8, 8.4, 9.3
$(7.8 + 8.4 + 9.3) \div 3 = 25.5 \div 3 = 8.5$

Algebra Chapter 8 Lesson 4
Review 1 A
Solve the Word Problems for Decimals

1)In class there were 35 students. Students who can swim are 0.2 of the whole class. How many students can swim?

2) A book has 60 pages. A student reads 0.5 of the whole book. How many pages does he read?

3) Andrew has 15 stamps. He gave his brother 0.2 of all of the stamps. How many stamps did the boy gave to his brother?

4) A tailor prepared 8 dresses. He sewed 0.5 of all of the dresses. How many dresses did he sew?

5) A kindergarten has 36 children. Girls made up 0.25 of all of the children. How many girls were in the kindergarden?

6) A notebook has 24 pages. A girl wrote in 0.625 of the whole notebook. How many pages remain?

7) A book cost $18, and it equals 0.3 of my whole money. How much money do I have?

8) Anna ate 6 plums; that equals 0.6 of all of the plums. How many plums did I have?

9) Briana read 10 pages; that equals 0.4 of all of the pages in the book. How many pages are in the book?

10) In the class, 8 students can dance; this is 0.25 of the whole class. How many students are in class?

Algebra Chapter 8 Lesson 4
Review 1 B
Solve the Word Problems

1) Lucy ate 5 of 10 apples. What part of apples did she eat? Write ratio as a decimal.

2) April read 20 pages. What part of the book did she read, if the book has 50 pages? Write ratio as a decimal.

3) Ben has 24 pencils. He gave his friend 6 pencils. What part of pencils did he gave to his friend? Write ratio as a decimal.

4) Lisa has $36. She spent $9. What part of the money did she spends? Write ratio as a decimal.

5)The shop had 900 lbs. of sugar. On the first day, they sold 0.3 of this sugar, and on the second day, they sold the same. How many pounds did they sell in the two days together?

6) A book has 80 pages. David read 0.25 of the whole book. How many pages more does he have left to read?

7) A class had 40 students. The girls made up 0.625 of the whole class. How many boys were in the class?

8)Mother has $10. She used 0.6 of her whole money to buy 4 loaves of bread. What much is one bread worth?

9) Find the average for the numbers: 1.5, 2.2, 3.2?

10) Find the average for the numbers: 4.5, 5.2, 6.2, 8.1?

Algebra Chapter 8 Lesson 5

Word Problems on Fractions

In word problems in fraction each word has special meanings
"Base" (B) represents "Whole Quantity",
"Amount" (A) usually represents "Portion of Whole Quantity"

Finding Amount by Known Base

Amy had whole money \$40. She spent $\frac{3}{5}$ of her whole money. How much amount money did she spend? To find how much money did she spent, multiply the base, \$40, by the fraction $\frac{3}{5}$:

$\frac{3}{5}$ of \$40 $\quad\quad$ **$\$40 \times \frac{3}{5} = \$24$**

To find amount by the known base, multiply the base by a fraction or divide the base by denominator; after this, multiply the result at the numerator: $\$40 \div 5 \times 3 = \$8 \times 3 = \$24$

Finding Base by Known Amount

Amy spent $\frac{3}{5}$ of her whole money; this is equals \$24. How much whole money did she have before? To find how much whole money she had, divide amount of the money, \$24, by the fraction $\frac{3}{5}$:

$\frac{3}{5}$ equivalent \$24 $\quad\quad$ **$\$24 \div \frac{3}{5} = 24 \times \frac{5}{3} = 24 \div 3 \times 5 = \$40$**

To find the base by known amount, divide amount by the fraction or multiply amount by the reciprocals for the fraction.

Find the Ratio

Amy has \$40. She spent \$24. What part of the whole money did she spent? To find the ratio between the money she spent to the whole of her money, divide the money she spent, \$24, by her whole money, \$40. Write ratio as a reduced fraction.

\$24 from \$40 $\quad\quad$ **$\$24 \div \$40 = \frac{24}{40} = \frac{3}{5}$**

To find ratio between two numbers, divide the numbers.

Algebra Chapter 8 Lesson 5
Review 1 A
Solve the Word Problems for Fractions

1) A class has 35 students. Students, who can swim make up $\frac{1}{5}$ of the whole class. How many students can swim?

2) A book has 60 pages. Student read $\frac{1}{2}$ of the whole book. How many pages did he read?

3)Andrew has 15 stamps. He gave his brother $\frac{1}{5}$ of all of the stamps. How many stamps did Andrew give his brother?

4) A tailor prepared 8 dresses. He sewed $\frac{1}{2}$ of all of the dresses. How many dresses did he sew?

5) A kindergarten has 36 children. The girls make up $\frac{1}{4}$ of the whole children. How many girls are in the kindergarden?

6) A notebook has 24 pages. Amanda wrote in $\frac{5}{8}$ of the whole notebook. How many pages remain?

7) A shop has 900 lb of sugar. On the first day, they sold $\frac{3}{10}$ of all sugar. On the second day, they sold the same amount. How many pounds did they sell in the two days together?

8) A book has 80 pages. Mike read $\frac{1}{4}$ of the whole book. How many pages more does he have left to read?

9) A class has 40 students. The boys' make-up $\frac{5}{8}$ of the whole class. How many girls are in the class?

10) Mother has \$10. She used $\frac{3}{5}$ of her whole money to buy 4 loaves bread. How much is one bread worth?

Algebra Chapter 8 Lesson 5
Review 1 B
Solve the Word Problems for Fractions

1) A book cost \$18; this it equals $\frac{3}{10}$ of the whole money. How much is the whole money?

2) Sara ate 6 plums; this is equals $\frac{3}{5}$ of all of the plums. How many plums did she have?

3) Brandon read 10 pages; this it equals $\frac{1}{10}$ of the whole pages of the book. How many pages were in the book?

4) In the class, 10 students can dance; this is equals $\frac{1}{4}$ of the whole class. How many students are in the class?

5) Casey ate 5 of 10 apples. What part of apples did she eat? Write ratio as a reduced fraction.

6) Kristy read 20 pages. What part of the book did she read, if the book has 50 pages? Write ratio as a reduced fraction.

7) Lillian has 24 pencils. She gave her son 6 pencils. What part of the pencils did she give to her son? Write ratio as a reduced fraction.

8) Claire had \$36. She spent \$9. What part of her money did she spend? Write ratio as a reduced fraction.

9) Find the average for the numbers: $1\frac{1}{2}$, $2\frac{1}{5}$, $3\frac{1}{5}$.

10) Find the average for the numbers: $4\frac{1}{2}$, $5\frac{1}{5}$, $6\frac{1}{5}$, $8\frac{1}{10}$.

Algebra Chapter 8 Lesson 6

Word problems on Direct Proportionality
Directly Proportional Quantities

If first quantity increases or decreases, then the second quantity increases or decreases in the same way, then both quantities are called directly proportional.

a) If 4 apples sell for $1.32. How much do 11 apples cost?

$132 \div 4 = 33$ c $\qquad$ $33 \times 11 = 363$ c = \$ 3.63

First Property of Direct Proportionality Quantities

If the ratio corresponding value of two quantities is a constant number then these quantities are directly proportional.

$132 \div 4 = 33$ c $\qquad$ $363 \div 11 = 33$ c

Second Property of Directly Proportional Quantities

If two quantities are directly proportional, then the ratios between corresponding value of two quantities is equal.

$132 \div 4 = 33$ c $\qquad$ $363 \div 11 = 33$ c

$132 \div 4 = 363 \div 11$

b) Pineapple costs $12 for 4 lb. At this rate, how much would 3 lb of pineapple cost? $\quad 12 \div 4 = \$ 3 \qquad 3 \times 3 = \$ 9$

c) A car can travel 264 mi on 11 gallons of gas. At the same rate, how many gallons of gas would be needed to travel 528 mi?

$264 \div 11 = 24$ mi/gal $\qquad$ $528 \div 24 = 22$ gal

d) A candle will burn 2 in for 40 min. How many inches of the candle will burn in 1 hour? $\qquad$ 1 hour = 60 minutes

$40 \div 2 = 20$ min/in $\qquad$ $60 \div 20 = 3$ in

e) Ashley has lost 2 lb in 4 weeks. At this rate, how long will it take her to lose 5 lb?

$2 \div 4 = 0.5$ lb/week $\qquad$ $5 \div 0.5 = 10$ week

Algebra Chapter 8 Lesson 6
Review 1 A
Solve the Word Problems on Direct Proportionality

1) How many would 3 pens cost, if 5 cost $ 2.50?

2) How many would 5 notebooks cost, if 3 notebooks cost 12 c ?

3) How many details can a worker do for 4 hours, if he does 28 etails for 2 hours?

4)How many pillowcases can a woman sew for 6 days, if she can sew 12 pillowcases in 2 days?

5)A worker can make 15 details in 3 hours. How much time does he need to make 40 details?

6)A studio can sew 12 dresses in 3 days. How many days would they need if they want to sew 20 dresses?

7) Amy can type 18 pages in 6 hours. How many page can she type in 4 hours?

8) A carpenter can make 36 frames for windows in 3 days. How many frames can he make in 2 days?

9) Anna can bind 160 books in 20 days. How many books can she bind in 10 days?

10) A gardener can cut down 80 trees in 5 days. How many trees can he cut down in 3 days?

Algebra Chapter 8 Lesson 6
Review 1 B
Solve the Word Problems

1) A factory got 632 lb of starch from 4 tons of potatoes.
How much starch can a factory get from 6 tons of potatoes?

2) A team made 750 keys in 3 hours. How many keys can the team make in 2 hours?

3) In a year, 10,000 m^2 of forest gives oxygen to 300 people.
How many people can get oxygen from 900,000m^2 of forest?

4) A factory got 16 lb of apple puree from 20 lb of apples.
How much apple puree can the factory get from 45 lb of apples?

5) A factory got 5 lb of oil from 20 lb of seed.
How much oil can the factory get from 7 lb of seed?

6) Brain paid $1.75 for 2.5 lb of cookies.
How much of the same cookie can the boy buy for $2.45?

7) Linda paid $1.8 for 200 gram of candy.
How much does 1 kg of the same candy cost? 1 kg = 1000 g

8) David paid $60 for 15 balls. How many balls can he buy for $20?

9)A girl paid $20 for 4 books. How much 6 of the same would books cost?

10) Vicki paid $24 for 6 months for a newspaper. How many months can the woman get the same newspaper for $ 16?

Algebra Chapter 8 Lesson 7

Word Problems on Uniform Motion

The motion of a moving object whose speed and direction do not change is called uniform motion.

Word problem at moving content three quantities

Distance - d   Time - t   Rate – r (Speed)

Rate – Distance passed for time unit (for 1 hour, 1 minute)

Movement units

Distance (d) measure mi, m, km

Time (t) measure in min, hrs.

Rate (r) measure in mi/hrs., m/min, km/hrs.

Distance d

To find distance, multiply rate by time: $d = rt$

What distance did a man runs in 2 hours, if his rate was 5 mi/hr?

$d = 5 \times 2 = 10$ **mi**

If men drive at the speed of 70mph for 2 hours, how far will he travel?

$d = 70\times2=140$ **mi**

Time t

To find time, divide distance by speed: $t = d \div r$

How much time needed if a man went 8 mi with a rate of 4 mi/hrs.?

$t = 8 \div 4 = 2$ **hrs.**

If a man drives at a speed of 52 mph for 91 mi, how long will he travel?

$t = d \div r = 91 \text{ mi} \div 52 \text{ mph} = 7 \div 4 = 1.75$ **hours**

Rate (Speed) r

To find the speed, divide distance by time: $r = d \div t$

How much rate does a girl have, if she runs 6 mi in 2 hours.

$r = 6 \div 2 = 3$ **mi / hr.**

Woman drove a distance of 252 mi in $4\frac{2}{3}$ hours. What was her speed?

$r = d \div t = 252 \div 4\frac{2}{3} = \frac{252}{1} \times \frac{3}{14} = 18\times3 = 54$ **mph**

Algebra Chapter 8 Lesson 7
Review 1 A
Find the Unknowns of Uniform Motion

		Distance	Time	Speed
1) a	Pedestrian	x mi	4 hrs.	5 mi/hr.
1) b	Tractor	x mi	2 hrs.	30 mi/hr.
1) c	Train	x mi	3 hrs.	120 mi/hr.
1) d	Plane	x mi	2 hrs.	1000 mi/hr.

2) a	Skier	26 mi	x hrs.	13 mi/hr.
2) b	Train	240 mi	x hrs.	60 mi/hr.
2) c	Car	160 mi	x hrs.	80 mi/hr.
2) d	Plane	5000 mi	x hrs.	2500 mi/hr.

3) a	Bicycle	28 mi	2 hrs.	x mi/hr.
3) b	Car	210 mi	3 hrs.	x mi/hr.
3) c	Plane	15000 mi	3 hrs.	x mi/hr.

4) a	Eagle	x mi	1 hr.	120 mi/hr.
4) b	Starling	x mi	2 hrs.	80 mi/hr.
4) c	Swallow	x mi	3 hrs.	74 mi/hr.

5) a	Deer	x mi	3 hrs.	10 mi/hr.
5) b	Shark	72 mi	x hrs.	36 mi/hr.
5) c	Ship	280 mi	4 hrs.	x mi/hr.

6) a	Sparrow	120 mi	2 hrs.	x mi/hr.
6) b	Zebra	45 mi	x hrs.	15 mi/hr.
6) c	Boat	x mi	4 hrs.	24 mi/hr.

7) a	Car	The same	2 hrs.	60 mi/hr.
7) b	Car	The same	3 hrs.	x mi/hr.

Algebra Chapter 8 Lesson 7

Find the unknowns

	Distance	Time	Rate
8) a	x miles	2 hours	60 mi/hr.
8) b	120 miles	x hours	60 mi/hr.
8) c	120 miles	2 hours	x mi/hr.

9) a	x mi	2 hrs.	40 mi/hr.
9) b	24 mi	x hrs.	6 mi/hr.
9) c	300 mi	5 hrs.	x mi/hr.

10) a	70 m	7 secs	The same
10) b	90 m	x secs	The same

Review 1 B

1) A man runs with a speed of 5 mi/h and spent 4 hours. What distance did he pass?
2) A motorcyclist drove 2 hours with a speed of 60 km/h. What distance did he drive?
3) A boy runs 6 meters in 1 second. What distance can he run in 15 seconds?
4) A sound runs 330 meters in the air for 1 second. What distance can the sound run for 3 seconds?
5) Skier skies 3 hours with a speed of 12 km/h. What distance did he run?
6) A plane flies 675 km in one hour. What distance can the plane fly in 3 hours?
7) What distance does a boat travel in 7 hours, if it has a speed of 35 km/h?
8) A train travels 60 km in an hour. What distance does it travel in $\frac{3}{4}$ hour?
9) A bus drove 90 km at a speed of 45 km/h. How much time it spent?
10) What speed does a man have, if he runs 12 mi in 3 hours?

Algebra Chapter 8 Lesson 8

Word Problems on Percent

Converting Percent and Decimals, Fraction, or Mixed Number

Percent $\frac{1}{100} = 1\%$ **and** $0.01 = 1\ \%$

Converting a Decimal, Fraction or Mixed Number to a Percent
If a fraction, decimal or mixed number needs to convert to a percent, multiply the fraction, decimal or mixed number by 100%.

A decimal to a percent: $0.75 = 0.75 \times 100\% = 75\%$

A fraction to a percent: $\frac{3}{8} = \frac{3}{8} \cdot 100\% = \frac{75}{2}\% = 37\frac{1}{2}\% = 37.5\%$

A mixed number to a percent: $1\frac{4}{5} = 1.8 \times 100\% = 180\%$ $\quad 1 = 100\%$

If a percent needs to convert to a decimal, fraction, or mixed number divide the percent by 100%

To a decimal $50\% = \frac{50\%}{100\%} = 0.5$ **To a fraction** $25\% = \frac{25\%}{100\%} = \frac{1}{4}$

To a mixed number $150\% = \frac{150\%}{100\%} = \frac{3}{2} = 1\frac{1}{2}$ $\quad 100\% = 1$

Percent Proportion: $\frac{A}{B} = \frac{R}{100}$ $\quad 100A = BR$

In word problems in percent each word has special meanings
"Amount" (A) represents R %, "Base" (B) represents 100%,
"is" means =, "of" means multiplication, "what number" means x

Percent Formulas: Amount: $A = \frac{BR}{100}$ $\quad \underline{A = B \cdot \frac{R}{100}}$ (1)

Base: $B = \frac{100A}{R}$ $\quad B = A \cdot \frac{100}{R}$ $\quad \underline{B = A \div \frac{R}{100}}$ (2)

Rate: $R = \frac{100A}{B}$ $\quad \underline{R = \frac{A}{B} \cdot 100\%}$ (3)

If a percent is less than 100% then the Amount is less than the Base
$R < 100\%$ $\quad A < B$

If a percent is more than 100% then the Amount is greater than the Base: $R > 100\%$ $\quad A > B$

Algebra Chapter 8 Lesson 8

Solve Percent Problems by Equations ("is" and "of")

1) Find A: What numbers is 25% of 40? $A = 0.25 \cdot 40$ $\underline{A = 10}$

2) Find B: 10 is 25% of what number?

$10 = \frac{25}{100}B$ $\frac{10}{0.25} = \frac{0.25B}{0.25}$ $\frac{1000}{25} = B$ $\underline{B = 40}$

3) Find R: What percent of 40 is 10?

$\frac{R}{100} \times 40 = 10$ $100 \times \frac{R}{100} \times 40 = 10 \times 100$ $40R = 1000$ $\underline{R = 25\%}$

Solve Percent Problems by Proportion

1) Find A: Michael solves 80% of 20 problems in test. How many problems he solved?

$\frac{A}{B} = \frac{R}{100}$ $\frac{A}{20} = \frac{80}{100}$ $\frac{A}{20} = \frac{8}{10}$ $10A = 160$ $A = \underline{16 \text{ problems}}$

2) Find B: David solved 16 problems in test, it is equal 80% of all test. How many problems were in test?

$\frac{A}{B} = \frac{R}{100}$ $\frac{16}{B} = \frac{80}{100}$ $\frac{16}{B} = \frac{8}{10}$ $8B = 160$ $\underline{B = 20 \text{ problems}}$

3) Find R: What percent of 20 is 16?

$\frac{A}{B} = \frac{R}{100}$ $\frac{16}{20} = \frac{R}{100}$ $\frac{4}{5} = \frac{R}{100}$ $5R = 400$ $\underline{R = 80\%}$

Solve Percent Problems by Formulas

1) Find A: Olivia had $ 30, she spent 70% of all her money. How much money she spent? To find the amount, multiply the base by the percent, which converted to decimal.

Formula 1 $\underline{A = B \cdot \frac{R}{100}}$ $A = 30 \cdot \frac{70}{100}$ $A = 30 \cdot 0.7$ $A = \underline{\$\,21}$

2)Find B: Anna spent $ 21, it is 70% of all her money. How much money she had? To find the base, divide the amount by the percent, which converted to decimal.

Formula 2 $\underline{B = A \div \frac{R}{100}}$ $B = 21 \div \frac{70}{100}$ $B = 21 \div 0.7 = 210 \div 7 = \underline{\$\,30}$

3)Find R: Vicki had $ 30, she spent $ 21. What percent of money she spent? To find a percent, divide the numbers and multiply the result by 100%. Formula 3 $\underline{R = \frac{A}{B} \cdot 100\%}$ $R = \frac{21}{30} \cdot 100\%$ $R = \frac{7}{10} \times 100\%$ $R = \underline{70\%}$

To solve the percent problems, choose one of three ways that is easy for you.

Algebra Chapter 8 Lesson 8
Review 1 A
Solve the Word Problems on Percent

1) Find 20 % of 20?

2) Find 25 % of 200?

3) A book has 80 pages. The boy read 40% of the book. How many pages does he have left to read?

4) Of 40 children, 62.5% were boys. How many were girls?

5) The book has 240 pages. I read 40% of the book in the first day, then in the second day, I read 14 pages more than the first day. How many pages remain for me to read?

6)In the garden were 1720 trees: 25% of all trees were cherry and 15% of all trees were apple. How many cherry and apple trees total were in the garden?

7) Find the number if 20% of this number is 15?

8) Find the number, if 25% of this number is 6?

9) Find the number, if 50% of this number is 14?

10) Find the number, if 75% of this number is 30?

Algebra Chapter 8 Lesson 8

Review 1 A
Solve the Word Problems

11) On a test, Mike had 70% of the problems right. If he had 14 problems correct, how many questions were on the test?

12) A factory has 5 new machines and 80% were old machine. How many machines are in the factory?

13) From a box, Sam took 30 pages of paper. How many pages of paper were in the box, if in the box there remains 85% of all pages?

14) What percent is $7 from $8?

15) What percent is 1 lb from 8 lb?

16) What percent is 2 from 5?

17) What percent is 4 from 8?

18) In 50 minute of lesson, students solve an equation in 10 minutes. What percent of lesson time did students use to solve the equation?

19) Of 30 students, 6 students got an "A" on a test. What percent of students got an "A"?

20) There are 160 ml of acid in 800 ml of a solution of acid and water. What percent of the solution is acid?

Algebra Chapter 8 Lesson 8
Review 1 B
Solve the Word Problems

1)A bookstore manager determines that a markup rate of 30% is necessary to make a profit. What is the markup on a book that costs the bookstore $ 10?

2)A disc player that regularly sells for $ 400 is selling for 25% off the regular price. What is the discount?

3)Amy sold $ 9000 of Jewelry this week. Find her commission for the week if she receives a commission rate of 5.6%?

4)How much commission will Jack make on the sale of a $ 325,000 house if he receives 2% of the selling price?

5)A family paid $ 25,000 as a down payment for a home. If this represents 20% of the price of the home, find the price of the home.

6)A student spent $ 390 for books, if its 15% of cost for last semester at her community college. How much cost of last semester is at her community college?

7) The sales tax is $ 96 on a stereo sound system purchase of $ 1600. Find the sales tax rate.

8) The sales tax is $ 108 on a necklace purchase of $ 1200. Find the sales tax rate.

9) The college bookstore is giving a discount of $8 on calculator that normally sell for $ 32. What is the discount rate?

10) An electric grill that regularly sells for $ 120 is selling for $ 36 off the regular price. What is the discount rate?

Algebra Chapter 8 Lesson 8
Review 1 B
Solve the Word Problems

11) Amy was making a wage of $ 14.50 an hour before receiving a 12% in hourly pay. What Anna's new hourly wage?

12) A clothing store bought leather suit for $ 84 and used a markup rate of 45%. What is the selling price?

13) A $ 400 fax machine is on sale for 25% off.
Find the amount of discount and the sale price.

14) A $ 4000 designer dress is on sale for 40% off. Find the amount of discount and the sale price.

15) The original price of a shirt is $ 60. What is a new price of a shirt with a 40% discount.

16) A lawn mower is on sale for 35% off the regular price of $ 1200. Find the sale price.

17) The whole price of coffee increased from $ 2 per pound to $ 3 per pound in one year. What was the percent increase in the price of 1-pound coffee?

18) The price of gasoline rose from $ 1.5 to $ 1.8 on five months. What was the percent increase in the price of gasoline?

19) A company with 1500 employees reduces the number of employees to 1200. What is the percent decrease?

20) The stereo television that regularly sells for $ 200 is on sale for $ 150. Find the discount rate.

Algebra Chapter 8 Lesson 9
Table 11 Word of Problems

1) Decimals

a) To find the amount by known base, multiply the base by a decimal:

Find the amount 0.6 of 40 $40 \times 0.6 = 24$

b) To find the base by known amount, divide the amount by a decimal:

Finding the base: 0.6 is equivalent $ 24 $24 \div 0.6 = 40$

c) To find the ratio between two numbers, divide the numbers.

Find the ratio 24 from 40 $24 \div 40 = 0.6$

2) Fractions

a) To find the amount by known base, multiply the base by a fraction.

Finding the amount: $\frac{3}{5}$ of 40 $40 \times \frac{3}{5} = \$24$

b) To find the base by known amount, divide amount by a fraction.

Finding the base: $\frac{3}{5}$ equivalent 24 $24 \div \frac{3}{5} = \$40$

c) To find the ratio between two numbers, divide the numbers.

Find the ratio: 24 from $40 $24 \div 40 = \frac{24}{40} = \frac{3}{5}$

3) Uniform Motion: Distance - d Time - t Rate – r (Speed)

a) To find distance, multiply rate by time: d = rt

What distance did a man go for 2 hours if his rate was 5 mi/hr?

$d = 5 \cdot 2 = 10$ mi

b) To find time, divide distance by speed. $t = d \div r$

How much time does a man needs if he runs 8 mi with a rate 4 mi/hr?

$t = 8 \div 4 = 2$ hr

c) To find speed, divide distance by time. $r = d \div t$

What is the rate a girl has, if she runs 6 mi for 2 hours

$r = 6 \div 2 = 3$ mi / hr

4) Directly proportional quantities

If 4 apples sell for $1.32. How much do 11 apples cost?

$132 \div 4 = 33$ c $33 \times 11 = 363$ c = $3.63

Algebra Chapter 8 Lesson 9 Table 12 Percent Problems

Solve Percent Problems by Equations ("is" and "of")

1) Find A: What numbers is 25% of 40? $A = 0.25 \cdot 40$ $\underline{A = 10}$

2) Find B: 10 is 25% of what number?

$10 = \frac{25}{100}B$ $\frac{10}{0.25} = \frac{0.25B}{0.25}$ $\frac{1000}{25} = B$ $\underline{B = 40}$

3) Find R: What percent of 40 is 10?

$\frac{R}{100} \times 40 = 10$ $100 \times \frac{R}{100} \times 40 = 10 \times 100$ $40R = 1000$ $\underline{R = 25\%}$

Solve Percent Problems by Proportion

1) Find A: Michael solves 80% of 20 problems in test.
How many problems he solved?

$\frac{A}{B} = \frac{R}{100}$ $\frac{A}{20} = \frac{80}{100}$ $\frac{A}{20} = \frac{8}{10}$ $10A = 160$ $A = \underline{16 \text{ problems}}$

2) Find B: David solved 16 problems in test, it is equal 80% of all test.
How many problems were in test?

$\frac{A}{B} = \frac{R}{100}$ $\frac{16}{B} = \frac{80}{100}$ $\frac{16}{B} = \frac{8}{10}$ $8B = 160$ $\underline{B = 20 \text{ problems}}$

3) Find R: What percent of 20 is 16?

$\frac{A}{B} = \frac{R}{100}$ $\frac{16}{20} = \frac{R}{100}$ $\frac{4}{5} = \frac{R}{100}$ $5R = 400$ $\underline{R = 80\%}$

Solve Percent Problems by Formulas

1) Find A: Olivia had $ 30, she spent 70% of all her money. How much money she spent? To find the amount, multiply the base by the percent, which converted to decimal.

Formula 1 $\underline{A = B \cdot \frac{R}{100}}$ $A = 30 \cdot \frac{70}{100}$ $A = 30 \cdot 0.7$ $A = \underline{\$\,21}$

2) Find B: Anna spent $ 21, it is 70% of all her money. How much money she had? To find the base, divide the amount by the percent, which converted to decimal.

Formula 2 $\underline{B = A \div \frac{R}{100}}$ $B = 21 \div \frac{70}{100}$ $B = 21 \div 0.7 = 210 \div 7 = \underline{\$\,30}$

3)Find R: Vicki had $ 30, she spent $ 21. What percent of money she spent? To find a percent, divide the numbers and multiply the result by 100%. Formula 3 $\underline{R = \frac{A}{B} \cdot 100\%}$ $R = \frac{21}{30} \cdot 100\%$ $R = \frac{7}{10} \times 100\%$ $R = \underline{70\%}$

To solve the percent problems, choose one of three ways which is easy for you.

Algebra Chapter 8 Lesson 9
Table 13 Percent Application

1. Sales Tax Formula

Amount of sales tax = Cost of merchandise × Tax rate

Example: A graphing calculator cost \$ 100. The sales tax rate is 5.5%. Compute the total cost.

5.5% = 0.55 100×0.055 =\$5.5 100 + 5.5 = \$105.5

2. Percent of commission

Amount of commission = Dollar sales × Percent of commission

Example: Diana earned 5% on a house she sold for \$ 100,000 Find her dollar commission. 100000× 0.05 = \$ 5000

3. Percent of decrease

Amount of decrease = original value - new value

Amount of decrease = original value × percent of decrease

Percent of decrease = amount of decrease ÷ original value ×100%

Example: A paper shredder was decreased price from \$ 80 to \$ 60 What is the percent decreased in price?

80 – 60 = 20 20 ÷ 80 × 100% = 0.25 × 100% = 25%

4. Percent of discount

Amount of decrease = original value - new value

Amount of decrease = original value × percent of decrease

Percent of decrease = amount of decrease ÷ original value ×100%

Example: The Company put suits that usually sell for \$ 250 on sale for \$ 170. What the percent discount?

250– 170 = \$ 80 80 ÷ 250 × 100% = 0.32 × 100% = 32%

5. Percent of increase

Amount of increase = new value – original value

Amount of increase = original value × percent of increase

Percent of increase = amount of increase ÷ original value × 100%

Example: The sales in one year Honda Civic increased from 400,000 to 408,000. Compute the percent increase.

408000 – 400000=8000 8000÷400000 ×100 = 0.2×100=20%

6. Percent markup

Amount of markup = new value – original value

Amount of markup = original value × percent of markup (rate)

Percent of markup = amount of markup ÷ original value × 100%

Example: A college bookstore marks up of math book is 30%. What is mark up for math book that has a price of \$ 70.

\$70 × 0.3 = \$ 21 70 + 21 = \$ 91

Algebra Chapter 8 Lesson 9
Table 14 Common Factor
Distributive laws of multiplication

1) Whole Numbers (Put A before parentheses)

AB + AC = A (B + C)   AB – AC = A (B - C)

If two products have a common factor, put the common factor before parentheses. AB + AC   A – common factor

In parentheses, put results of division products by common factor

AB ÷ A = B   AC ÷ A = C

$16 \cdot 5 + 4 \cdot 5 = 5\ (16 + 4) = 5\ (20) = 100$

$22 \cdot 5 - 2 \cdot 5 = 5\ (22 - 2) = 5\ (20) = 100$

2) Decimals

a) (B + C) A = AB + AC $(0.4 + 0.1)5 = 0.4 \times 5 + 0.1 \times 5 = 2 + 0.5 = 2.5$

AB + AC = A(B + C) $0.4 \times 5 + 0.1 \times 5 = 5\ (0.4 + 0.1) = 5(0.5) = 2.5$

b) A(B - C) = AB – AC $5\ (0.4 - 0.1) = 5 \times 0.4 - 5 \times 0.1 = 2 - 0.5 = 1.5$

AB – AC = A(B – C) $5 \times 0.4 - 5 \times 0.1 = 5\ (0.4 - 0.1) = 5(0.3) = 1.5$

3) Negative Numbers

If first term negative, put (- A) before parentheses and change each sign inside parentheses: -AB – AC = - A (B + C)   - AB + AC = - A(B - C)

a) $-16 \cdot 5 - 4 \cdot 5 = -5\ (16 + 4) = -5\ (20) = -100$

$-22 \cdot 5 + 2 \cdot 5 = -5\ (22 - 2) = -5\ (20) = -100$

b) $-7.5 \times 5 - 2.5 \times 5 = -5(7.5 + 2.5) = -5(10) = -50$

$-12.5\ (5) + 2.5(5) = -5(12.5 - 2.5) = -5(10) = -50$

4) To solve problems work by order of operations or change division for multiplication then put A before parentheses.

AB + AC = A (B + C)   AB – AC = A (B - C)

a) $2\frac{5}{8} \div \frac{1}{8} + 1\frac{3}{8} \div \frac{1}{8} = 2\frac{5}{8} \times 8 + 1\frac{3}{8} \times 8 = 8(2\frac{5}{8} + 1\frac{3}{8}) = 8(4) = 32$

b) $2\frac{5}{8} \div \frac{1}{8} - 1\frac{3}{8} \div \frac{1}{8} = 2\frac{5}{8} \times 8 - 1\frac{3}{8} \times 8 = 8(2\frac{5}{8} - 1\frac{3}{8}) = 8(1\frac{2}{8}) = 8(\frac{10}{8}) = 10$

5) To solve problems work by order of operations or change division for multiplication then put (- A) before parentheses

- AB - AC = - A (B + C)   - AB + A C = - A (B - C)

a) $-2\frac{5}{8} \div \frac{1}{8} - 1\frac{3}{8} \div \frac{1}{8} = -2\frac{5}{8} \times 8 - 1\frac{3}{8} \times 8 = -8\ (2\frac{5}{8} + 1\frac{3}{8}) = -8\ (4) = -32$

b) $-2\frac{5}{8} \div \frac{1}{8} + 1\frac{3}{8} \div \frac{1}{8} = -2\frac{5}{8} \times 8 + 1\frac{3}{8} \times 8 = -8\ (2\frac{5}{8} - 1\frac{3}{8}) = -8\ (1\frac{2}{8}) = -8\ (\frac{10}{8}) = -10$

Algebra Chapter 8 Lesson 9
Solve Equations
Mental Math 17

1) $x + 6 = 10$
2) $7 + x = 10$

3) $x - 9 = 5$
4) $x - 8 = 9$

5) $x - 7 = 8$
6) $12 - y = 3$

7) $13 - y = 8$
8) $14 - y = 5$

9) $15 - y = 7$
10) $16 - y = 9$

11) $8 \cdot x = 72$
12) $x \cdot 9 = 63$

13) $x \div 8 = 7$
14) $x \div 7 = 6$

15) $x \div 6 = 9$
16) $72 \div z = 9$

17) $72 \div z = 8$
18) $63 \div z = 9$

19) $56 \div z = 7$
20) $54 \div z = 9$

Algebra Chapter 8 Lesson 9
Mental Math 18

Table of Exponents 2

$$a^1 = a$$

1) 1^3	11) 2^1
2) 2^3	12) 2^2
3) 3^3	13) 2^3
4) 4^3	14) 2^4
5) 5^3	15) 2^5
6) 6^3	16) 2^6
7) 7^3	17) 2^7
8) 8^3	18) 2^8
9) 9^3	19) 2^9
10) 10^3	20) 2^{10}

Algebra Chapter 8 Lesson 10
Laws for operations
Distributive laws of multiplication

(A + B) C = AC + BC **(A - B) C = AC – B C**

Example a) $\frac{5}{6}\times 3=\frac{5}{6}\times\frac{3}{1}=\frac{5}{2}=2\frac{1}{2}=2.5$

Example b) $\frac{5}{3}\times 6=6\times\frac{5}{3}=6\div 3\times 5=2\times 5=10$

Example c) $\left(1\frac{1}{3}-\frac{5}{6}\right)\times 3=\frac{4}{3}\times\frac{3}{1}-\frac{5}{6}\times\frac{3}{1}=4-\frac{5}{2}=4-2.5=1.5$

Worksheet 6 A
Evaluate by using laws

1) $9(1-\frac{4}{9})$

2) $(\frac{2}{3}-\frac{3}{5})\times 15$

3) $9(3-\frac{5}{9})$

4) $(1+\frac{5}{6})\times 6$

5) $7(2+\frac{3}{7})$

6) $(1-\frac{1}{6})\times 6$

7) $6(1\frac{1}{2}-\frac{2}{3})$

8) $(\frac{5}{7}+\frac{4}{7})\times 7$

9) $7(\frac{5}{7}-\frac{4}{7})$

10) $(3+\frac{2}{3})\times 3$

11) $3(2+\frac{1}{6})$

12) $(2\frac{1}{4}-\frac{5}{8})\times 4$

13) $3(1\frac{5}{6}-1\frac{1}{3})$

14) $(3\frac{1}{2}-\frac{1}{4})\times 4$

15) $3(1\frac{2}{3}-\frac{5}{6})$

16) $(1\frac{2}{7}-\frac{1}{3})\times 21$

17) $6(2\frac{1}{2}+1\frac{1}{3})$

18) $(4+\frac{5}{8})\times 2$

19) $3(1\frac{1}{3}-\frac{5}{12})$

20) $(1\frac{1}{4}-\frac{5}{8})\times 4$

Algebra Chapter 8 Lesson 10

Solve by finding common factor

Common Factor (Put C before parentheses)

AC + BC = C (A + B) **AC – B C = C (A - B)**

Example $1\frac{1}{4}\times\frac{3}{4}+\frac{3}{4}\times 2\frac{3}{4}=\frac{3}{4}(1\frac{1}{4}+2\frac{3}{4})=\frac{3}{4}\times\frac{4}{1}=3$

Worksheet 6 B

Evaluate by using laws

1) $8\times 1\frac{3}{4}-8\times\frac{5}{8}$

2) $3\frac{2}{9}\times\frac{3}{5}+\frac{3}{5}\times 6\frac{7}{9}$

3) $7\times 1\frac{1}{7}-7\times\frac{5}{14}$

4) $4\times 2\frac{1}{4}-4\times 1\frac{7}{8}$

5) $2\frac{6}{7}\times\frac{7}{10}+2\frac{1}{7}\times\frac{7}{10}$

6) $5\frac{5}{6}\times\frac{3}{4}-1\frac{5}{6}\times\frac{3}{4}$

7) $4\frac{4}{5}\times\frac{5}{8}+\frac{5}{8}\times 3\frac{1}{5}$

8) $3\frac{2}{3}\times 1\frac{1}{2}-1\frac{2}{3}\times 1\frac{1}{2}$

9) $\frac{5}{6}\times 5\frac{1}{5}-\frac{5}{6}\times 2\frac{1}{5}$

10) $3\frac{1}{3}\times 1\frac{1}{5}+1\frac{1}{5}\times 1\frac{2}{3}$

11) $1\frac{3}{5}\times 2\frac{1}{2}+1\frac{3}{5}\times 7\frac{1}{2}$

12) $4\frac{2}{3}\times\frac{3}{4}-3\frac{1}{3}\times\frac{3}{4}$

13) $4\frac{3}{5}\times\frac{5}{6}-\frac{5}{6}\times 1\frac{3}{5}$

14) $4\frac{1}{2}\times\frac{1}{6}+\frac{1}{6}\times 1\frac{1}{2}$

15) $3\frac{3}{4}\times 1\frac{1}{6}-2\frac{1}{4}\times 1\frac{1}{6}$

16) $\frac{1}{3}\times 15-\frac{1}{5}\times 15$

17) $1\frac{3}{4}\times 8-1\frac{1}{2}\times 8$

18) $1\frac{1}{2}\times 6-1\frac{1}{6}\times 6$

19) $2\frac{1}{2}\times 6-1\frac{1}{2}\times 6$

20) $1\frac{3}{4}\times 12-1\frac{1}{4}\times 12$

21) $1\frac{3}{4}+1\frac{1}{4})\div\frac{1}{4}$

22) $(1\frac{3}{5}-1\frac{2}{5})\div\frac{2}{5}$

23) $(4\frac{3}{5}+1\frac{2}{5})\div\frac{1}{5}$

24) $(1\frac{3}{4}-1\frac{1}{4})\div\frac{1}{8}$

25) $(2\frac{1}{2}+1\frac{1}{2})\div\frac{1}{6}$

26) $(3\frac{2}{3}-1\frac{2}{3})\div\frac{2}{3}$

27) $(1\frac{5}{6}+1\frac{1}{6})\div\frac{1}{6}$

28) $(4\frac{3}{5}-1\frac{3}{5})\div\frac{1}{5}$

29) $(1\frac{2}{3}+1\frac{1}{3})\div\frac{1}{3}$

30) $(4\frac{5}{6}-2\frac{1}{6})\div\frac{1}{6}$

Algebra Chapter 8 Lesson 11
Reviews for Chapter 8
Review 8 A

Date **Name**

Read directions carefully. **Answers must be in indicated place.**

Show all steps for full credits. **Each problem is worth 5 points.**

Do only 1 operation per 1 step. **Solve 1 problem for 5 minutes.**

1) A book has 80 pages. David read 0.15 of the whole book. How many pages more does he have left to read?

2)In the class, 5 students can dance; this is $\frac{1}{4}$ of the whole class. How many students are in the class? How many students cannot dance?

3)Kristy read 10 pages. What part of the book did she read, if the book has 50 pages? Write the ratio as a reduced fraction. How many pages more does she have left to read?

4) Find the average for the numbers: $\frac{1}{2}, \frac{1}{5}, \frac{1}{5}, \frac{1}{10}$

5) A girl paid $24 for 2 books. How much would 3 of the same books cost?

6) A bus drive 42 mi at a speed of 15 mi/h. How much time it spent?

7) What speed does a boy have, if he drove 45 mi in 4 hours?

8) Of 40 children, 37.5% were boys. How many were girls?

9) A factory has 2 new machines and 80% were old machine. How many machines are in the factory?

10) In 50 minutes of lesson, students solve an equation in 10 minutes. What percent of lesson time did students use to solve the equation?

Algebra Chapter 8 Lesson 11
Review 8 B

Date	**Name**
Read directions carefully.	**Answers must be in indicated place.**
Show all steps for full credits.	**Each problem is worth 5 points.**
Do only 1 operation per 1 step.	**Solve 1 problem for 5 minutes.**

1) A book has 60 pages. David read 0.55 of the whole book. How many pages more does he have left to read?

2) In the class, 8 students can dance; this is $\frac{2}{5}$ of the whole class. How many students are in the class? How many students cannot dance?

3)Kristy read 8 pages. What part of the book did she read, if the book has 40 pages? Write the ratio as a reduced fraction. How many pages more does she have left to read?

4) Find the average for the numbers: $1\frac{1}{2}$, $1\frac{1}{5}$, $3\frac{1}{5}$, $3\frac{1}{10}$

5)A girl paid $12 for 2 books. How much would 3 of the same books cost?

6) A bus drove 36 mi at a speed of 15 mi/h. How much time it spent?

7) What speed does a boy have, if he drove 61 mi in 4 hours?

8) Of 24 children, 37.5% were boys. How many were girls?

9) A factory has 4 new machines and 60% were old machine. How many machines are in the factory?

10) In 75 minutes of lesson, students solve an equation in 15 minutes. What percent of lesson time did students use to solve the equation?

Algebra Chapter 8 Lesson 12
Tests for Chapter 8
Test 8 A

Date **Name**

Read directions carefully. **Answers must be in indicated place.**
Show all steps for full credits. **Each problem is worth 5 points.**
Do only 1 operation per 1 step. **Solve 1 problem for 5 minutes.**

1) A book has 80 pages. David read 0.25 of the whole book. How many pages more does he have left to read?

1) ______

2) In the class, 6 students can dance; this is $\frac{1}{5}$ of the whole class.

How many students are in the class? How many students cannot dance?

2) ______

3) Kristy read 20 pages. What part of the book did she read, if the book has 50 pages? Write the ratio as a reduced fraction. How many pages more does she have left to read?

3) ______

4) Find the average for the numbers: $\frac{1}{2}, \frac{1}{5}, 1\frac{1}{5}, 1\frac{1}{10}$

4) ______

5) A girl paid $15 for 3 books. How much would 4 of the same books cost?

5) ______

Algebra Chapter 8 Lesson 12
Test 8 A

Date **Name**

Read directions carefully. **Answers must be in indicated place.**
Show all steps for full credits. **Each problem is worth 5 points.**
Do only 1 operation per 1 step. **Solve 1 problem for 5 minutes.**

6) A bus drove 54 mi at a speed of 20 mi/h.
How much time it spent?

6) ______

7) What speed does a boy have, if he drove 49 mi in 4 hours?

7) ______

8) Of 40 children, 62.5% were boys. How many were girls?

8) ______

9) A factory has 3 new machines and 80% were old machine.
How many machines are in the factory?

9) ______

10) In 50 minutes of lesson, students solve an equation in 15 minutes.
What percent of lesson time did students use to solve the equation?

10) ______

Algebra Chapter 8 Lesson 12
Test 8 B

Date **Name**

Read directions carefully. **Answers must be in indicated place.**
Show all steps for full credits. **Each problem is worth 5 points.**
Do only 1 operation per 1 step. **Solve 1 problem for 5 minutes.**

1) A book has 60 pages. David read 0.65 of the whole book. How many pages more does he have left to read?

1) ______

2) In the class, 9 students can dance; this is $\frac{3}{5}$ of the whole class. How many students are in the class? How many students cannot dance?

2) ______

3) Kristy read 16 pages. What part of the book did she read, if the book has 40 pages? Write the ratio as a reduced fraction. How many pages more does she have left to read?

3) ______

4) Find the average for the numbers: $3\frac{1}{2}$, $3\frac{1}{5}$, $1\frac{1}{5}$, $1\frac{1}{10}$

4) ______

5) A girl paid $12 for 3 books. How much would 4 of the same books cost?

5)______

Algebra Chapter 8 Lesson 12
Test 8 B

Date **Name**

Read directions carefully. **Answers must be in indicated place.**
Show all steps for full credits. **Each problem is worth 5 points.**
Do only 1 operation per 1 step. **Solve 1 problem for 5 minutes.**

6) A bus drove 46 mi at a speed of 20 mi/h.
How much time it spent?

6) ______

7) What speed does a boy have, if he drove 65 mi in 4 hours?

7) ______

8) Of 24 children, 62.5% were boys. How many were girls?

8) ______

9) A factory has 6 new machines and 60% were old machine.
How many machines are in the factory?

9) ______

10) In 75 minutes of lesson, students solve an equation in 30 minutes.
What percent of lesson time did students use to solve the equation?

10) _____

Algebra Answers Chapter 8

Lesson 1 1) 2.3 hr 2) 16.25 mi/hr 3) 9 girls 4) 15 machines 5) 40%

Lesson 2 Word problems on Operations

Review 1 A

1) 59 flowers 2) 89 roses 3) 69 cents 4) 58 birds 5) 88 cents
6) 2 pencils 7) 16 balloons 8) 11 roses 9) 24 hammers 10) 40 cents

Review 1 B

1) 80 hours 2) 280 seats 3) 225 words 4) 120 seats 5) 40 hours
6) 9 packages 7) 3 hours 8) 4 floors 9) 7 people 10) 9 students

Review 1 C

1) 46 markers 2) 69 nickels 3) 5 pens 4) 22 cups 5) 50 seats
6) 50 players 7) 60 sheets 8) 6 trees 9) 4 hours 10) 5 pages

Lesson 3 Word problems on Worth

Review 1 A

1) 10 c 2) 20 c 3) 30 c 4) 90 c 5) 72 c
6) $ 60 7) $ 50 8) 20 plants 9) 8 liters 10) 12 lb

Review 1 B

1) 8 buttons 2) 20 meters 3) $ 240 4) 90 c 5) $ 12
6) $ 4 7) 7 c 8) 20 envelopes 9) $ 9 10) 4 times

Review 1 C

1) 7 books 2) 6 tickets 3) 3 boys 4) 6 pencils 5) 3 bouquets
6) 8 logs 7) 2 bunches 8) $ 24 9) 4 packs 10) $ 280

Lesson 4 Word Problems on Decimals

Review 1 A

1) 7 students 2) 30 pages 3) 3 stamps 4) 4 dresses 5) 9 girls
6) 9pages 7) $ 60 8) 10 plums 9) 25 pages 10) 32 students

Review 1 B

1) 0.5 2) 0.4 3) 0.25 4) 0.25 5) 540 lb
6) 60 pages 7) 15 boys 8) $ 1.5 9) 2.3 10) 6

Lesson 5 Word Problems on Fractions

Review 1 A

1) 7 students 2) 30 pages 3) 3 stamps 4) 4 dresses 5) 9 girls
6) 9 pages 7) 540 lb 8) 60 pages 9) 15 girls 10) $ 1.5

Review 1 B

1) $ 60 2) 10 plums 3) 100 pages 4) 40students 5) $\frac{1}{2}$
6) $\frac{2}{5}$ 7) $\frac{1}{4}$ 8) $\frac{1}{4}$ 9) 2.3 10) 6

Algebra Answers Chapter 8

Lesson 6 Word Problems on Direct Proportionality

Review 1 A

1) \$1.50 2) 20 c 3) 56 details 4) 36 cases 5) 8 hr
6) 5 days 7) 12 pages 8) 24 frames 9) 80 books 10) 48 trees

Review 1 B

1) 948 lb 2) 500 keys 3) 27000 people 4) 36 lb 5) 1.75 lb
6) 3.5lb 7) \$9 8) 5 balls 9) \$30 10) 4 mo

Lesson 7 Uniform Motion

Review 1 A

1a) 20 mi 1b) 60 mi 1c) 360 mi 1d) 2000 mi

2a) 2 hrs. 2b) 4 hrs. 2c) 2 hrs. 2d) 2 hrs.

3a) 14 mi/hr. 3b) 70 mi/hr. 3c) 5,000 mi/hr.

4a) 120 mi 4b) 160 mi 4c) 222 mi

5a) 30 mi 5b) 2 hrs. 5c) 70 mi/hr.

6a) 60 mi/hr. 6b) 3 hrs. 6c) 96 mi

7a) 120 mi 7b) 40 mi/hr.

8a) 120 mi 8b) 2 hrs. 8c) 60 mi/hr.

9a) 80 mi 9b) 4 hrs. 9c) 60 mi/hr.

10a) 10 m/sec. 10b) 9 secs.

Review 1 B

1) 20 mi 2) 120 km 3) 90 m 4) 990 m 5) 36 mi

6) 2,025 mi 7) 245 mi 8) 45 mi 9) 2 hrs. 10) 4 mi/hr.

Algebra Answers Chapter 8

Lesson 8 Word Problems on Percent

Review 1 A

1) 4	2) 50	3) 60 pages	4) 15 girls	5) 34 pages
6) 688 trees	7) 75	8) 24	9) 28	10) 40
11) 20 questions	12) 25 machines	13) 200 pages	14) 87.5%	15) 12.5%
16) 40%	17) 50%	18) 20%	19) 20%	20) 20%

Review 1 B

1) $A = 10 \times 0.3 = \$\ 30$ 2) $A = 400 \times 0.25 = \$100$

3) $A = 9000 \times 0.056 = \$\ 504$ 4) $A = 325{,}000 \times 0.02 = 6{,}500$

5) $B = 25000 \div 0.2 = \$\ 125{,}000$ 6) $B = 390 \div 0.15 = \$2{,}600$

7) $R = 96 \div 1600 \times 100\% = 6\ \%$ 8) $R = 108 \div 1200 \times 100\% = 9\%$

9) $R = 8 \div 32 \times 100\% = 25\%$ 10) $R = 36 \div 120 \times 100\% = 30\%$

11) $A = 14.50 \times 0.12 = \$\ 1.74$ $14.50 + 1.74 = \$\ 16.24$

12) $A = 84 \times 0.45 = \$\ 37.8$ $84 + 37.8 = \$\ 121.8$

13) $A = 400 \times 0.25 = \$\ 100$ $400 - 100 = \$\ 300$

14) $A = 4000 \times 0.25 = \$\ 1{,}000$ $4000 - 1000 = \$\ 3{,}000$

15) $A = 60 \times 0.4 = \$\ 24$ $60 - 24 = \$\ 36$

16) $A = 1200 \times 0.35 = \$\ 420$ $1200 - 420 = \$\ 780$

17) $3 - 2 = 1$ $R = \frac{1}{2} \times 100\% = 50\%$

18) $1.8 - 1.5 = 0.3$ $R = 0.3 \div 1.5 \times 100\% = 20\%$

19) $1500 - 1200 = 300$ $R = 300 \div 1500 \times 100\% = 20\%$

20) $200 - 150 = 50$ $R = 50 \div 200 \times 100\% = 25\%$

Algebra Answers Chapter 8

Lesson 9 Mental Math 17

1) 4	2) 3	3) 14	4) 17	5) 15
6) 9	7) 5	8) 9	9) 8	10) 7
11) 9	12) 7	13) 56	14) 42	15) 54
16) 8	17) 9	18) 7	19) 8	20) 6

Lesson 9 Mental Math 18

1) 1	2) 8	3) 27	4) 64	5) 125
6) 216	7) 343	8) 512	9) 729	10) 1000
11) 2	12) 4	13) 8	14) 16	15) 32
16) 64	17) 128	18) 256	19) 512	20) 1024

Lesson 10 Worksheet 6 A

1) 5	2) 1	3) 22	4) 11	5) 17
6) 5	7) 5	8) 9	9) 1	10) 11
11) 6.5	12) 6.5	13) 1.5	14) 13	15) 2.5
16) 20	17) 23	18) 9.25	19) 2.75	20) 2.5

Lesson 10 Worksheet 6 B

1) 9	2) 6	3) 5.5	4) 1.5	5) 3.5
6) 3	7) 5	8) 3	9) 2.5	10) 6
11) 16	12) 1	13) 2.5	14) 1	15) 1.75
16) 2	17) 2	18) 2	19) 6	20) 6
21) 12	22) 0.5	23) 30	24) 4	25) 24
26) 3	27) 18	28) 15	29) 9	30) 16

Lesson 11 Review 8 A

1) 68 pages 2) 15 students 3) $\frac{1}{5}$, 40 pages 4) 0.25 5) $ 36

6) 2.8 hr 7) 11.25 mi/hr 8) 25 girls 9) 10 machines 10)20%

Lesson 11 Review 8 B

1) 27 pages 2) 12 students 3) $\frac{1}{5}$, 32 pages 4) 2.25 5) $ 18

6) 2.4 hr 7) 15.25 mi/hr 8) 15 girls 9) 10 machines 10)20%

Lesson 12 Test 8 A

1) 60 pages 2) 24 students 3) $\frac{2}{5}$, 30 pages 4) 0.75 5) $20

6) 2.7 hr 7) 12.25 mi/hr 8) 15 girls 9) 15 machines 10)30%

Lesson 12 Test 8 B

1) 21 pages 2) 6 students 3) $\frac{2}{5}$, 24 pages 4) 2.25 5)$16

6) 2.3 hr 7) 16.25 mi/hr 8) 9 girls 9) 15 machines 10) 40%

Geometry

Chapter 9

Introductory Geometry

Students!

Students, you are smarter than a calculator!
Students, you are smarter than a computer!
Students, you need to believe in yourself.
Students, be patient and do your best.

Do not be afraid to make mistakes.
Tell yourself positive words.
I am young! I am smart!
I am strong! I can do anything!

Smart students use their brain!
Your brain is always with you!
A calculator is not always with you.
Mathematics is work for the brain.

Be open to new ideas in math!
Work hard and try your best!
Check answers and ask questions!
Practice! Practice! Practice!

Geometry Chapter 9 Lessons
Introductory Geometry
Lessons 1 – 12
Advanced Problems

Geometry Chapter 9 Contents
Introductory Geometry

Contents have topics and exercises, which students need to solve before tests.1)Work with lesson 1, read rules and examples, 2) Solve the problems from contents, 3) Check answers in the end of chapter 1, 4) Work with the next lessons 2, 3, 4, 5, …, 5) Solve Test 1 A, 6) Grading Scale without minuses
A=90–100% B+=85–89% B=80–84% C+=75–79%
C=70–74% D+=65–69% D=60–64 % F=0 - 59%
7)If you solve 10 problems correctly, you get an A, 8) If have an A, you can move on to next chapter, 9) If your grade is lower than an A, work with the mistakes, then solve Test 1B, 10) When you finish all of the chapters, solve Final Exam in last chapter.
Lesson 1 Work with Table 10 of Geometry. Solve the Pretest # 1–#5
Lesson 2 Segments
Solve the problems Review 1 A # 6-# 10. Review 1B #6-#10
Lesson 3 Angles
Solve the problems Review 1 A # 6-# 10. Review 1B#6-# 10
Lesson 4 Triangles
Solve the problems Review 1 A # 6–# 10. Review1B#6–# 10
Lesson 5 Right Triangles
Solve the problems Review 1 A , Review1B#16–# 25
Lesson 6 Perimeter
Solve the problems Review 1 A # 1 – # 20 odd
Lesson 7 Area
Solve the problems Review 1 A # 1 – # 20 odd
Lesson 8 Circle
Solve the problems Review 1 A # 6-# 10. Review1B#6-#10

Lesson 9 Solid Geometry
Solve the problems Review A # 1, 3, 5, 7, 9, 11, 13, 15, 17, 19
Solve the problems Review B # 1, 3, 5, 7, 9, 11, 13, 15, 17, 19
Lesson 10 Work with Table 9, Table 10
Solve the problems Mental Math 15, 16 odd
Lesson 11 Reviews, Solve the problems Review 9 A # 1 – # 10 all
Solve the problems Review 9 B # 1 – # 10 all
Lesson 12 Tests, Solve the problems Test 9 A # 1 – # 10 all
Solve the problems Test 9 B # 1 – # 10 all

Geometry Chapter 9 Lesson 1

Table 10 Plane Geometry

Right triangles

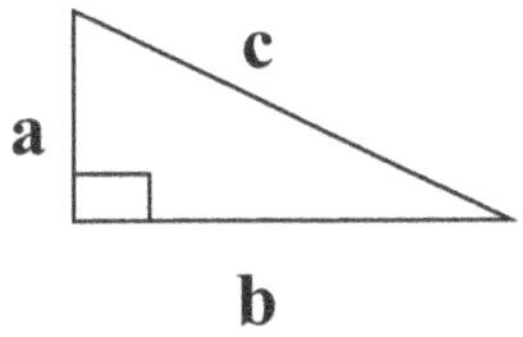

1) Pythagorean Theorem 1) $c^2 = a^2 + b^2$

2) Area 2) $A_\Delta = 0.5ab$

Triangles

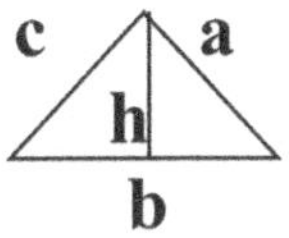

3) Perimeter 3) $P = a + b + c$

4) Area 4) $A_\Delta = 0.5bh$

Parallelogram

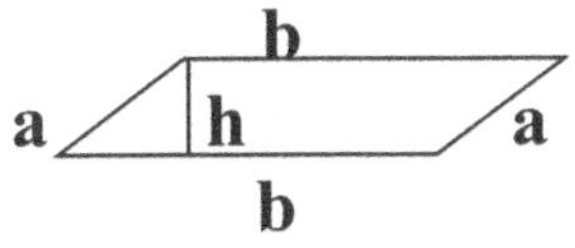

5) Perimeter 5) $P_{par} = 2a + 2b$

6) Area 6) $A_{par} = b\ h$

Rhombus

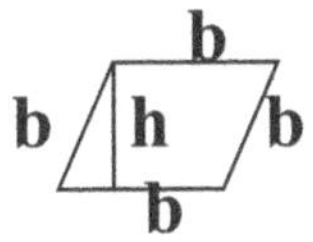

7) Perimeter 7) $P_{rh} = 4b$

8) Area 8) $A_{rh} = b \cdot h$

Rectangle

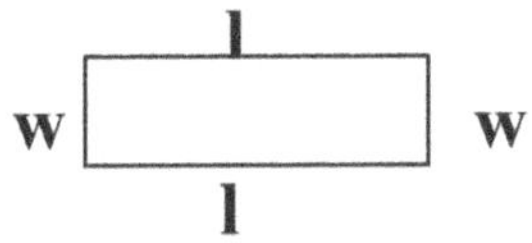

9) Perimeter 9) $P_{rec} = 2l + 2w$

10) Area 10) $A_{rec} = l \cdot w$

Square

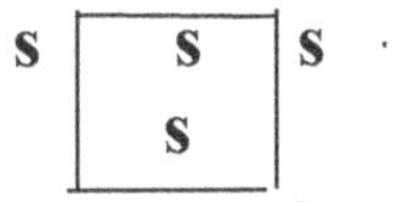

11) Perimeter 11) $P_{\blacksquare} = 4s$

12) Area 12) $A_{\blacksquare} = s^2$

Trapezoid

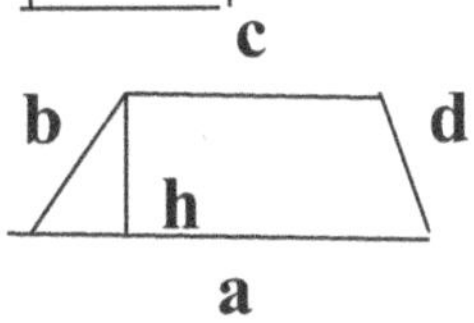

13) Perimeter 13) $P_{tr} = a + b + c + d$

14) Area 14) $A_{tr} = 0.5\ h\ (a + c)$

Circle

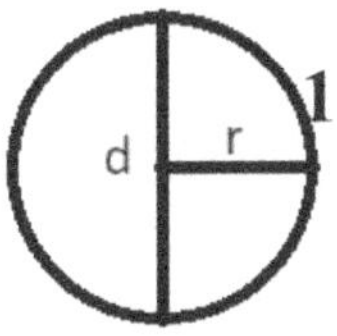

15) Number π in fraction $\approx \frac{22}{7}$ 16) Number π in decimal ≈ 3.14

17) Diameter $d = 2r$ 18) Circumference 1 (d) $C = \pi d$

19) Circumference 2 (r) $C = 2\pi r$ 20) Area $A_c = \pi r^2$

Solid Geometry

Cube 1) $V = s^3$ 2) $SA = 6s^2$ 3) $d = s\sqrt{3}$ Sphere 4) $V = \frac{4}{3}\pi r^3$ 5) $SA = 4\pi r^2$

Rectangular Solid:6) $V = l \cdot w \cdot h$ 7) $SA = 2lw + 2lh + 2wh$ 8) $d^2 = l^2 + w^2 + h^2$

Regular Prism 9) $V = \text{Base} \cdot h$ Regular Pyramid 10) $V = \frac{1}{3}\text{Base} \cdot h$

Right Cylinder 11) $V = \pi r^2 h$ Right Cone 12) $V = \frac{1}{3}\pi r^2 h$

Memorize Formulas of Geometry

Geometry Chapter 9 Lesson 1

Date **Name**

Pretest

Solve the problems

1) Find the area of a trapezoid, if its height is $3\frac{1}{3}$ cm, and the bases are $9\frac{3}{4}$ cm and $8\frac{1}{4}$ cm. Draw the picture.

1) ______

2) Find the area of a circle if the radius is 14 in and $\pi = \frac{22}{7}$.

Draw the picture.

2) ______

3) Find the area of a ring of a circle if its radii are 6.5 in and 5.5 in and $\pi \approx 3.14$. Draw the picture.

3) ______

4) Find the volume of a rectangular solid if its length is 12.5 in, its width is 5 in and its height is 4 in

4) ______

5) Find the exact value volume of a sphere if the radius r = 6 ft.

5) ______

Geometry Chapter 9 Lesson 2

Segments

Geometry is the study of properties of figures with different shapes and sizes. Geometry has two parts: plane and solid geometry.
Plane geometry is the study of properties of flat figures in the plane.
A plane is a flat surface like a desk or a blackboard.
A flat figure is contained in a plane. Solid geometry is the study of properties of figures having three dimensions.
A point is like a tiny dot and represented by capital letter A, B, C, …
A straight line goes off in both directions, the line is endless.
A straight line divides a plane by semi planes.
A straight line represented by two capital letters or one small letter.
A straight line constructed with a ruler and formed by two points.

Figure 1

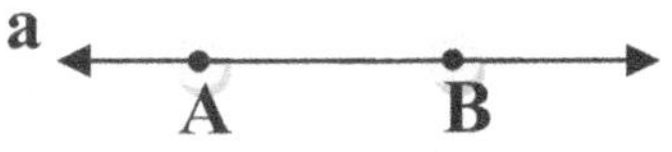

Two lines with one common point called intersection lines.

Figure 2

Two lines, Rays, Segments

Two lines on the plane called parallel if they have no common point intersection. Symbol of parallel lines is a || b

Figure 3

Part of the line limited by one point called a ray.
A ray represented by two capital letters OC.
The first letter is the origin of the ray; the second letter is any point on the ray. Figure 4

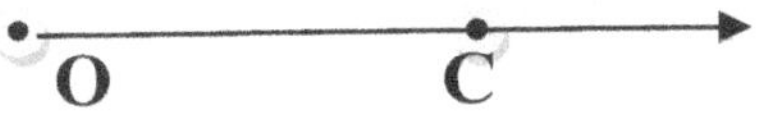

The part of a line limited by two points and all points between the two points called a straight segment. A segment has an origin and a point that called an end. A segment is represented by two capital letters and has a positive length AB > 0

Figure 5

Geometry Chapter 9 Lesson 2

Measurement

Segment measures in units of length in the American system:
in, ft, yd, mi

1 inches = 1 in   1 foot = 1 ft   1 yard = 1 yd   1 mile = 1 mi
1 ft = 12 in   1 yd = 3 ft   1 yd = 36in   1 mi = 5280 ft

Segment measures in units of length in the Metric system:
mm, cm, dm, m, and km.

1cm = 10mm   1dm = 10cm   1m = 10dm   1dm = 100mm
1m = 100cm   1m = 1000mm   1km = 1000m

Segments measured and constructed with a ruler that has origin and units. Two segments are parallel if they lie on the parallel lines.

a
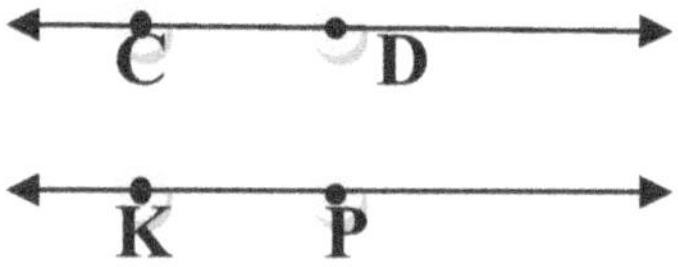

b

Figure 6   CD = 2 cm   a || b
KP = 20 mm   CD ||KP

Operations with segments

Two figures are congruent, if put over of each other, they coincide. Congruent segments have the same lengths. Smallest segment has smallest length.

To compare two segments means to find which segments are more or less than another or maybe its equals.

Figure 7   AB = 4 cm   AB = CD
CD = 40 mm   MN< AB
MN = 3 cm   CD > MN

When a point divides a given segment into two segments, then the length of given segment is equal to the sum of the lengths of the two segments. AC = AB + BC Segments can add and subtracted.

AB + BC = AC   AC – AB = BC   AB = 8 in   BC = 2 in

Figure 8

$AC = AB + BC = 8\text{ in} + 2\text{ in} = 10\text{ in}$
$BC = AC - AB = 10\text{ in} - 8\text{ in} = 2\text{ in}$

When a point divides a segment into two equal segments or by half, this point called the middle of segment. AB = BC= 5 in

Segments can multiply or divide by number:AB×2=AC AC÷2=AB

Figure 9

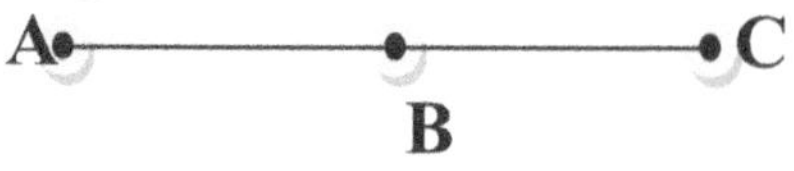

$AC = AB \times 2 = 5 \times 2 = 10\text{ in}$
$AB = AC \div 2 = 10 \div 2 = 5\text{ in}$

Geometry Chapter 9 Lesson 2
Segments
Review 1 A
Solve the problems
Draw the pictures

1) If AB = 6cm BC = 4cm MN = 40mm
Compare segments
a) AB and BC b) BC and AB c) BC and MN

2) Add and subtract segments, if AB = 6cm BC = 4cm
a) AB + BC b) AB – BC

3) Find AB + BC if AB = 8 cm BC = 4cm Draw a picture

4) Find DE + EF if DE = 6 in EF = 2 in Draw a picture

5) Find AB – BC if AB = 7in BC = 5in

6) Find DE – EF if DE = 8 in EF = 3 in

7) Find AB × 3 if AB = 2cm Draw a picture

8) Find 4 × DE if DE = 2 in

9) Find AB ÷ 3 if AB = 12cm Draw a picture

10) Find DE ÷ 2 if DE = 14 in

Geometry Chapter 9 Lesson 2
Review 1 B
Solve the problems
Draw the pictures

1) Find the sum of two segments if the lengths of segments equal 3 in and 7 in

2) Find the sum of two segments if the lengths of two segments equal, 1 dm and 5 cm   1 dm = 10 cm

3) How much longer is segment AB than segment CD if AB = 10 in and CD = 4 in?

4) How much shorter is segment CD than segment AB if AB = 10 cm and CD = 4 cm?

5)Find the second segment if the length of the first segments is 5 in and the length of the second segment is 2 in more than first segment.

6)Find the second segment if the length of the first segment is 8in and the length of the second segment is 3 in less than the length of first segment.

7)How much longer is the first segment than the second segment if the lengths of the segments are 1 dm and 4 cm? 1 dm = 10cm

8)How much shorter is the second segment than the first segment if the length of segments are 6 cm and 1 dm. 1 dm = 10 cm

9)Find the sum of two segments if the lengths of the two segments are 5 in and 3 in

10)Find the difference of two segments if the lengths of two segments are 6 in and 3 in

Geometry Chapter 9 Lesson 3
Angles

An angle in a figure formed when two rays meet at a common point. The common point called the vertex. The rays called the sides of the angle. An angle represented by three capital letters or one letter on vertex. Any angle divides a plane by two parts. An arc denotes the inner part of the angle. Figure 1

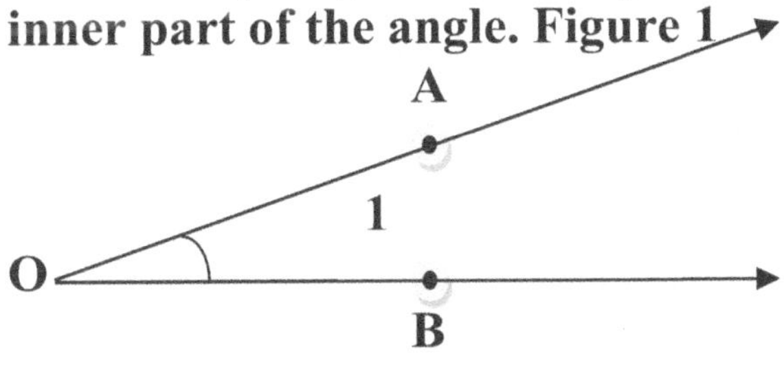

$\angle AOB$ $\angle BOC$

$\angle O$ $\angle 1$

Angles measured in degrees, minutes, and seconds:

1 degree = 60minutes $1^0 = 60'$ 1 minute = 60 seconds $1' = 60''$

A protractor is an instrument used for measurement and construction of angles. Each angle has a positive measurement: $\angle ABC > 0$ Congruent angles have the same measurement:

Figure 2

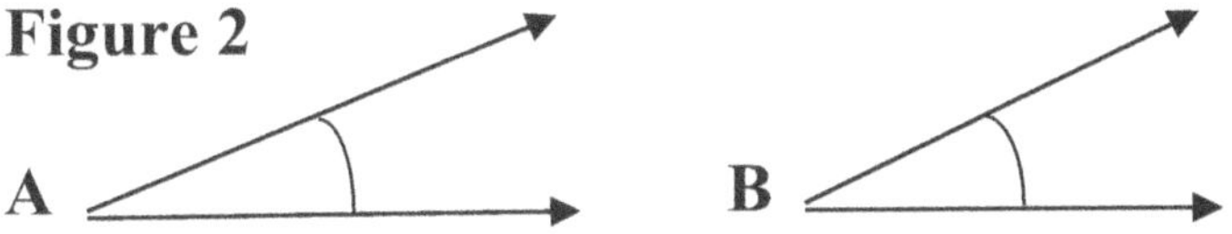

$\angle A = 30^0$

$\angle B = 30^0$

$\angle A = \angle B$

Operations with angles

A ray starting from the vertex of an angle and dividing this angle into two equal parts called the bisector of an angle and is denoted: OB = l. Angles can add and subtracted: $\angle AOB + \angle BOC = \angle AOC$, $\angle AOC - \angle AOB = \angle BOC$ Angles can multiply and divide by number:

$2 \times \angle AOB = \angle AOC$ $\angle AOC \div 2 = \angle AOB$

Figure 3

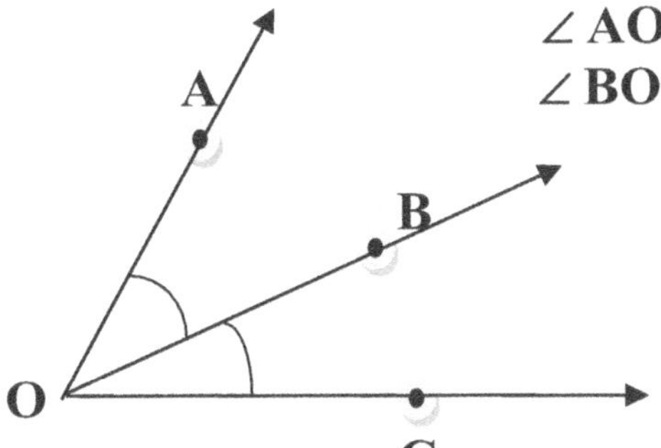

$\angle AOB = 30^0$ $\angle BOC = 30^0$

$\angle AOC > \angle AOB$ $\angle BOC < \angle AOC$

$\angle AOC = \angle AOB + \angle BOC = 30^0 + 30^0 = 60^0$

$\angle BOC = \angle AOC - \angle AOB = 60^0 - 30^0 = 30^0$

$\angle AOC = 2 \times \angle AOB = 2 \times 30^0 = 60^0$

$\angle AOB = \angle AOC \div 2 = 60^0 \div 2 = 30^0$

Geometry Chapter 9 Lesson 3

Different kind of angles

An angle larger than 0^0 but smaller than 90^0 is an acute angle:

$0^0 \leq \angle A < 90^0$ **$\angle A = 30^0$**

Figure 4

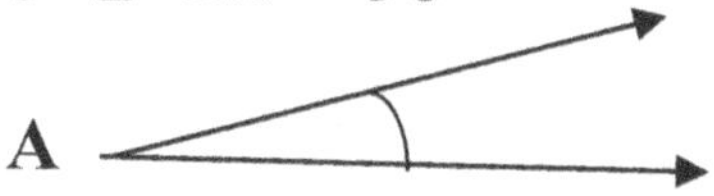

An angle larger than 90^0 but smaller than 180^0 called an obtuse angle: $90^0 < \angle A < 180^0$ **$\angle A = 120^0$**

Figure 5

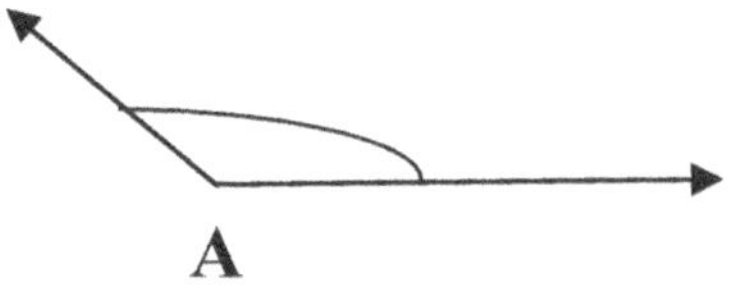

A straight angle is an angle that measures 180^0. A straight angle formed by straight lines.

$\angle A = 180^0$

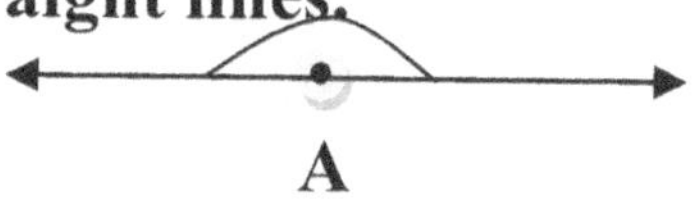

Figure 6

A right angle is half of a straight and equal 90^0 **$\angle A = 90^0$**

Figure 7

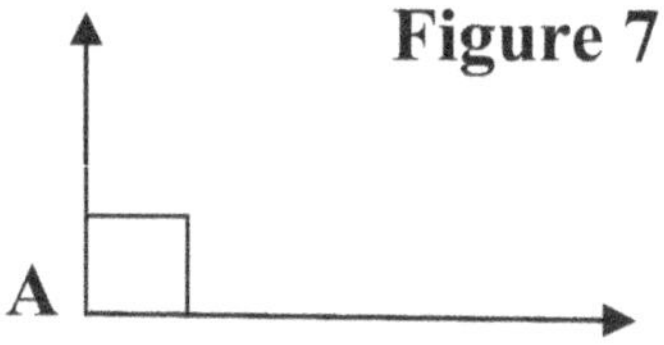

A circle is a full one revolution and equals 360^0

Figure 8

$\angle O = 360^0$

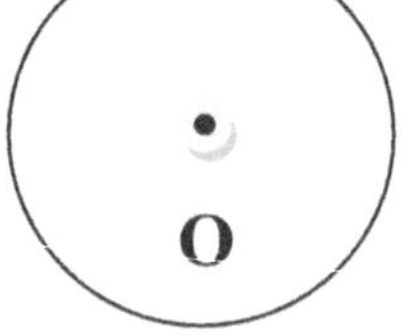

Geometry Chapter 9 Lesson 3

Pairs of Angles

Two angles are complementary if the sum of their measures is 90^0:

$\angle 1 + \angle 2 = 90^0$ $\quad 30^0 + 60^0 = 90^0 \quad 45^0 + 45^0 = 90^0$

Figure 9

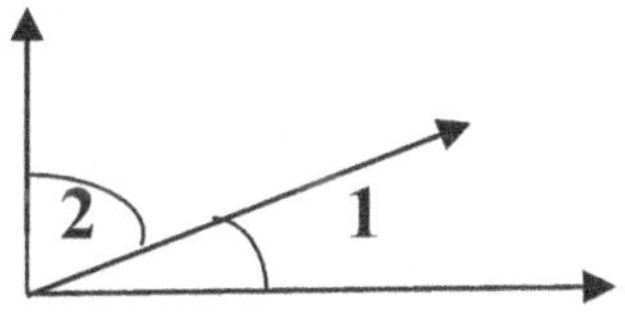

Two angles whose sum is equals 180^0 are supplementary angles. Two angles with one common side and another two sides formed straight line called supplementary. $\angle 1 + \angle 2 = 180^0 \quad 60^0 + 120^0 = 180^0$

Figure 10

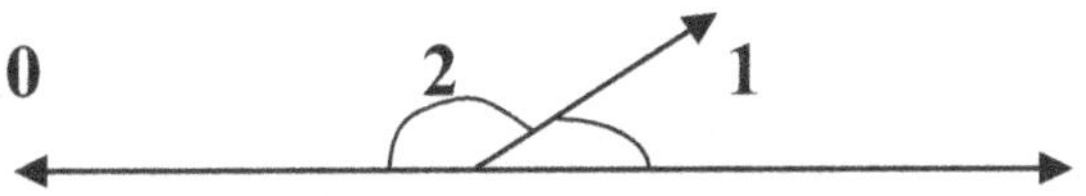

Two intersection lines formed four angles and two pair of verticals angles. Vertical angles are congruent. $\angle 1 = \angle 3$ **and** $\angle 2 = \angle 4$ **are verticals** $\angle 1 = \angle 3 = 30^0 \quad \angle 2 = \angle 4 = 150^0$

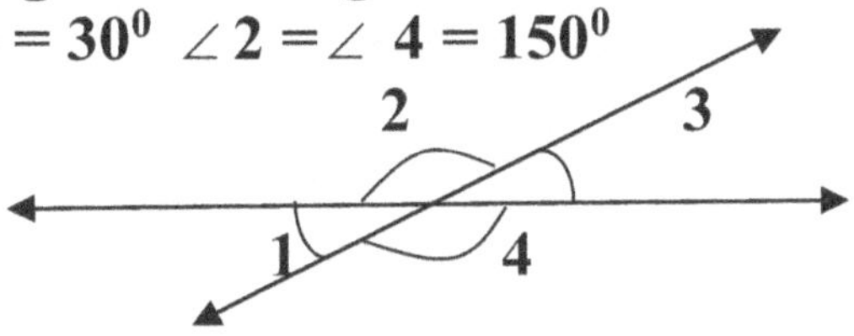

Figure 11

Two lines are perpendicular if the angle between them is 90^0. Segments that lay on perpendicular lines called perpendicular segments:Figure 12 b

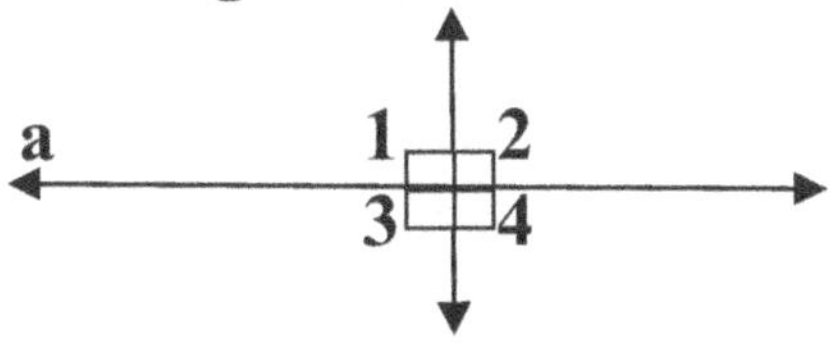

$\angle 1 = \angle 2 = \angle 3 = \angle 4 = 90^0$

$a \perp b$

When two parallel lines are crossed by a transversal, eight angles are formed. c Figure 13 $a \parallel b$ **transversal c**

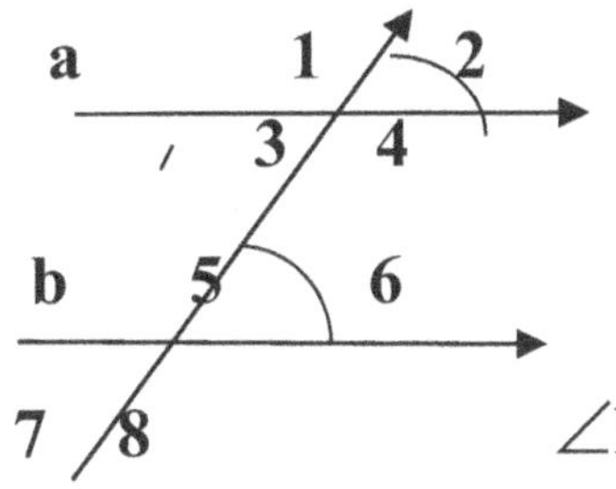

$\angle 2 = 30^0 \quad \angle 1 = 180^0 - 30^0 = 150^0$

$\angle 3 \cong \angle 6 \quad \angle 4 \cong \angle 5$ (interior)

$\angle 1 \cong \angle 8 \quad \angle 2 \cong \angle 7$ (exterior)

$\angle 1 \cong \angle 5 \quad \angle 2 \cong \angle 6$ (corresponding)

Geometry Chapter 8 Lesson 3

Problems with angles

Example 1 Second angle is 0.5 of the first angle of supplementary angles. Find both angles. The first angles is represented by x degree. The second angle represents 0.5 x. The sum of the supplementary angle is 180^0. Equation: $x + 0.5x = 180$ $1.5x = 180$

$x = 180 \div 1.5 = 1800 \div 15 = 120$ $0.5 \cdot 120 = 60^0$

First angle is 120^0. Second angle is 60^0

Figure 14

Example 2

Supplementary angles have ratio like of 9 to 3. Find the angles. Let's represent 1 part for x. The first angle is 9x, the second angle is 3x.

Equation: $9x + 3x = 180$ $12x = 180$ $x = 180 \div 12$ $x = 15^0$

First angle is $9 \cdot 15 = 135^0$ Second angle is $3 \cdot 15 = 45^0$

Figure 15

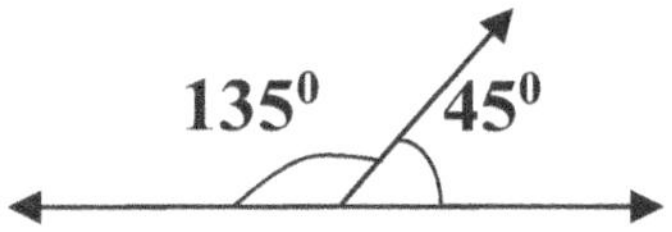

Example 3

Find the supplementary angles if the first angle is 30 more than the second angle. Let's represent the second angle as x. The first angle is x + 30 Equation: $x + 30 + x = 180$ $2x + 30 = 180$

$2x + 30 - 30 = 180 - 30$ $2x = 150$ $x = 150 \div 2$ $x = 75^0$

$75^0 + 30^0 = 105^0$ Figure 16

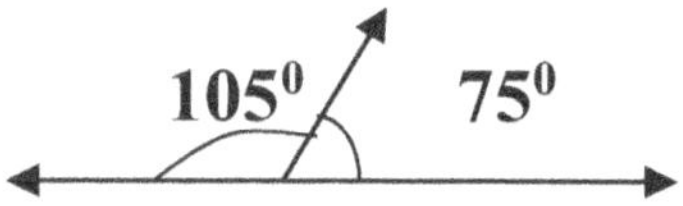

Example 4

Find the supplementary angles if the second angle is 30 less than the first angle. Find angles. The first angle is x. Second angle is x – 30.

Equation: $x + x - 30 = 180$ $2x - 30 = 180$ $2x - 30 + 30 = 180 + 30$

$2x = 210$ $x = 210 \div 2$ $x = 105^0$ $105 - 30 = 75^0$

Figure 17

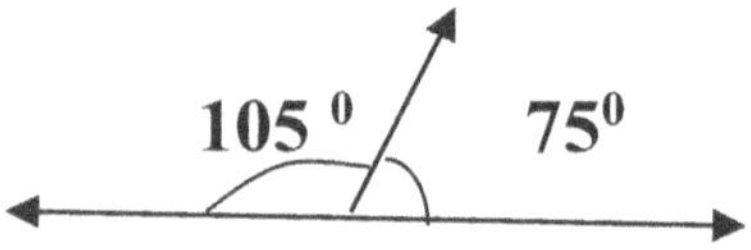

Geometry Chapter 9 Lesson 3

Angles
Review 1 A
Draw the pictures
Solve the problems

1) Given: $\angle 2 = 120^0$ $\angle 1 = 45^0$ Find: $\angle 2$ - $\angle 1$

2) Given: $\angle 2 = 90^0$ $\angle 1 = 45^0$ Find: $\angle 1 + \angle 2$

3) Given: $\angle 2 = 150^0$ $\angle 1 = 60^0$ Find: $\angle 2 - \angle 1$

4) Given: $\angle 1 = 30^0$ Find: $5\angle 1$

5) Given: $\angle 2 = 135^0$ Find: $\angle 2 \div 3$

6) Given: $\angle 2 = 120^0$ Find: $\angle 2 \div 2$

7) Find one of the complementary angles if another angle equals 60^0

8) Find one of the supplementary angles if another angle equals 60^0

9) Find one of the supplementary angles if another angle equals 150^0

10) Find one of the supplementary angles if another angle equals 135^0

Geometry Chapter 9 Lesson 3

Review 1 B

Solve the problems
Draw the pictures

1) One angle is 0.2 of the second angle of supplementary angles. Find both angles.

2) One angle is 0.8 of the second angle of supplementary angles. Find both angles.

3) Supplementary angles have a ratio of 9 to 11. Find the angles.

4) Supplementary angles have a ratio of 7 to 8. Find the angles.

5) Find the second angle, which is 9 times more than the first angle of supplementary angles.

6) One of the supplementary angles is 5 times less than the second angle. Find the biggest angle.

7) Find the supplementary angles if one angle is 20 more than the other angle.

8) Find the supplementary angles if one angle is 40 less than the other angle.

9) One of the supplementary angles is 4 time less than the second angle. Find the biggest angle.

10) Supplementary angles have a ratio of 4 to 5. Find the angles.

Geometry Chapter 9 Lesson 4

Triangles

A triangle is a geometric figure that contains three points and three segments connecting its points.

ΔABC Triangle has 3 vertexes, 3 sides, 3 angles.

The three points of a triangle has vertexes A, B, C.

The three segments of the triangle have sides called AB, BC, AC or a, b, c. The three angles formed by the sides of the triangle called the inner angles of triangles: $\angle A, \angle B, \angle C$

Each side of a triangle is less the sum of the other two sides:

$a < b + c$	$b < a + c$	$c < a + b$
$6 < 12+8$	$12 < 6 + 8$	$8 < 6 + 12$

Figure 1

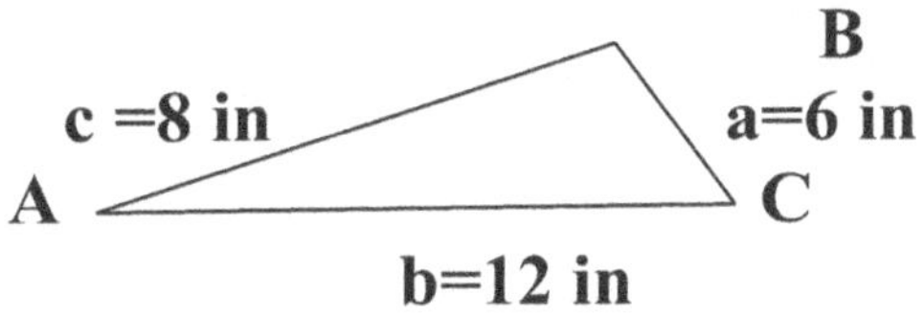

If three sides and three inner angles for one triangle accordingly equal three sides and three angles for another triangle then two triangles are congruent.

Figure 2

Different kind of triangles

An acute triangle contains three acute angles (Figure 3a)

$0^0 < A < 90^0$ $\quad$ $0^0 < B < 90^0$ $\quad$ $0^0 < C < 90^0$

A right triangle contains one right angle and two acute angles (Figure 3b) $C = 90^0$

An obtuse triangle contains one obtuse angle and two acute angles (Figure 3c) $90^0 < C < 180^0$ Sum of inner angles of a triangles equals 180^0. $A + B + C = 180^0$ Example $50^0 + 60^0 + 70^0 = 180^0$

Figure 3

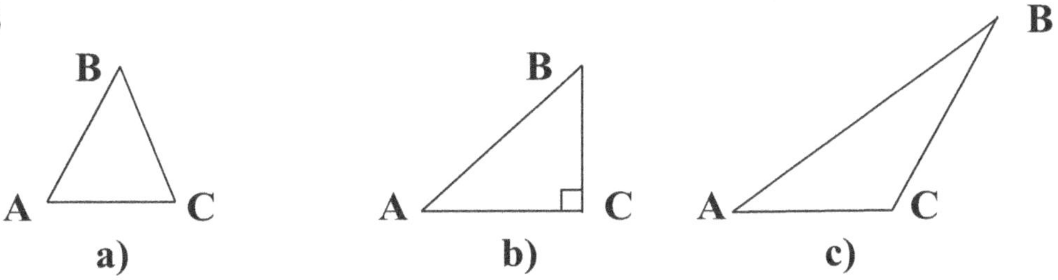

Geometry Chapter 9 Lesson 4
Different elements of triangles

A triangle has vertexes, sides, angles and bisectors, medians, and heights. A ray starting from the vertex of the angle, which divides this angle by two equal parts, is called the bisector and denotes OB = l.

The segment of the bisector of the angle of the triangle connecting the vertex of the triangle with a point on the opposite side called the bisector of the triangle.

A triangle has three bisectors, which intersect at one point inside the triangle: ΔABC $AB = c$ $AC = b$ $BC = a$
$\angle 1 = \angle 2$ $\angle 3 = \angle 4$ $\angle 5 = \angle 6$ $AN = l_a$ $BD = l_b$ $CM = l_c$

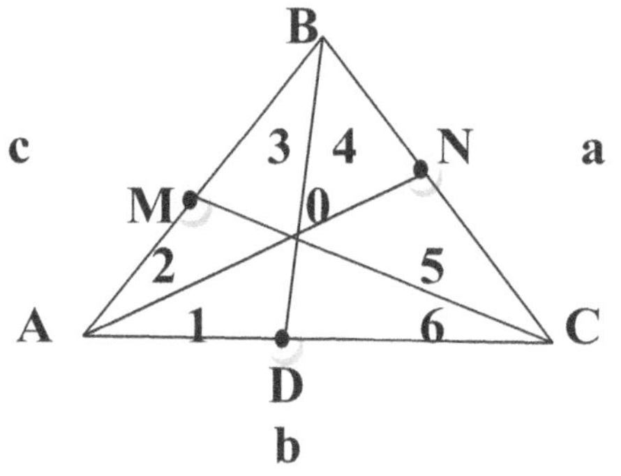

Figure 4

A segment, connecting the vertex to the middle of the opposite side called the median of the triangle.

A triangle has 3 medians, which intersect at one point inside the Triangle
ΔABC $AM = MB$ $BN = NC$ $AD = DC$ $AN = m_a$ $BD = m_b$ $CM = m_c$

Figure 5

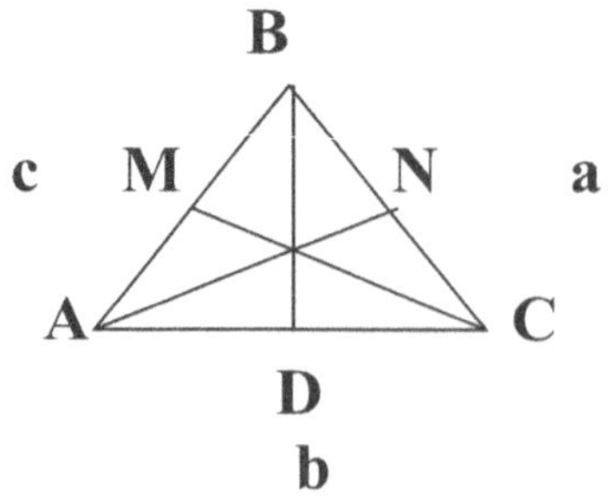

Geometry Chapter 9 Lesson 4

Height of triangle

Perpendicular from vertex of triangle to straight line contains opposite side, called height of triangle Δ ABC

A triangle has 3 heights, which intersect in one point.

In an acute triangle, the 3 heights intersect inside of the triangle Point O: (Figure 6a)

In a right triangle, the 3 heights intersect at the vertex of the right-angle Point C: (Figure 6b) $CM \perp AB$

In an obtuse triangle, the 3 heights intersect outside of the triangle – Point O: (Figure 6c)

$\Delta ABC \quad AN \perp BC \quad BD \perp AC \quad CM \perp AB \quad AN = h_a \; BD = h_b \; CM = h_c$

(Figure 6 a and Figure 6 c)

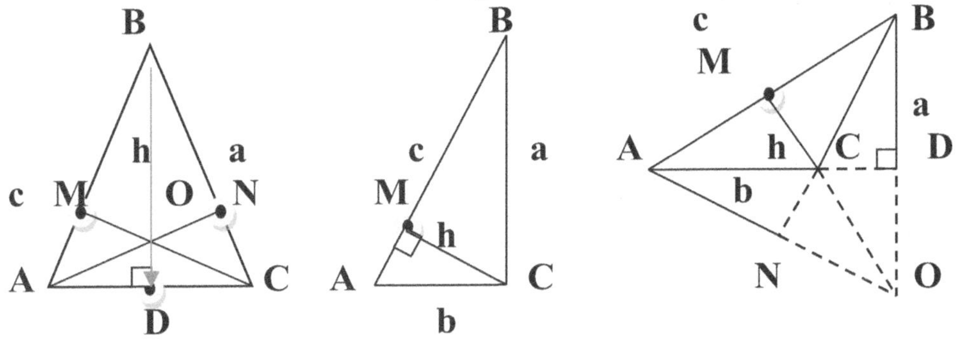

a) b) c)

Figure 6

Geometry Chapter 9 Lesson 4

Special triangles

The sum of the three lengths of the sides of a triangle called the perimeter. A scalene triangle has three different sides: P=a+b+c

Figure 7

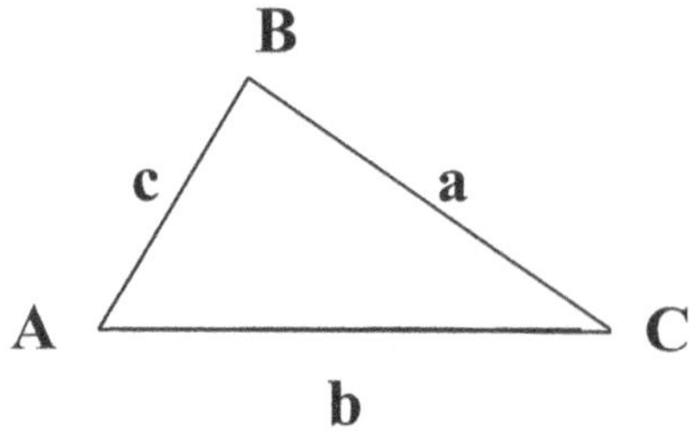

An isosceles triangle is a triangle with two equal sides.
The two equal sides are called legs a and a, the third side is called base b: If AB = BC = a and AC = b than P = 2a + b
Angles on the base isosceles triangle are congruent
If in $\triangle ABC$ AB = BC than $\angle A = \angle C$
In an isosceles triangle, the bisector from the vertex to the base is the median and height.

Figure 8

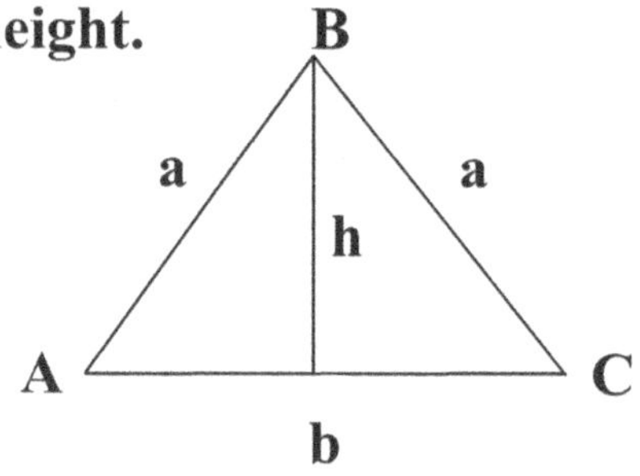

Equilateral Triangle

An equilateral triangle is a triangle with three equal sides:
If AB = BC = AC = b то P = 3b. If in $\triangle ABC$ AB = BC = AC then $\angle A = \angle B = \angle C$ and $\angle A + \angle B + \angle C = 180^0$
$\angle A = \angle B = \angle C = 180^0 \div 3 = 60^0$
In an equilateral triangle, all three bisectors are medians and heights:

Figure 9

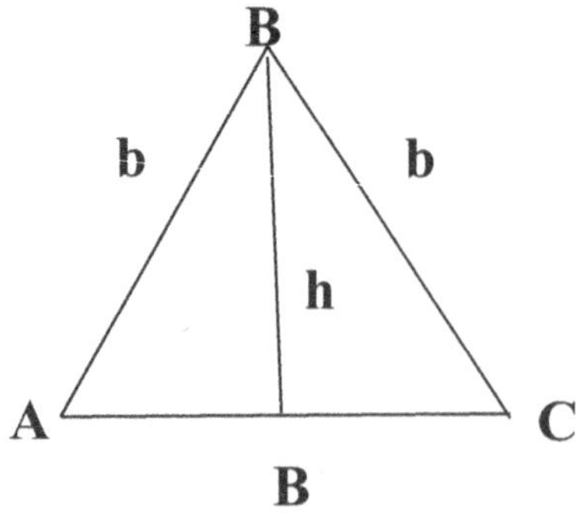

Geometry Chapter 9 Lesson 4

Triangles
Review 1 A
Solve the problems
Draw the pictures

1) Find the third angle of a triangle if the two other angles are 40^0 and 50^0.
2) Find the third angle of a triangle if the two other angles are 55^0 and 60^0.

3) Find the angle of a right triangle if one of the angles is 45^0.

4) Find the angle of a right triangle if one of the angle is 30^0.

5) Find the perimeter of a triangle if its sides equal 5, 6, and 7 cm.

6) Find the third side of a triangle if two of its sides are 8 m and 9m, and the perimeter is 27 m.

7) Find the perimeter of an isosceles triangle if the base is 16 in and one of legs is 10 in

8) Find the legs of an isosceles triangles if the base is 8 ft and the perimeter is 18 ft

9) Find the base of an isosceles triangle if one the leg is 10 yd and perimeter is 36 yd

10) Find the perimeter of an equilateral triangle if its side is 6 ft.

Geometry Chapter 9 Lesson 4

Review 1 B
Solve the problems
Draw the pictures

1) Find the side of an equilateral triangle if its perimeter is 24 in.

2) Find the perimeter of an isosceles triangle if one of legs is 20 cm and the base is 10 cm

3) Find one leg of an isosceles triangle if its base is 1.3 in and its perimeter is 2.5 in.

4) The first side of a triangle is 17 cm of a triangle. The second side of the triangle is double of the first side. Third side of the triangle less the second side of the triangle on 10 cm. Find the perimeter of the triangle.

5) Find the base of an isosceles triangle if its perimeter 3.4 ft and it's one of legs is 1.3 ft.

6) The perimeter of an isosceles triangle is 40 cm. The leg is twice than the base of the triangle. Find the sides of the triangle.

7) The perimeter of an isosceles triangle is 26 in. Its base is 6 times less than its leg. Find the sides of the triangle.

8) The perimeter of an isosceles triangle is 20 in. Its leg is 2 in less than its base. Find the sides of the triangle.

9) The perimeter of an isosceles triangle is 40 in. Its leg is 5 in more than its base. Find the sides of the triangle.

10) The perimeter of an isosceles triangle is 22 cm. Its base is 2 cm less than its leg. Find the sides of the triangle.

Geometry Chapter 9 Lesson 5

Right triangles

A right triangle contains one right angle and two acute angles: $C = 90^0$
For example, special right triangles: 30^0, 60^0, 90^0 and 45^0, 45^0, 90^0

The sides of a right triangle that form the right angle are called legs a and b
The side of a right triangle that is opposite the right angle is called hypotenuse (c)

The perimeter of a triangle is equal to the sum of its sides.
If hypotenuse AB = c, legs AC = b and BC = a then $P = a + b + c$

In a right triangle, the hypotenuse is more than the any other legs

$C = 90^0$	$A < 90^0$	$B < 90^0$
$A < C$	$a < c$	$c > a$
$B < C$	$b < c$	$c > b$

The sum of two acute angles in a right triangle is equal to 90^0
$A + B + C = 180^0$ $\quad$ $C = 90^0$ $\quad$ $A + B = 90^0$

The property for the sides of a right triangle called the Pythagorean Theorem.
In a right triangle, the square of the hypotenuse equals the sum of the square of the legs.

$c^2 = a^2 + b^2$ or $a^2 + b^2 = c^2$ $\quad$ $a^2 = c^2 - b^2$ or $b^2 = c^2 - a^2$
Figure 1

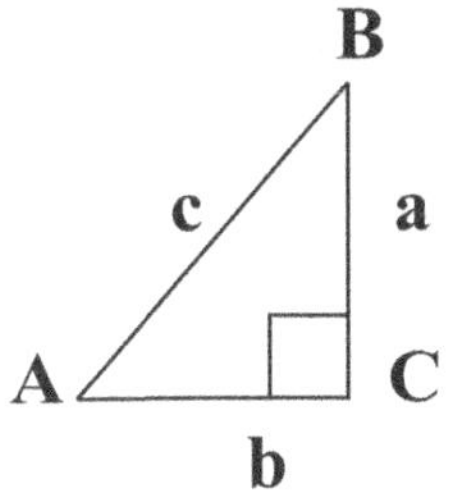

Geometry Chapter 9 Lesson 5

Pythagorean Theorem

Triangle with sides 3, 4, 5 people knew before proven Pythagorean theorem $3^2 + 4^2 = 5^2$ $9 + 16 = 25$

Right triangles, in which sides expressed in natural numbers, called Pythagorean triples $6^2 + 8^2 = 10^2$ $36 + 64 = 100$

Figure 2

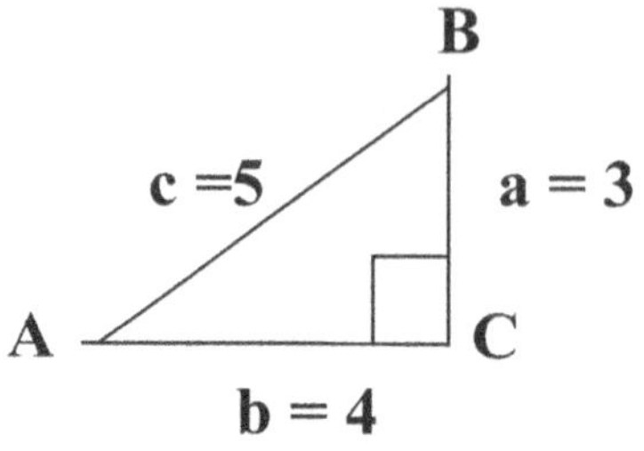

Examples for Right Triangle ABC (Figure 2)

a) $c = 17$ ft $b = 8$ ft $a = ?$ $a^2 + b^2 = c^2$
$a^2 + 8^2 = 17^2$ $a^2 + 64 = 289$ $b^2 = 289 - 64$ $b^2 = 225$
$b = 15$ft

b) $a = 20$ yd $c = 29$ yd $b = ?$ $c^2 = a^2 + b^2$
$29^2 = 20^2 + b^2$ $841 = 400 + b^2$ $441 = b^2$ $b = 21$ yd

c) $a = 52$ m $b = 39$ m $c^2 = a^2 + b^2$ $c^2 = 52^2 + 39^2$
$c^2 = 2704 + 1521$ $c^2 = 4225$ $c = 65$ m

Rectangle

d) **Find the length of a rectangle, if the diagonal is 15 in and the width is 9 in.**

$d = 15$ in $w = 9$ in $l^2 + w^2 = d^2$ $l^2 + 9^2 = 15^2$
$x^2 + 81 = 225$ $x^2 = 225 - 81$ $x^2 = 144$ $x = 12$ in

Figure 3

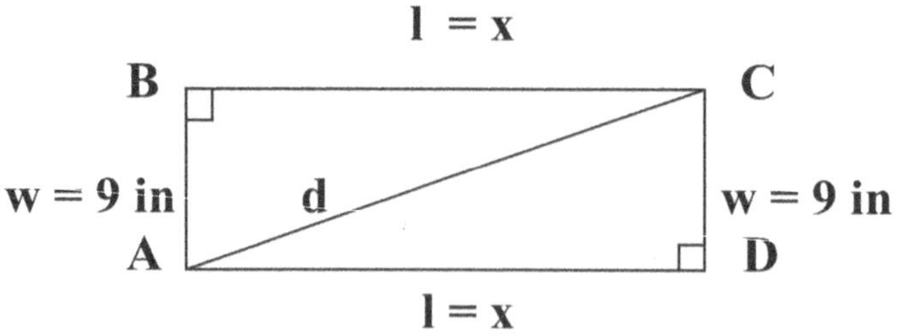

Geometry Chapter 9 Lesson 5

Right triangle

Review 1 A

Draw the right triangles
Find the missing side of each right triangle
$a^2 + b^2 = c^2$

	a	b	c		a	b	c
1)	?	4	5	2)	6	?	10
3)	9	12	?	4)	?	16	20
5)	15	?	25	6)	18	24	?
7)	?	28	35	8)	24	?	40
9)	27	36	?	10)	?	40	50
11)	33	?	55	12)	36	48	?
13)	?	52	65	14)	42	?	70
15)	45	60	?	16)	?	64	80
17)	51	?	85	18)	54	72	?
19)	?	76	95	20)	60	?	100
21)	5	12	?	22)	?	24	26
23)	15	?	39	24)	20	48	?
25)	?	60	65				

Geometry Chapter 9 Lesson 5

Review 1 B

Draw the right triangles

Find the missing side of each right triangle

$$a^2 + b^2 = c^2$$

	a	b	c		a	b	c
1)	3	?	7.8	2)	0.7	2.4	?
3)	?	4.8	5	4)	2.1	?	7.5
5)	2.8	9.6	?	6)	?	2.1	2.9
7)	4	?	5.8	8)	0.8	1.5	?
9)	?	3	3.4	10)	2.4	?	5.1
11)	0.9	4	?	12)	?	8	8.2
13)	1.1	?	6.1	14)	1.3	8.4	?
15)	?	6.3	8.7	16)	3	?	$7\frac{4}{5}$
17)	$1\frac{2}{5}$	4.8	?	18)	?	$7\frac{1}{5}$	$7\frac{1}{2}$
19)	$2\frac{4}{5}$	?	10	20)	4	$4\frac{1}{5}$	?
21)	?	$1\frac{1}{2}$	1.7	22)	$1\frac{3}{5}$	?	3.4
23)	$2\frac{2}{5}$	4.5	?	24)	?	8	$8\frac{1}{5}$
25)	1.3	?	$8\frac{1}{2}$				

Geometry Chapter 9 Lesson 6

Perimeter

A polygon means a many-sided figure.
A polygon has three or more sides of the figure.

Points called vertexes and segments called sides.
A triangle has 3 vertexes, 3 sides, and 3 inner angles (Figure 1a)

A quadrilateral has 4 vertexes, 4 sides, and 4 inner angles (Figure 1b)
A pentagon has 5 vertexes, 5 sides, and 5 inner angles (Figure 1c)

A hexagon has 6 vertexes, 6 sides, and 6 angles (Figure 1d)
Two vertexes of polygon, which belong to one side, called adjacent.

A segment connecting any two nonadjacent vertexes called diagonal (d)
Figure 1 b

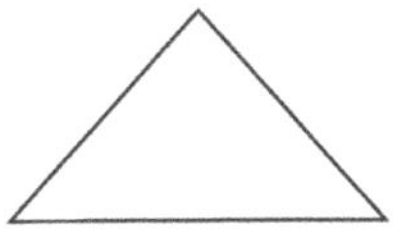

Figure 1 a

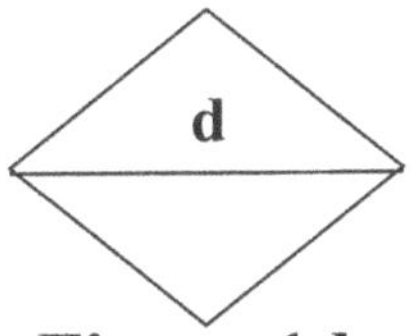

Figure 1 b

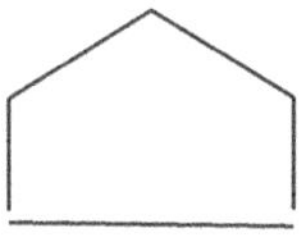

Figure 1 c

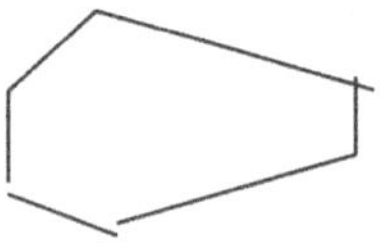

Figure 1 d

Quadrilateral
A quadrilateral has 4 vertexes, 4 sides, and 4 inner angles.
Two nonadjacent vertexes or sides of a quadrilateral are called opposite vertexes or sides.

The perimeter is the distance around a figure or the sum of the length of all sides. The measurement of the perimeter is in linear units:
1 in, 1 ft, 1 yd, 1 mi, 1 mm, 1cm, 1dm, 1m, 1 km

Geometry Chapter 9 Lesson 6

Parallelogram

A parallelogram is a quadrilateral in which the opposite sides are parallel. If ABCD is a parallelogram, then AD || BC and AB || DC

Figure 2

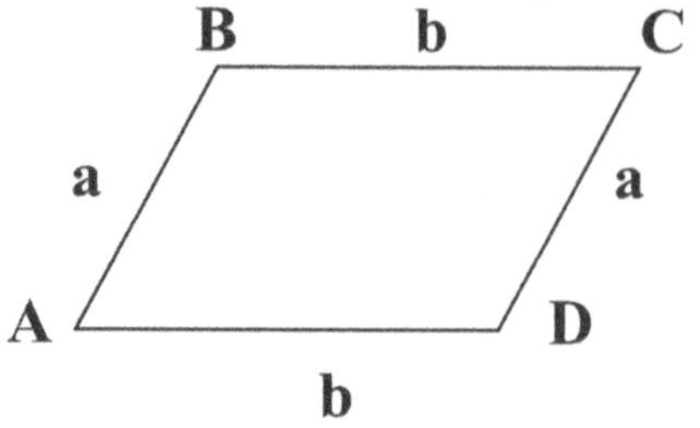

The sum of inner angles for any parallelogram is equal 360^0

$\angle A + \angle B + \angle C + \angle D = 360^0$

The sum of angles adjacent to one side of a parallelogram is equal to 180^0. Figure 2

If AB || DC then $\angle A + \angle D = 180^0$ **and** $\angle B + \angle C = 180^0$

If AD || BC then $\angle A + \angle B = 180^0$ **and** $\angle C + \angle D = 180^0$

Opposite angles of a parallelogram are equal:

If ABCD is a parallelogram, then $\angle A = \angle C$ **and** $\angle B = \angle D$

The opposite sides of a parallelogram are equal:

If ABCD is a parallelogram, then AB = CD and AD = BC

Parallelogram ABCD, where AB = a AD = b

The perimeter of a parallelogram with sides a and b can use any of the following formulas:

$P_{par} = a + b + a + b$ $\quad P_{par} = 2a + 2b$ $\quad P_{par} = 2(a + b)$

Rhombus

A rhombus is a parallelogram in which measures of the four sides are equal. Figure 3

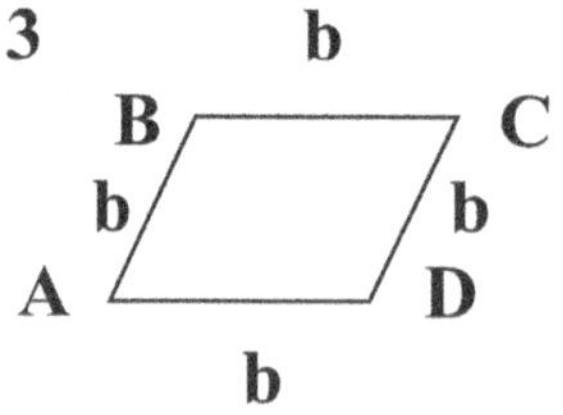

If AB = BC = CD = AD = b then P = b + b + b + b $\quad$ **P = 4b**

All properties of parallelogram are correct for a rhombus:

AB || DC $\quad$ **AD || BC** $\quad$ $\angle A = \angle C$ **and** $\angle B = \angle D$

Geometry Chapter 9 Lesson 6

Rectangle

A rectangle is parallelogram in which all four angles are right angles.

Figure 4

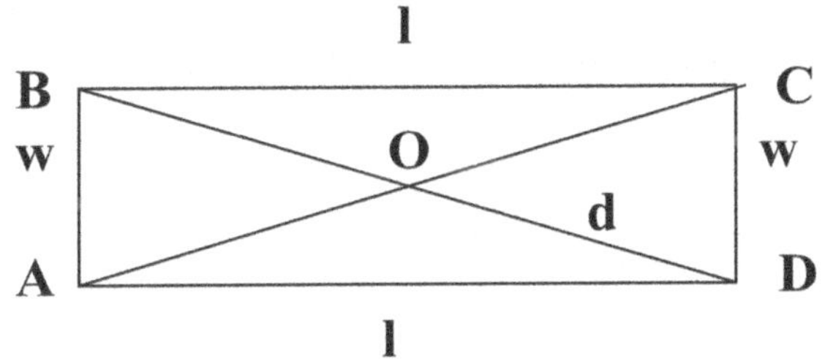

If ABCD is a rectangle, then $\angle A = \angle B = \angle C = \angle D = 90^0$

The dimensions of a rectangle are length l and width w.
The perimeter of a rectangle is equal to the sum of its sides.

If ABCD is a rectangle $AD = BC = l$ $AB = CD = w$

$P = l + w + l + w$ $P = 2l + 2w$ $P = 2(l + w)$

All properties of a parallelogram are the same for a rectangle
A rectangle has its own property: the diagonals of a rectangle are congruent. If ABCD is a rectangle, then $AC = BD = d$

Square

A square is a rectangle in which measures of all four sides are equal.
A square is a rhombus in which all four angles are right angles.
A square is a parallelogram in which all four sides are equal and all angles are right angles.

Figure 5

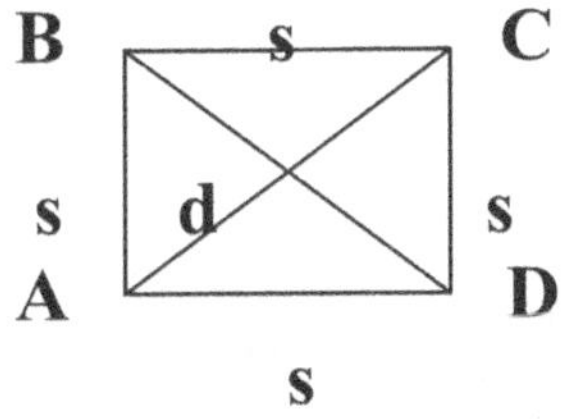

All properties for a rectangle, rhombus, and parallelogram are the same for a square. The perimeter of a square is equal to the sum of the four sides: **If $AB = BC = CD = AD = s$** $P = s + s + s + s$ $P = 4s$

$\angle A = \angle B = \angle C = \angle D = 90^0$ $AC = BD = d$

Geometry Chapter 9 Lesson 6

Trapezoid

A trapezoid is a quadrilateral that has exactly two opposite sides that are parallel. The two parallel sides called bases and the two non-parallel sides called legs.

Figure 6

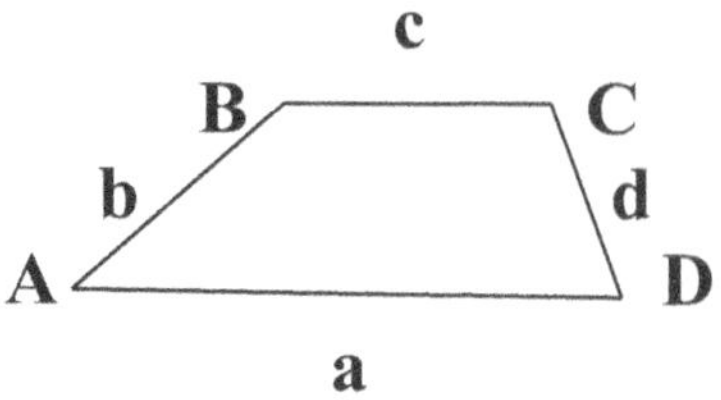

AD || BC   AD and BC are bases   AB and CD are legs
If ABCD is a trapezoid, AD = a   AB = b   BC = c   CD = d
P = a + b + c + d

A trapezoid called right trapezoid if it has one right angle:
If ABCD is a right trapezoid and ∠A = 90⁰ then ∠B = 90⁰

Figure 7

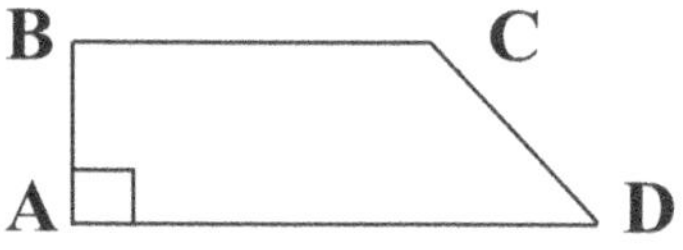

A trapezoid is called an isosceles trapezoid if it has measures of legs are equals. If ABCD is an isosceles trapezoid and AD || BC than AB = CD

Figure 8

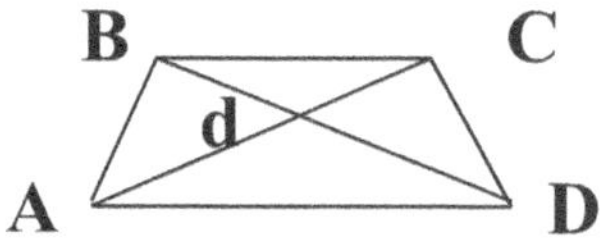

An isosceles trapezoid has diagonals whose length are equal.
AC = BD = d

Geometry Chapter 9 Lesson 6

Kite

A kite is quadrilateral in which there are two pairs of equal adjacent sides. If ABCD is a kite, then AB = AD and CB = CD

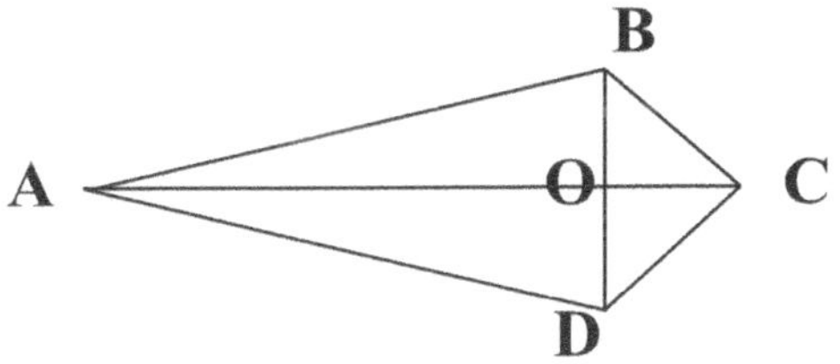

Figure 9

Regular polygons

A polygon called regular which measures of all sides and all angles are equal. A regular triangle is equilateral triangle and its perimeter: P=3 s.
A regular quadrilateral is the square and its perimeter: P=4 s
A regular pentagon is a polygon that measures of five equal sides: P=5 s. (Figure 10a)
A regular hexagon is a polygon that measures of six equal sides: P=6 s (Figure 10 b)

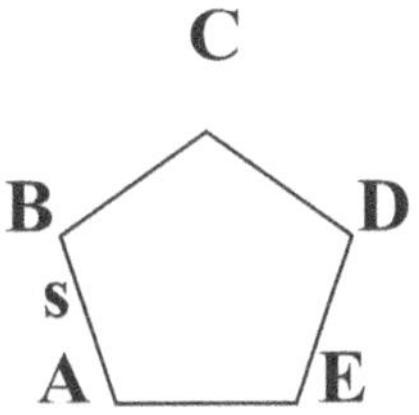

Figure 10 a

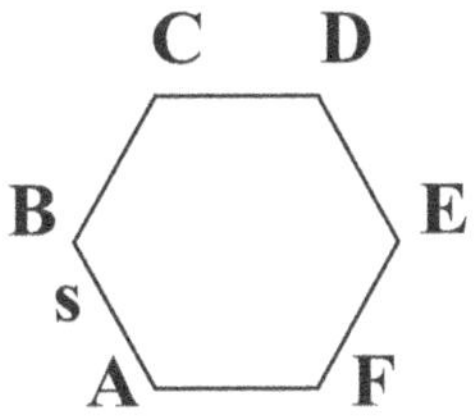

Figure 10 b

Formulas

The perimeter of a triangle is the sum of the length of all three sides.
$P = a + b + c$
The perimeter of a square is four times the length of the sides. $P = 4\ s$
The perimeter of a rhombus is four times the length of the sides. $P=4b$
The perimeter of a rectangle is the sum of twice of the length and twice of the width: $P = 2\ l + 2\ w$
The perimeter of a parallelogram is the sum of twice of the adjacent sides. $P = 2\ a + 2\ b$

Geometry Chapter 9 Lesson 6

Perimeter
Review 1 A
Solve the problems
Draw the pictures

1) Find the perimeter of a triangle, if its sides equal 6cm, 8cm, 10 cm.
2) Find the perimeter of an equilateral triangle, if a side equals 5 in.
3) Find a side of an equilateral triangle, if the perimeter equals 30 ft.
4) Find the perimeter of an isosceles triangle, if its sides equal 10cm, 10 cm, and 8cm.
5) Find the perimeter of a square, if its side equal 5 yd.

6) Find the side of a square, if its perimeter equals 40 in.
7) Find the perimeter of a rhombus, if its sides equal 10 dm.
8) Find the side of a rhombus, if its perimeter equals 24 cm.
9) Find the perimeter of a rectangle, if its length equals 6 cm, and the width is 4 cm
10) Find the length of a rectangle if its perimeter is 44 in, and its width is 10 in.

11) Find the width of a rectangle if its perimeter is 88 ft and its length is 30 ft
12) Find the perimeter of a parallelogram, if the length of its sides are 10m and 8m
13) Find the side of a parallelogram if its perimeter is 60 ft and one of its side 10 ft
14) Find the perimeter of a trapezoid, if its sides are 10cm, 8cm, 4cm, and 6cm
15) Find the perimeter of regular pentagon, if each side equals 10 ft.

16) Find the side of a regular pentagon if its perimeter is 50 yd.
17) Find the perimeter of a hexagon, if its side equal 4m, 6m, 4m, 2m, 6m, and 2m.
18) Find the perimeter of a regular hexagon, if each side equals 5 ft.
19) Find the side of a regular hexagon if its perimeter is 60 m.
20) Find perimeter of a kite, if its sides are equals 6 in and 10 in.

Geometry Chapter 9 Lesson 7

Area

The area is the number of square units that covers a surface.
Congruent polygons have the same areas: if $F_1 = F_2$ then $A_1 = A_2$
If a polygon formed by two different figure, then the area of the polygon equals the sum of the areas for each figure:
if $F = F_1 + F_2$ then $A = A_1 + A_2$
Measurement Area is Square Unit
1 mm^2, 1 cm^2, 1 dm^2, 1 m^2, 1 km^2 and 1 in^2, 1 ft^2, 1 yd^2, 1 mi^2

1 cm = 10 mm	**1 ft = 12 in**	**1 yd = 3 ft**
1 sq.cm = 100 sq.mm	**1 ft^2 = 144 in^2**	**1 yd^2 = 9 ft^2**

Square

The area of a square is equal to the square of its sides:
ABCD is square $A = s \cdot s$ $\qquad A_{\square} = s^2$
Example: s = 6 in. $A_{\square} = ?$ $\qquad A_{\square} = s^2 = 6^2 = 6 \cdot 6 = 36$ in ^{2}

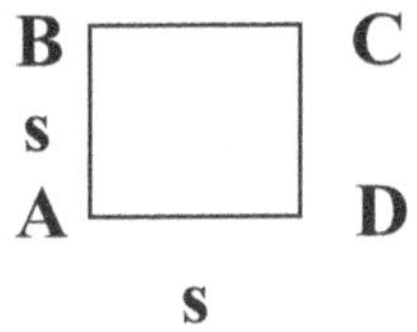

Figure 1

Rectangle

The area of a rectangle is equal to the product, length by its width.
If ABCD is a rectangle, AD = l, AB = w then $A_{rec} = l \cdot w$
Example: l = 10 m $\qquad$ w = 5 m $\quad A_{rec} = l \cdot w = 10\ m \cdot 5\ m = 50\ m^2$

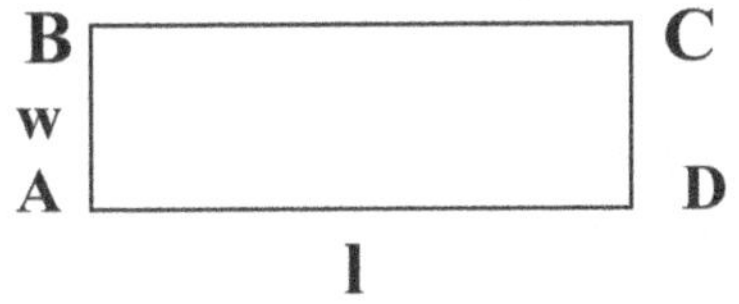

Figure 2

Rectangle and Square

Find the area of a figure that contains a rectangle and a square together. Example: ABCD a rectangle, DCMN a square
if l = 10 in, w = 5 in, s = 5 in. $A = A_1 + A_2 = 10 \cdot 5 + 5^2 = 50 + 25 = 75$ in^2

Figure 3

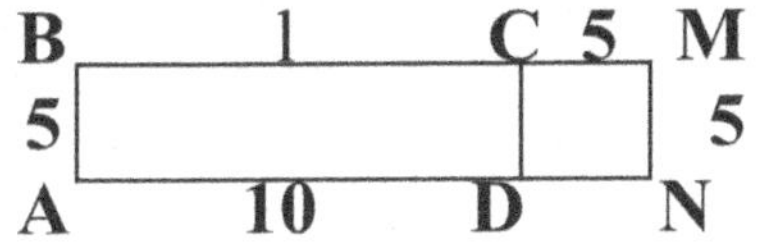

Geometry Chapter 9 Lesson 7

Rhombus

The area of a rhombus is equal to the product, base by its height.
If ABCD is a rhombus, AD = b, BM = h then $A_{rh} = b \cdot h$

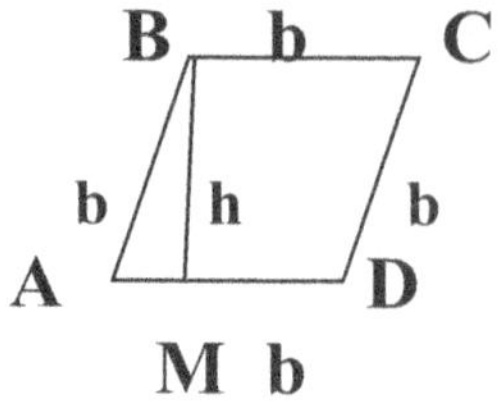

Example: $b = 10$ cm $h = 5$ cm
$A_{rh} = bh$ $A = 10 \cdot 5 = 50\ cm^2$

Figure 4

Parallelogram

One sides of a parallelogram called the base.
Perpendicular from any point on the opposite side to the line containing the base called the height of the parallelogram.
The area of a parallelogram is equal to the product, base by its height: If ABCD is a parallelogram, AD = b BM = h then
$A_{par} = b\,h$

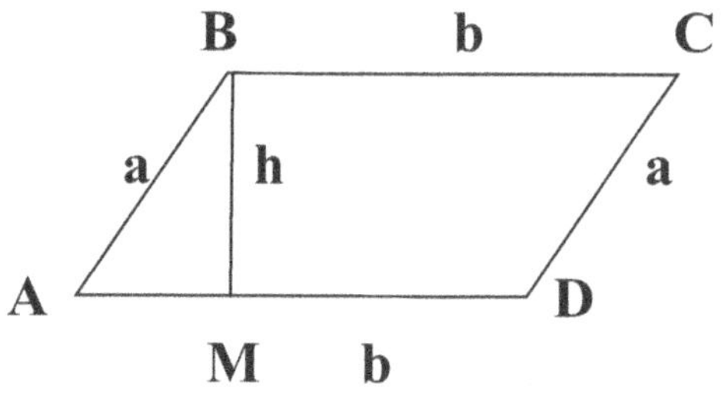

Example:
$b = 20$ m $h = 10$ m
$A_{par} = b\,h = 20 \cdot 10 = 200\ m^2$

Figure 5

Trapezoid

Perpendicular from any point of the upper base to the line containing the lower base called the height of the trapezoid. The area of a trapezoid is equal product half of height by sum of bases. If ABCD is a trapezoid, AD = a BC = c BM = h then $A_{tr} = 0.5h\,(a + c)$

Figure 6

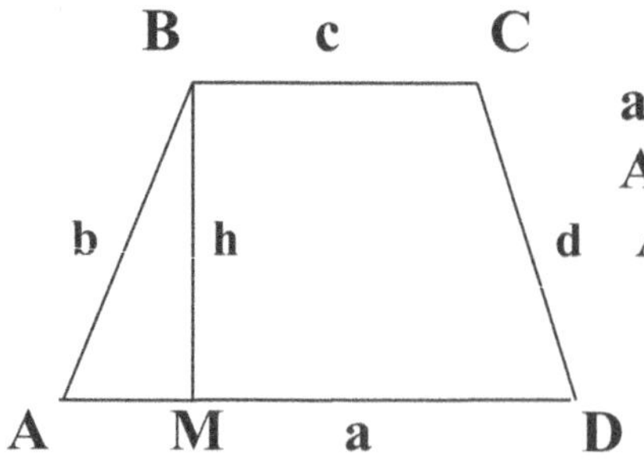

Example:
$a = 12$ m $c = 8$ m $h = 6$ m
$A_{tr} = 0.5\,h\,(a + c)$
$A_{tr} = 0.5 \cdot 6\,(12 + 8) = 3 \cdot 20 = 60\ m^2$

Geometry Chapter 9 Lesson 7

Formula for the Area of a Triangle

One of sides of a triangle called the base.
Perpendicular from any vertex to the line containing the base, is called the height of the triangle.
The area of a triangle equal half of product, base by height.

If ΔABC $AC = b$ $BM = h$ then $A_\Delta = 0.5bh$

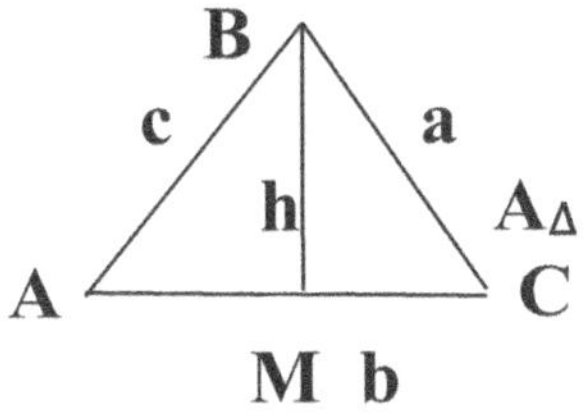

Example:

$b = 5$ cm $h = 4$ cm

$A_\Delta = 0.5\ bh = 0.5 \cdot 5 \cdot 4 = 0.5 \cdot 20 = 10\ cm^2$

Figure 7

Hero's Formula for Area of a Triangle

A triangle has sides a, b, c and a semi perimeter $s = (a + b + c) \div 2$

Area of triangle is equal $A_\Delta = \sqrt{s(s - a)(s - b)(s - c)}$

Figure 8

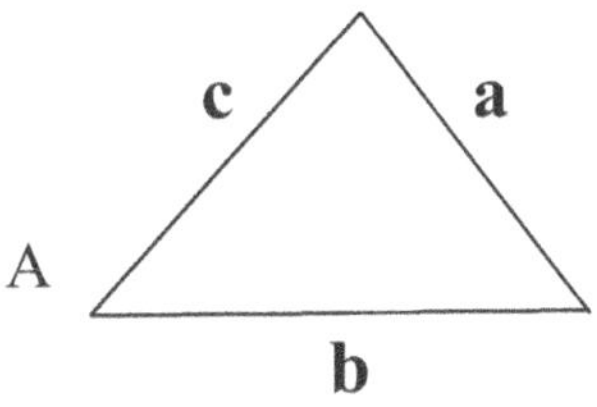

Example:

Find the area of a triangle with sides 3m, 4m, 5m.

$s = (3 + 4 + 5) \div 2 = 12 \div 2 = 6$

$A_\Delta = \sqrt{6(6 - 3)(6 - 4)(6 - 5)} = \sqrt{6 \cdot 3 \cdot 2 \cdot 1} = \sqrt{36} = 6\ \text{м}^2$

Right triangle

The area of a right triangle equal half of the product of its legs.

If ΔABC $C = 90^0$ $AC = b$ $BC = a$ than $A_\Delta = 0.5ab$

Example: $a = 3$ ft $b = 4$ ft $A_\Delta = 0.5\ ab = 0.5 \cdot 3 \cdot 4 = 6\ ft^2$

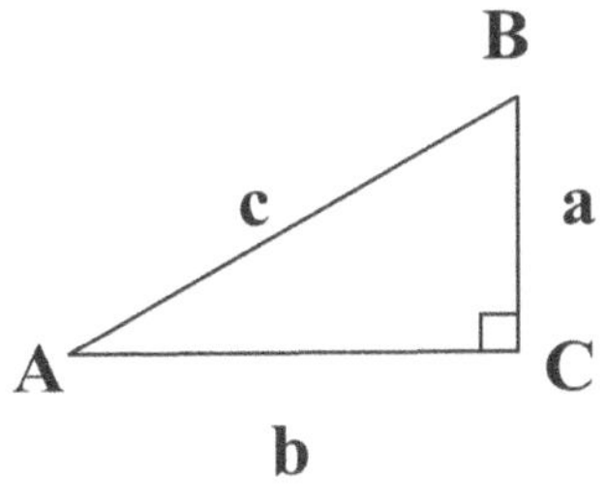

Figure 9

Geometry Chapter 9 Lesson 7

Area
Review 1 A
Solve the problems
Draw the pictures

1) Find the area of a square, if its side is 8 in.
2) Find the length of a square if the area of the square is 25 cm^2.

3) Find the area of a rectangle, if its dimensions are 9yd and 4 yd.
4) Find the length of a rectangle, if its area is 72 sq.cm, and its width is 8cm.

5) Find the width of a rectangle, if its area is 132 ft^2, and its length is 12ft.
6) Find the base of a rhombus if its height is 5m and its area is 60 m^2.

7) Find the height of a parallelogram if its base is 15in and its area is 150 in^2.
8) Find the height of a trapezoid, if its bases are 10m and 8m, and its area is 90 m^2.

9) Find the big base of trapezoid, if $c = 10$ in, $h = 6$ in, $A = 66$ in^2
10) Find the small base of a trapezoid, if $a = 10$ ft, $h = 4$ ft, $A = 36$ ft^2

11) Find the area of a triangle, if its base is 10m, and its height is 8m.
12) Find the base of a triangle, if its height is 6m, and its area is 60 m^2.

13) Find the area of a right triangle, if its legs are 10in and 8 in.
14) Find one of leg of a right triangle, if its other leg is 12m, and its area is 60 m^2.

15) Find the area of a triangle with sides 3, 4, and 5 м.
16) Find the area of a triangle with sides 6, 8, and 10in.
17) Find the area of a triangle with sides 5, 12, and 13ft.
18) Find the area of a triangle with sides 9, 12, and 15 cm.
19) Find the area of a triangle with sides 7, 24, and 25 yd.
20) Find the area of a triangle with sides 12, 16, and 20 dм.

Geometry Chapter 9 Lesson 8

Circle

A circle is a geometric figure, made up of all points, and ordered by a given distance from a given point. The given point called the center of the circle. A segment connecting the center with any point on the circle called the radius of the circle. The given distance is a radius of the circle. All radii are equal in one circle. A segment passing through the center of the circle called the diameter of circle. O – Center AO=OB=r
AB is diameter AB = d (Figure 1) The diameter of a circle is twice the radius: d = 2r. The degree of measurement of a full circle is 360 degrees.

Figure 1

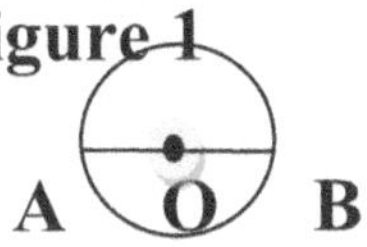

Arc and chord

Two points on a circle divide the circle into two parts, each part formed by arc of circle. The arc of a circle measured in degrees.

Figure 2

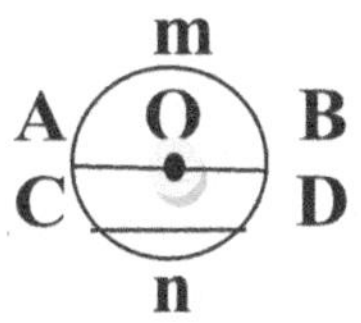

Circles can draw with compasses or patterns. A segment connecting two points of a circle called a chord. CD is a chord and $\cup C\,n\,D$ is an arc (Figure 2) The measure of two arcs together with the same ends is 360^0.
$\cup AmB + \cup A\,n\,B = 360^0$
An arc called a semicircle if the segment connects the ends of the diameter. A semicircle is half of a circle. $\cup AmB = \cup AnB = 180^0$
Arcs can be compared, added, and subtracted: (Figure 3)
$\cup AB = \cup BC = \cup AC$ $\quad\cup AB < \cup ACB$ $\quad\cup ACB > \cup AB$
$\cup AB + \cup BC = \cup ABC$ $\quad\cup ABC - \cup AB = \cup BC$
Arc can be multiplied or divided by number:
$\cup AB \cdot 2 = \cup ABC$ $\quad\cup A\,BC \div 2 = \cup AB$
An angle with the vertex in the center of a circle called a central angle. Degree measured of a central angle of a circle is equal to the degree measured for an arc. $\angle AOB = \cup A\,n\,B$ (Figure 4)

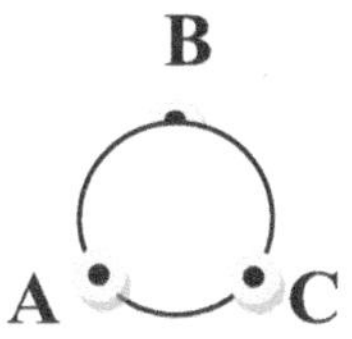

Figure 3

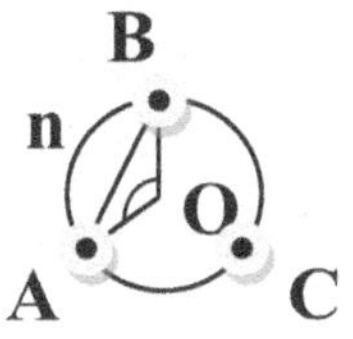

Figure 4

Geometry Chapter 9 Lesson 8

Circumference

The circumference is the distance around a circle. Two formulas can used for finding the circumference. First formula of circumference

$C = \pi d \quad \frac{C}{d} = \pi \quad d \times \frac{C}{d} = \pi \times d \quad C = \pi d$ **where** $\pi \approx 3.14$ **or** $\pi \approx 3\frac{1}{7}$

$\pi \approx \frac{22}{7}$ **Example 1: Find the circumference if d = 7 in, use** $\frac{22}{7}$ **for** π.

$C = \pi d = \frac{22}{7} \cdot 7 \approx 22$ **in Second formula of circumference** $C = 2\pi r$

If $d = 2r \qquad C = \pi d \qquad C = \pi \cdot 2r \qquad C = 2\pi r$

Example 2: Find the circumference if r = 10 cm, use ≈ 3.14 **for** π

$C = 2\pi r = 2 \cdot 3.14 \cdot 10 \approx 62.8$ **cm**

Example 3: Find the exact answer of the circumference in terms of π **if r = 7 in.** $C = 2\pi r = 2\pi \cdot 7 = 2 \cdot 7\,\pi = 14\pi$ **in**

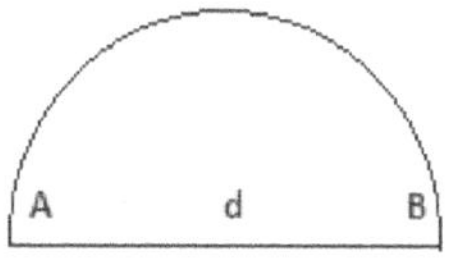

Circumference of a semicircle is

$C = 0.5\,(2\pi r) = \pi r$ **or** $C = 0.5\pi d$

Area of the Circle

The area of a circle is equal to the product of number π **and the square of radius r:** $A_c = \pi r^2$. **The area of semicircle** $A = 0.5\,\pi r^2$ **Example 4: Find the area of the circle if radius r = 10 cm** $\pi = 3.14$ $A_c = \pi r^2 = \pi \times 10^2 = 100\pi = 100 \times 3.14 \approx 314$ **cm²** **Example 5: Find the area of the circle if d = 14 cm**

$\pi \approx \frac{22}{7}$ **d = 14 in r = 7 in** $A_c = \pi r^2 = \frac{22}{7} \cdot 7^2 = \frac{22}{7} \times \frac{49}{1} = 22 \cdot 7 \approx 154$ **in²**

Example 6: Find the exact answer of the area in terms of π **if r = 8 in.**

$A = 0.5\,\pi r^2 = 0.5\pi \cdot 8^2 = 0.5 \cdot 64\,\pi = 32\pi$ **in²**

Concentric circles

Two circles with the same center but different radii called concentric circles. Concentrically circles form a ring. The thickness of the ring is the equal the difference between its radii: h = R – r The area of a ring can be found by the formula: $A_{ring} = \pi R^2 - \pi r^2 = \pi\,(R^2 - r^2)$

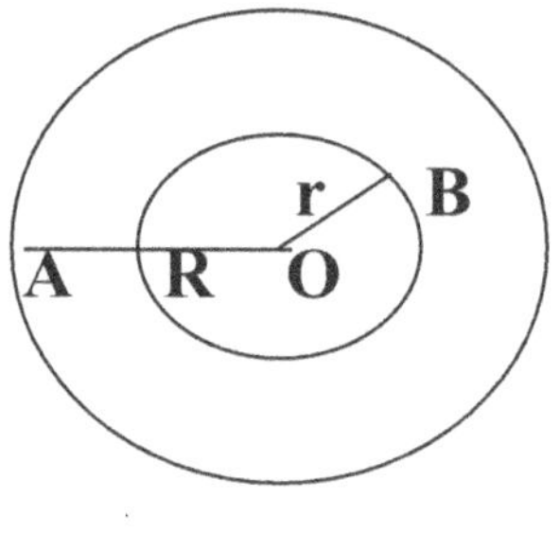

Example 7: Figure 5

Find the area of a ring if its radii are

R = 10 in r = 5 in $\pi \approx 3.14$

$A_{ring} = \pi\,(10^2 - 5^2) = \pi\,(100 - 25) = 75\,\pi$

$= 75(3.14) \approx 235.5$ **in²**

Geometry Chapter 9 Lesson 8
Circles
Review 1 A
Solve the problems
Use 3.14 for π

1) Find the diameter of a circle, if its radius equals 15 ft.

2) Find the radius of a circle, if its diameter equals 20 in.

3) Find the circumference of a circle, if its radius equals 5 yd.

4) Find the circumference of a circle, if its diameter equals 10 in.

5) Find the diameter of a circle if its circumference equals 314m.

6) Find the radius of a circle, if its circumference equals 628 cm.

7) Find the circumference of a semicircle, if its diameter is 5 in.

8) Find the circumference of a semicircle, if its radius is 10 ft.

9) Find the diameter of a semicircle if its circumference is $C=4\pi$ in

10) Find the radius of a semicircle if its circumference is $C=4\pi$ in

Geometry Chapter 9 Lesson 8

Review 1 B
Solve the problems
Use 3.14 for π

1) Find the area of a circle, if its radius equals 10 yd.

2) Find the area of a circle, if its diameter is 20 m.

3) Find the radius of a circle if the area equals 314 in^2.

4) Find the diameter of a circle if the area equals 1256 m^2.

5) Find the area of a circle, if its circumference equals 20π in.

6) Find the circumference of a circle if its area equals 100π ft^2.

7)Find the exact value of the area of a ring, if its big radius is 2.5 cm and its small radius is 1.5 cm.

8) Find the big radius of a ring, if its area is 4π yd, and the small radius is 1.5 yd.

9) Find the small radius of a ring, if its area is 4π cm, and the big radius is 2.5 cm.

10) Find the circumference of a circle, if its diameter is 7 ft and $\pi \approx \frac{22}{7}$

Geometry Chapter 9 Lesson 9

Solid Geometry
Unit for volume

1 cm = 10 mm	$1\ cm^3 = 1000\ mm^3$	
1 yd = 3 ft	$1\ yd^3 = 27\ ft^3$	s is a side
SA is a surface area	V is a volume	d is a diagonal

Cube

The volume of a cube is equal cube of the its side: $V = s^3$

Surface Area of Cube $SA = 6s^2$

Diagonal of Cube is $B_1D = d$ $d = s\sqrt{3}$

Example 1
Find the volume, the surface of area, and the diagonal of cube, if s = 4 ft.

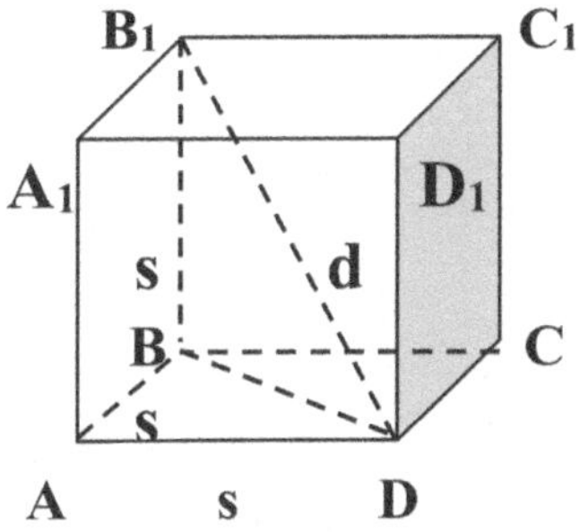

Figure 1

$V = s^3 = 4^3 = 64\ ft^3$

$SA = 6 \cdot 4^2 = 6 \cdot 16 = 96\ ft^2$

$d = s\sqrt{3} = 4\sqrt{3}$ ft

Rectangular Solid

The volume of a rectangular solid is equal to the product length, width, and height. A rectangular solid has length l, width w, and height h.

Volume of Rectangular Solid $V = lwh$
Surface Area of Rectangular Solid $SA = 2lw + 2lh + 2wh$
Diagonal of Rectangular Solid $d^2 = l^2 + w^2 + h^2$

Geometry Chapter 9 Lesson 9
Rectangular solid

Example 2

Find the volume, the surface of area, and the diagonal of rectangular solid, if l = 10 in, w = 6 in, h = 8 in.

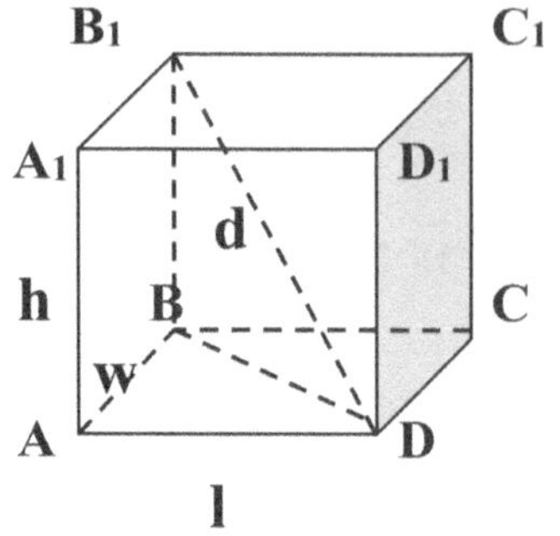

Figure 2

$V = lwh = 10 \cdot 6 \cdot 8 = 60 \cdot 8 = 480 \text{ in}^3$

$SA = 2lw + 2lh + 2wh = 2 \cdot 10 \cdot 6 + 2 \cdot 10 \cdot 8 + 2 \cdot 6 \cdot 8 =$
$20 \cdot 6 + 20 \cdot 8 + 12 \cdot 8 = 120 + 160 + 96 = 376 \text{ in}^2$

$d^2 = l^2 + w^2 + h^2 = 10^2 + 6^2 + 8^2 = 100 + 36 + 64 = 100 + 100 = 200$
$d = \sqrt{200} = \sqrt{100 \cdot 2} = 10\sqrt{2}$ in

Sphere

The volume of a sphere is equal to the product four over three of the number π by cube of radius. An example of a sphere is any balls.

Volume of sphere is $V = \frac{4}{3}\pi r^3$ where $\pi \approx \frac{22}{7}$ or $\pi \approx 3.14$

Surface Area of Sphere $SA = 4\pi r^2$ $OB = r$

Figure 3

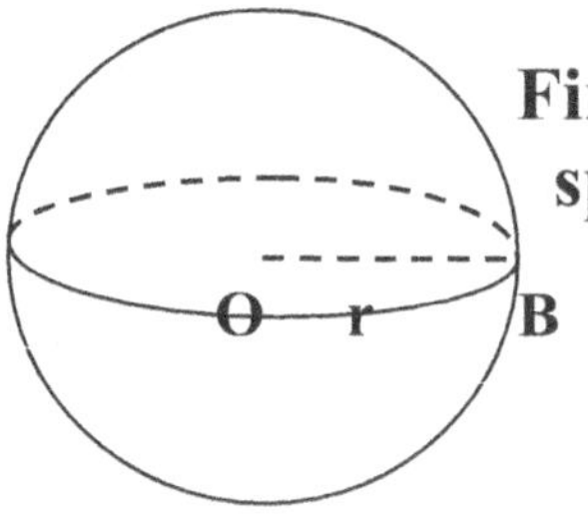

Example 3

Find exact volume and exact surface area of a sphere if the radius is 3 m.

$V = \frac{4}{3}\pi r^3 = \frac{4}{3}\pi \times 3^3 = \frac{4}{3} \times \frac{27}{1}\pi = 4 \times 9\pi = 36\pi \text{ m}^3$

$SA = 4\pi r^2 = 4\pi \times 3^2 = 4 \times 9\pi = 36\pi \text{ m}^2$

Geometry Chapter 9 Lesson 9

Regular Prism

The volume of a regular prism is equal to the product of the base and height: $V = Base \cdot h$ Base means area of triangles ABC or $A_1B_1C_1$

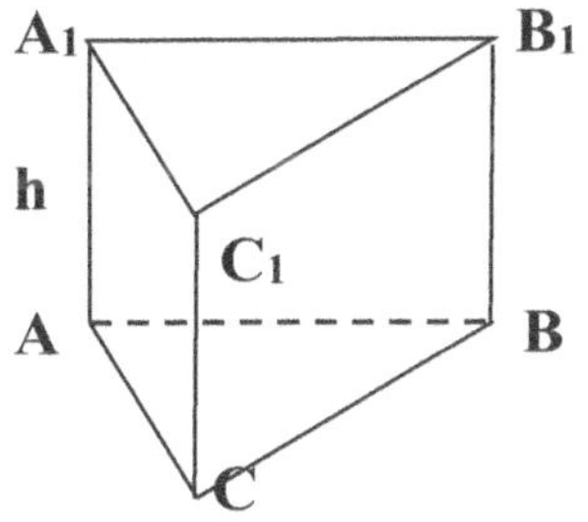

Figure 4

Example 4 Find the volume of a regular prism if the base is 25 m^2 and the height is 10 m. Base = Area of triangle ABC = 25 m^2

$h = 10$ m $V = B_{base} \cdot h = 25 \cdot 10 = 250\ m^3$

Right Circular Cylinder

The volume of a right circular cylinder is equal to the product of the number π, and the square of radius by height: $V = \pi r^2h$

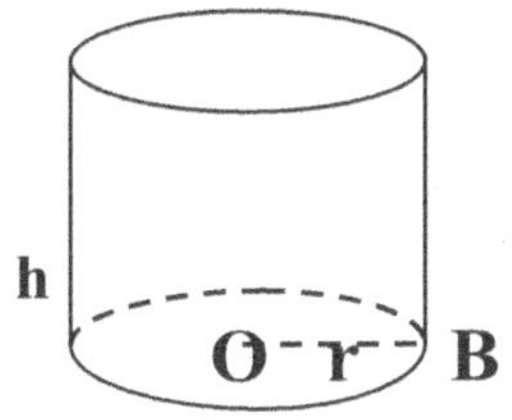

Figure 5

Example 5 Find volume of a right circular cylinder

if $r = 7$ in $h = 10$ in and $\pi \approx \frac{22}{7}$

$$V = \pi r^2h = \frac{22}{7} \cdot 7^2 \cdot 10 = \frac{22}{7} \cdot \frac{49}{1} \cdot 10 = 22 \cdot 7 \cdot 10 = 1540\ in^3$$

Geometry Chapter 9 Lesson 9

Regular Pyramid

The volume of a regular pyramid is equal to the product of one-third of base by height: $V = \frac{1}{3}\text{Base} \cdot h$ Base means area of triangles ABC

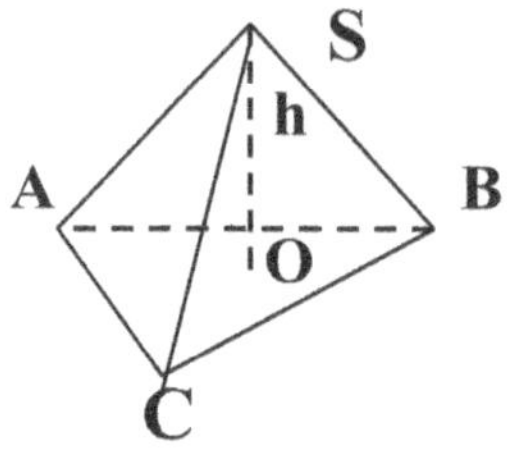

Figure 6

Example 6 Find the volume of a regular pyramid if the base is 120 m^2 and height is 5 m. The base is the area of the triangle ABC = 120 m^2

$h = 5$ m $V = \frac{1}{3}\text{Base} \cdot h = \frac{1}{3} \cdot 120 \cdot 5 = 40 \cdot 5 = 200\ m^3$

Right Circular Cone

The volume of a right circular cone is equal to the product of one-third of square radius by height: $V = \frac{1}{3}\pi r^2 h$ $\pi \approx 3.14$ or $\pi \approx \frac{22}{7}$

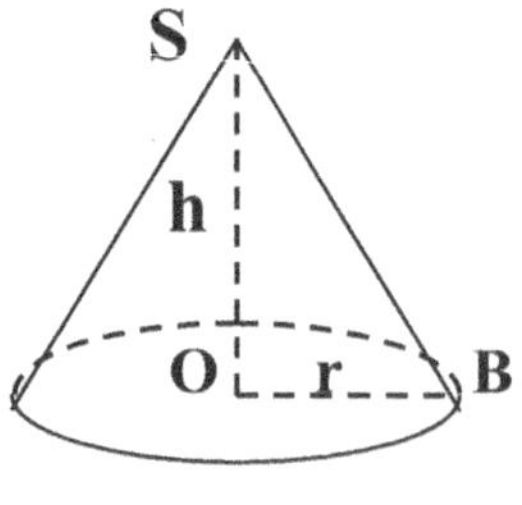

Figure 7

Example 7 Find the volume of a right circular cone if its height is 10 ft and radius is 3 ft: $h = 10$ ft $r = 3$ ft

$V = \frac{1}{3}\pi r^2 h = \frac{1}{3}\pi \cdot 3^2 \cdot 10 = \frac{1}{3} \cdot 9 \cdot 10\pi = 3 \cdot 10\pi = 30\pi = 30\ (3.14) = 94.2\ ft^3$

Geometry Chapter 9 Lesson 9

Solid Geometry
Review 1 A
Solve the problems

Cube		$V = s^3$		
1)	$s = 10$ cm	$V = ?$		
2)	$V = 343$ in^3	$s = ?$		
Cube		$SA = 6s^2$		
3)	$s = 10$ in	$SA=?$		
4)	$SA = 600$ cm^2	$s = ?$		
5)	$SA = 600$ ft^2	$V = ?$		
Cube		$d = s\sqrt{3}$		
6)	$s = 10$ ft	$d = ?$		
7)	$d = 20\sqrt{3}$ in	$s = ?$		
Rectangular solid		$V = lwh$		
8)	$l = 5$ ft	$w = 4$ ft	$h = 2$ ft	$V = ?$
9)	$l = 2.5$ in	$w = 0.8$ in	$V = 8$ in^3	$h = ?$
10)	$l = 1.25$ dm	$h = 0.8$ dm	$V = 4$ dm^3	$w = ?$
11)	$w = 1$ cm	$h = 2$ cm	$V = 10$ cm^3	$l = ?$

Geometry Chapter 9 Lesson 9

Review 1 A
Solve the problems

Rectangular solid		$SA = 2lw + 2\ lh + 2\ wh$		
12)	$l = 8$ ft	$w = 5$ ft	$h = 4$ ft	$SA = ?$
13)	$l = 8$ in	$w = 3$ in	$SA = 108$ in^3	$h = ?$
14)	$l = 6$ dm	$h = 4$ dm	$SA = 108$ dm^3	$w = ?$
15)	$w = 4$ cm	$h = 3$ cm	$SA = 108$ cm^3	$l = ?$
Rectangular solid		$d^2 = l^2 + w^2 + h^2$		
16)	$l = 6$ ft	$w = 3$ ft	$h = 2$ ft	$d = ?$
17)	$l = 6$ in	$w = 3$ in	$d = 7$ in	$h = ?$
18)	$w = 3$ dm	$h = 2$ dm	$d = 7$ dm	$l = ?$
19)	$l = 6$ cm	$h = 2$ cm	$d = 7$ cm	$w = ?$
20)	$l = 7$ in	$w = 6$ in	$h = 6$ in	$d = ?$

Geometry Chapter 9 Lesson 9

Solid Geometry
Review 1 B
Solve the problems

Sphere $V = \frac{4}{3}\pi r^3$ $\pi \approx 3.14$

1) $r = 3$ ft $V = ?$

2) $V = 113.04$ in^3 $r = ?$

3) $d = 6$ cm $V = ?$

4) $V = 113.04$ m^3 $d = ?$

Sphere $SA = 4\pi r^2$

5) $r = 3$ ft $SA = ?$

6) $SA = 1256$ in^2 $r = ?$

Regular Prism $V = \text{Base} \cdot h$

7) Base = 25 m^2 $h = 4$ m $V = ?$

8) Base = 125 in $V = 1000$ in^3 $h = ?$

9) $h = 10$ ft $V = 1000$ ft^3 Base = ?

Geometry Chapter 9 Lesson 9

Review 1 B
Solve the problems

Right Circular Cylinder $V = \pi r^2 h$ $\pi \approx 3.14$

10)	$r = 10$ in	$h = 20$ in	$V = ?$
11)	$r = 20$ ft	$V = 6280\ ft^3$	$h = ?$
12)	$h = 10$ cm	$V = 3140\ cm^3$	$r = ?$
13)	$d = 20$ m	$h = 20$ m	$V = ?$
14)	$h = 20$ yd	$V = 6280\ yd^3$	$d = ?$

Regular Pyramid $V = \frac{1}{3} \text{Base} \cdot h$

15)	$\text{Base} = 30\ in^2$	$h = 10$ in	$V = ?$
16)	$\text{Base} = 45\ ft^2$	$V = 300\ ft^3$	$h = ?$
17)	$h = 3$ cm	$V = 100\ cm^3$	$\text{Base} = ?$

Right Circular Cone $V = \frac{1}{3} \pi r^2 h$ $\pi \approx 3.14$

18)	$r = 10$ ft	$V = 314\ ft^3$	$h = ?$
19)	$h = 3$ cm	$V = 314\ cm^3$	$r = ?$
20)	$r = 3$ in	$h = 6$ in	$V = ?$

Geometry Chapter 9 Lesson 10

Table 9 Conditional Equations

Six Properties of True Equality: if $a = b$ then

$a + c = b + c$ $a - c = b - c$ $ac = bc$ $a \div c = b \div c$ $c \neq 0$

If $a + c = b + c$ $a = b$ $3y + 3 = 6 + 3$ $3y = 6$ $y = 2$

If $a - b = c$ $a = c + b$ $3y + 2 = 8$ $3y = 8 - 2$ $3y = 6$ $y = 2$

If $a + c = b + c$ $a = b$ $3y - 9 = 0$ $3y = 0+9$ $3y = 9$ $y = 3$

If $a + b = c$ $a = c - b$ $3y + 2 = 8$ $3y = 8 - 2$ $3y = 6$ $y = 2$

Conditional equation has one solution $2y = 8$ $y = 4$

Rule for Solving Equations

1) Open parentheses $a(b + c) = ab + ac$
2) Combine like terms
3) Eliminate number on left side
4) Eliminate letter on right side
5) Multiplication property

Equation with Whole Numbers

$-3(y - 1) = -2(y + 2) + 6$ $-3y + 3 = -2y - 4 + 6$ $-3y + 3 = -2y + 2$

$-3y + 2y = -3 - 2$ $-y = -5$ $y = 5$

Equations with Decimals

$3(y - 5.8) = 6.5(y + 1.2) - 12.25$ $3y - 17.4 = 6.5y + 7.8 - 12.25$

$3y - 17.4 = 6.5y - 4.45$ $3y = 6.5y + 12.95$

$-3.5y = 12.95$ $y = -3.7$

Equations with Fraction (× LCD)

$\frac{1}{6} - \frac{y}{4} = \frac{2-y}{3}$ $\frac{12}{1} \times \frac{1}{6} - \frac{12}{1} \times \frac{y}{4} = \frac{12}{1} \times \frac{2-y}{3}$ $2 - 3y = 4(2 - y)$

$2 - 3y = 8 - 4y$ $-3y + 4y = 8 - 2$ $y = 6$

Proportions

$12 \div 4 = 18 \div 6$ $a \div b = c \div d$ $\frac{12}{4} = \frac{18}{6}$ $3 = 3$ $\frac{a}{b} = \frac{c}{d}$

$12 \times 6 = 72$ $4 \times 18 = 72$ $12 \times 6 = 4 \times 18$ $ad = bc$

$12 \div 18 = 4 \div 6$ $6 \div 4 = 18 \div 12$ $6 \div 18 = 4 \div 12$

$\frac{12}{4} = \frac{18}{y}$ $12y = 4 \cdot 18$ $12y = 72$ $y = 72 \div 12$ $y = 6$

$\frac{12}{y} = \frac{18}{6}$ $18y = 12 \cdot 6$ $18y = 72$ $y = 72 \div 18$ $y = 4$

Geometry Chapter 9 Lesson 10

Table 10 Plane Geometry

Right triangles

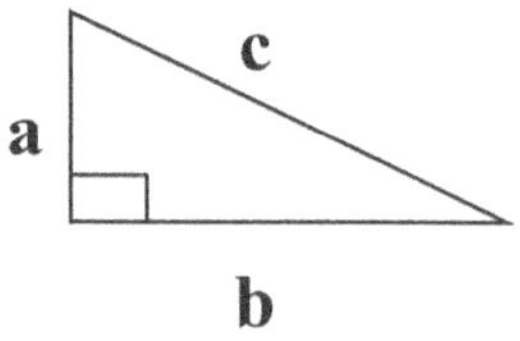

1) Pythagorean Theorem — 1) $c^2 = a^2 + b^2$

2) Area — 2) $A_{\Delta} = 0.5ab$

Triangles

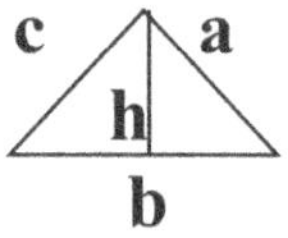

3) Perimeter — 3) $P = a + b + c$

4) Area — 4) $A_{\Delta} = 0.5bh$

Parallelogram

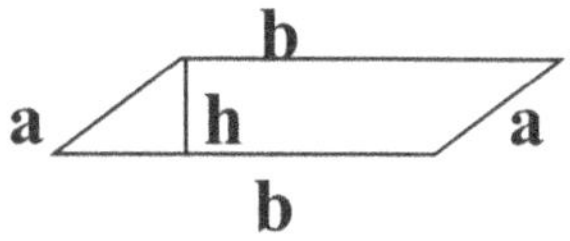

5) Perimeter — 5) $P_{par} = 2a + 2b$

6) Area — 6) $A_{par} = b\,h$

Rhombus

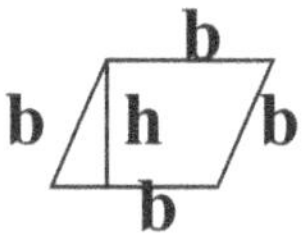

7) Perimeter — 7) $P_{rh} = 4b$

8) Area — 8) $A_{rh} = b \cdot h$

Rectangle

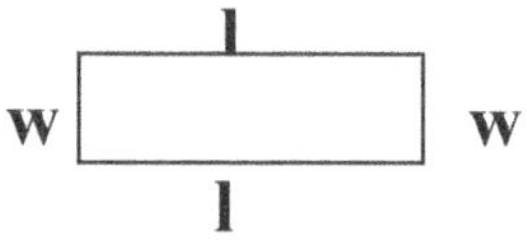

9) Perimeter — 9) $P_{rec} = 2l + 2w$

10) Area — 10) $A_{rec} = l \cdot w$

Square

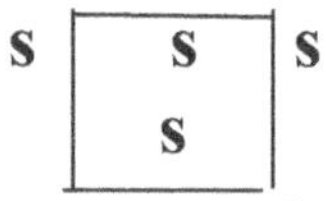

11) Perimeter — 11) $P_{\blacksquare} = 4s$

12) Area — 12) $A_{\blacksquare} = s^2$

Trapezoid

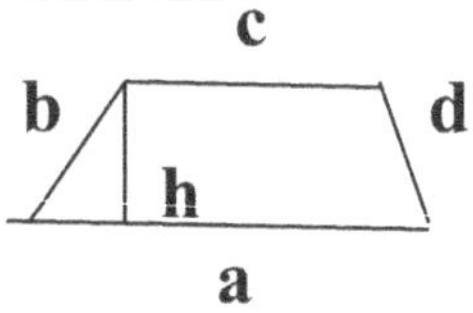

13) Perimeter — 13) $P_{tr} = a + b + c + d$

14) Area — 14) $A_{tr} = 0.5\,h\,(a + c)$

Circle

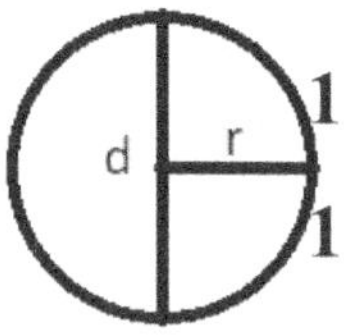

15) Number π in fraction $\approx \frac{22}{7}$ 16) Number π in decimal ≈ 3.14

17) Diameter $d = 2r$ 18) Circumference 1 (d) $C = \pi d$

19) Circumference 2 (r) $C = 2\pi r$ 20) Area $A_c = \pi r^2$

Solid Geometry

Cube 1) $V = s^3$, 2) $SA = 6s^2$ 3) $d = s\sqrt{3}$ Sphere 4) $V = \frac{4}{3}\pi r^3$ 5) $SA = 4\pi r^2$

Rectangular Solid: 6) $V = l \cdot w \cdot h$ 7) $SA = 2lw + 2lh + 2wh$ 8) $d^2 = l^2 + w^2 + h^2$

Regular Prism 9) $V = \text{Base} \cdot h$, Regular Pyramid 10) $V = \frac{1}{3}\,\text{Base} \cdot h$

Right Cylinder 11) $V = \pi r^2 h$, Right Cone 12) $V = \frac{1}{3}\pi r^2 h$

Memorize Formulas of Geometry

Mental Math 15
Law for Fractions

1) $\frac{1}{2}\times8+\frac{1}{2}\times2$

2) $\frac{4}{3}\times6-\frac{1}{3}\times6$

3) $\frac{1}{4}\times8+\frac{3}{4}\times8$

4) $\frac{6}{5}\times5-\frac{1}{5}\times5$

5) $\frac{1}{3}\times9+\frac{2}{3}\times9$

6) $\frac{3}{2}\times8-\frac{1}{2}\times8$

7) $\frac{1}{3}\times6+\frac{2}{3}\times6$

8) $\frac{5}{4}\times4-\frac{1}{4}\times4$

9) $\frac{1}{5}\times5+\frac{4}{5}\times5$

10) $\frac{7}{6}\times6-\frac{1}{6}\times6$

11) $\frac{1}{6}\times6+\frac{5}{6}\times6$

12) $\frac{5}{2}\times4-\frac{1}{2}\times4$

13) $\frac{3}{7}\times7+\frac{4}{7}\times7$

14) $\frac{7}{3}\times3-\frac{1}{3}\times3$

15) $\frac{3}{8}\times8+\frac{5}{8}\times8$

16) $\frac{5}{4}\times8-\frac{1}{4}\times8$

17) $\frac{4}{9}\times9+\frac{5}{9}\times9$

18) $\frac{7}{5}\times5-\frac{2}{5}\times5$

19) $\frac{3}{5}\times5+\frac{2}{5}\times5$

20) $\frac{7}{6}\times6-\frac{1}{6}\times6$

Geometry Chapter 9 Lesson 10

Mental Math 16
Exponents

1) 0^2	11) 10^2
2) 1^2	12) 20^2
3) 2^2	13) 30^2
4) 3^2	14) 40^2
5) 4^2	15) 50^2
6) 5^2	16) 60^2
7) 6^2	17) 70^2
8) 7^2	18) 80^2
9) 8^2	19) 90^2
10) 9^2	20) 100^2

Geometry Chapter 9 Lesson 11

Reviews for Chapter 9
Review 9 A

Date — Name

Read directions carefully. — Answers must be in indicated place.

Show all steps for full credits. — Each problem is worth 5 points.

Do only 1 operation per 1 step. — Solve 1 problem for 5 minutes.

1) Find one leg of a right triangle if its hypotenuse is 9.5 ft and another leg is 5.7 ft. Draw the picture.

1) ______

2) Find the area of a right triangle, if its legs are $5\frac{5}{6}$ m and $4\frac{4}{5}$ m. Draw the picture.

2) ______

3) Find the area of a triangle if it sides are 5 cm, 12 cm, and 13 cm. Draw the picture.

3) ______

4) Find the area of a square if it side is $1\frac{1}{2}$ ft. Draw the picture.

4) ______

5) Find the area of a rhombus if the base is $3\frac{1}{3}$ in and the height is $2\frac{1}{4}$ in. Draw the picture.

5) ______

Geometry Chapter 9 Lesson 11

Review 9 A

Date **Name**

Read directions carefully. **Answers must be in indicated place.**

Show all steps for full credits. **Each problem is worth 5 points.**

Do only 1 operation per 1 step. **Solve 1 problem for 5 minutes.**

Solve the problems

6) **Find the area of a trapezoid; if the height is $6\frac{2}{3}$ cm, and bases are $9\frac{1}{2}$ cm and $8\frac{1}{2}$ cm. Draw the picture.**

6) ______

7) **Find the circumference of circle that has a diameter is 28 in and $\pi \approx \frac{22}{7}$. Draw the picture.**

7) ______

8) **Find the area of a circle if its radius r = 10 m and $\pi \approx 3.14$ Draw the picture.**

8) ______

9) **Find the volume of a cube if its side is $1\frac{1}{2}$ in.**

9) ______

10) **Find the exact value volume of a sphere if radius $r = 1\frac{1}{2}$ ft.**

10) ______

Geometry Chapter 9 Lesson 11

Review 9 B

Date **Name**

Read directions carefully. **Answers must be in indicated place.**

Show all steps for full credits. **Each problem is worth 5 points.**

Do only 1 operation per 1 step. **Solve 1 problem for 5 minutes.**

1)Find one leg of a right triangle if its hypotenuse is 6.5 ft and the other leg is 5.2 ft. Draw the picture.

1) ______

2) Find the area of a right triangle, if it legs are $6\frac{1}{4}$ m and $4\frac{4}{5}$ m .

Draw the picture.

2) ______

3) Find the area of a triangle if its sides are 9 cm, 12 cm, and 15 cm Draw the picture.

3) ______

4) Find the area of a rectangle if its length is $3\frac{1}{3}$ ft and its width is $2\frac{2}{5}$ ft. Draw the picture.

4) ______

5) Find the area of a parallelogram if the base is $5\frac{3}{5}$ in and the height is $3\frac{3}{4}$ in. Draw the picture.

5) ______

Geometry Chapter 9 Lesson 11
Review 9 B

Date **Name**

Read directions carefully. **Answers must be in indicated place.**
Show all steps for full credits. **Each problem is worth 5 points.**
Do only 1 operation per 1 step. **Solve 1 problem for 5 minutes.**

Solve the problems

6) Find the area of a trapezoid, if its height is $5\frac{1}{3}$ cm, and bases are $6\frac{1}{2}$ cm and $5\frac{1}{2}$ cm. Draw the picture.

6) ______

7) Find the area of a circle if its radius 7 in and $\pi = \frac{22}{7}$.

Draw the picture.

7) ______

8) Find the area of a ring of a circle is its radii are 2.5 in and 1.5 in and $\pi \approx 3.14$. Draw the picture.

8) ______

9) Find the volume of a rectangular solid if its length is 4 in, width is 2.5 in and height is 0.5 in

9) ______

10) Find the exact value volume of a sphere if radius $r = 4\frac{1}{2}$ ft.

10) ______

Geometry Chapter 9 Lesson 12

Tests for Chapter 9
Test 9 A

Date | Name

Read directions carefully. | Answers must be in indicated place.

Show all steps for full credits. | Each problem is worth 5 points.

Do only 1 operation per 1 step. | Solve 1 problem for 5 minutes.

1) Find one leg of a right triangle if its hypotenuse is $7\frac{1}{2}$ ft and the other leg is $7\frac{1}{5}$ ft. Draw the picture.

1) ______

2) Find the area of a right triangle, if its legs are $6\frac{2}{3}$ m and $4\frac{1}{2}$ m. Draw the picture.

2) ______

3) Find the area of a triangle if its sides are 12 cm, 16 cm, and 20 cm. Draw the picture.

3) ______

4) Find the area of a square if its side is $2\frac{1}{2}$ ft. Draw the picture.

4) ______

5) Find the area of a rhombus if the base is $7\frac{1}{2}$ in and the height is $4\frac{2}{3}$ in. Draw the picture.

5) ______

Geometry Chapter 9 Lesson 12

Test 9 A

Date **Name**

Read directions carefully. **Answers must be in indicated place.**

Show all steps for full credits. **Each problem is worth 5 points.**

Do only 1 operation per 1 step. **Solve 1 problem for 5 minutes.**

Solve the problems

6) Find the area of a trapezoid, if the height is $2\frac{2}{3}$ cm, and the bases are $6\frac{1}{2}$ cm and $5\frac{1}{2}$ cm. Draw the picture.

6) ______

7) Find the circumference of a circle that has a diameter of 14 in and $\pi = \frac{22}{7}$. Draw the picture.

7) ______

8) Find the area of a circle if radius r = 20 m and $\pi \approx 3.14$ Draw the picture.

8) ______

9) Find the volume of a cube if its side is $2\frac{1}{2}$ in.

9) ______

10) Find the exact value volume of a sphere if radius r = 3 ft.

10) ______

Geometry Chapter 9 Lesson 12

Test 9 B

Date
Name
Read directions carefully.
Answers must be in indicated place.
Show all steps for full credits.
Each problem is worth 5 points.
Do only 1 operation per 1 step.
Solve 1 problem for 5 minutes.

1) Find one leg of a right triangle if its hypotenuse is $5\frac{1}{10}$ ft and the other leg is $2\frac{2}{5}$ ft. Draw the picture.

1) ______

2) Find the area of a right triangle, if its legs are $9\frac{3}{4}$ m and $5\frac{1}{3}$ m. Draw the picture.

2) ______

3) Find area of a triangle if its sides are 15 cm, 20 cm, and 25 cm
Draw the picture.

3) ______

4) Find the area of a rectangle if its length is $10\frac{1}{2}$ ft and its width is $5\frac{1}{3}$ ft. Draw the picture.

4) ______

5) Find the area of a parallelogram if the base is $7\frac{1}{2}$ in and the height is $6\frac{2}{3}$ in. Draw the picture.

5) ______

Geometry Chapter 9 Lesson 12

Test 9 B

Date **Name**

Read directions carefully. **Answers must be in indicated place.**

Show all steps for full credits. **Each problem is worth 5 points.**

Do only 1 operation per 1 step. **Solve 1 problem for 5 minutes.**

Solve the problems

6) Find the area of a trapezoid, if its height is $3\frac{1}{3}$ cm, and the bases are $9\frac{3}{4}$ cm and $8\frac{1}{4}$ cm. Draw the picture.

6) ______

7) Find the area of a circle if the radius is 14 in and $\pi \approx \frac{22}{7}$.

Draw the picture.

7) ______

8) Find the area of a ring of a circle if its radii are 6.5 in and 5.5 in and $\pi \approx 3.14$. Draw the picture.

8) ______

9) Find the volume of a rectangular solid if its length is 12.5 in, its width is 5 in and its height is 4 in

9) ______

10) Find the exact value volume of a sphere if the radius r = 6 ft.

10) ______

Geometry Answers Chapter 9

Lesson 1 1) 30 cm^2 2) 616 in^2 3) 37.68 m^2 4) 250 in^3 5) 288π ft^3

Lesson 2 Segments. Review 1 A

1) a) > b) < c) = 2) a) 10cm b) 2 cm 3) 12 cm 4) 8 in 5) 2 in
6) 5 in 7) 6 cm 8) 8 in 9) 4 cm 10) 7 in

Review 1B

1) 10 in 2) 15 cm 3) 6 in 4) 6 cm 5) 7 in
6) 5 in 7) 6 cm 8) 4 cm 9) 8 in 10) 3 in

Lesson 3 Angles. Review 1 A

1) 75^0 2) 135^0 3) 90^0 4) 150^0 5) 45^0
6) 60^0 7) 30^0 8) 120^0 9) 30^0 10) 45^0

Review 1 B

1) 150^0, 30^0 2) 100^0, 80^0 3) 81^0, 99^0 4) 84^0, 96^0 5) 18^0, 162^0
6) 30^0, 150^0 7) 80^0, 100^0 8) 70^0, 110^0 9) 36^0, 144^0 10) 80^0, 100^0

Lesson 4 Triangles. Review 1 A

1) 90^0 2) 65^0 3) 45^0 4) 60^0 5) 18 cm
6) 10 m 7) 36 in 8) 5 ft 9) 16 yd 10) 18 ft

Review 1 B

1) 8 in 2) 50 cm 3) 0.6 in 4) 75 cm 5) 0.8 ft
6)8, 16, 16 cm, 7)12, 12, 2 in, 8)6, 6, 8 in, 9)10, 15, 15 in, 10)8, 8, 6 cm

Lesson 5 Right Triangles. Review 1 A

1) 3 2) 8 3) 15 4) 12 5) 20
6) 30 7) 21 8) 32 9) 45 10) 30
11) 44 12) 60 13) 39 14) 56 15) 75
16) 48 17) 68 18) 90 19) 57 20) 80
21) 13 22) 10 23) 36 24) 52 25) 25

Review 1 B

1) 7.2 2) 2.5 3) 1.4 4) 7.2 5) 10
6) 2 7) 4.2 8) 1.7 9) 1.6 10) 4.5
11) 4.1 12) 1.8 13) 6 14) 8.5 15) 6
16) 7.2 17) 5 18) 2.1 19) 9.6 20) 5.8
21) 0.8 22) 3 23) 5.1 24) 1.8 25) 8.4

Lesson 6 Perimeter. Review 1 A

1) 24 cm 2) 15 in 3) 10 ft 4) 28 cm
5) 20 yd 6) 10 in 7) 40 dm 8) 6 cm
9) 20 cm 10) 12 in 11) 14 ft 12) 36 m
13) 20 ft 14) 28 cm 15) 50 ft 16) 10 yd
17) 24 cm 18) 30 ft 19) 10 m 20) 32 in

Geometry Answers Chapter 9

Lesson 7 Area. Review 1 A

1)	64 in^2	2)	5 cm	3)	36 yd^2	4)	9 cm
5)	11 ft	6)	12 m	7)	10 in	8)	10 m
9)	12 in	10)	8 ft	11)	40 m^2	12)	20 m
13)	40 in^2	14)	10 m	15)	6 m^2	16)	24in^2
17)	30ft^2	18)	54 cm^2	19)	84yd^2	20)	96dm^2

Lesson 8 Circle. Review 1 A

1) 30 ft 2) 10 in 3) 31.4 yd 4) 31.4 m 5) 100m
6) 100 cm 7) 7.85 in 8) 314 ft 9) 8 in 10) 4 in

Review 1 B

1) 314 yd^2 2) 314 m^2 3) 10 in 4) 40m 5) 314in^2
6) 62.8 ft 7) 4π cm^2 8) 2.5 yd 9) 1.5 cm 10) 22ft

Lesson 9 Solid Geometry. Review 1 A

1) 1000 dm^3 2) 7 in 3) 600 in^2 4) 10cm 5) 1000 ft^3
6) $10\sqrt{3}$ ft 7) 20 in 8) 40 ft^3 9) 4 in 10) 4 dm
11) 5 cm 12) 184 ft^2 13) 4 in 14) 3dm 15) 6 cm
16) 7 ft 17) 2 in 18) 6dm 19) 3cm 20) 11in

Review 1 B

1) 113.04 ft^3 2) 3in 3) 113.04 cm^3 4) 6 m 5) 113.04 ft^2
6) 10in 7) 100 m^3 8) 8 in 9) 100 ft^2 10) 6280 in^3
11) 5 ft 12) 10 cm 13) 6280 m^3 14) 20 yd 15) 100 in^3
16) 20 ft 17) 100 cm^2 18) 3 ft 19) 10 cm 20) 56.52 in^3

Lesson 11 Review 9 A

1) 7.6 ft 2) 14 m^2 3) 30 cm^2 4) 2.25 ft^2 5) 7.5 in^2
6) 60 cm^2 7) 88 in 8) 314 m^2 9) 3.375 in^3 10) 4.5π ft^3

Review 9 B

1) 3.9 ft 2) 15 m^2 3) 54 cm^2 4) 8 ft^2 5) 21 in^2
6) 32 cm^2 7) 154 in^2 8) 12.56 m^2 9) 5 in^3 10) 121.5π ft^3

Lesson 12 Test 9 A

1) 2.1 ft 2) 15 m^2 3) 96 cm^2 4) 6.25 ft^2 5) 35 in^2
6) 16 cm^2 7) 44 in 8) 1256 m^2 9) 15.625 in^3 10) 36π ft^3

Test 9 B

1) 4.5 ft 2) 26 m^2 3) 150 cm^2 4) 56 ft^2 5) 50 in^2
6) 30 cm^2 7) 616 in^2 8) 37.68 m^2 9) 250 in^3 10) 288π ft^3

Summary 1

Chapter 10

Basic

Math

Students!

**Students, you are smarter than a calculator!
Students, you are smarter than a computer!
Students, you need to believe in yourself.
Students, be patient and do your best.**

**Do not be afraid to make mistakes.
Tell yourself positive words.
I am young! I am smart!
I am strong! I can do anything!**

**Smart students use their brain!
Your brain is always with you!
A calculator is not always with you.
Mathematics is work for the brain.**

**Be open to new ideas in math!
Work hard and try your best!
Check answers and ask questions!
Practice! Practice! Practice!**

Summary 1 Chapter 10 Contents
Basic Math
Lessons 1 – 14

Contents have topics and exercises, which students need to solve before tests.1)Work with lesson 1, read rules and examples, 2) Solve the problems from contents, 3) Check answers in the end of chapter 1, 4) Work with the next lessons 2, 3, 4, 5, …, 5) Solve Test 1 A, 6) Grading Scale without minuses
A=90–100% B+=85–89% B=80–84% C+=75–79%
C=70–74% D+=65–69% D=60–64 % F=0 - 59%
7)If you solve 10 problems correctly, you get an A, 8) If have an A, you can move on to next chapter, 9) If your grade is lower than an A, work with the mistakes, then solve Test 1B, 10) When you finish all of the chapters, solve Final Exam in last chapter.

Summary 1 Chapter 10 Lesson 1

Basic Math

Date **Name**

College Placement Test (50 minutes)

Evaluate

1) $18000 \div (75^2 - 45^2)$

2) $5.6 \times 4 + 2.7 \times 4 - 8.3 \times 4$

3) $3\frac{3}{5} \times 15 - 2\frac{1}{5} \times 15$

4) $-3\frac{2}{3} \div \frac{5}{6} + 1\frac{1}{3} \div \frac{5}{6}$

5) Simplify $\frac{3}{7}(14x - 21y) - 15(\frac{2}{3}x - \frac{1}{3}y)$

6) Solve the equation $-2(3.1x - 1) + 3(1.2x + 1) = -14.5$

7)Find the perimeter of right triangle if it hypotenuse is $8\frac{1}{2}$ ft and another leg is 1.3 ft.

8) Find the area of trapezoid, if it height is $6\frac{2}{3}$ in, bases are $8\frac{1}{2}$ in and $9\frac{1}{2}$ in.

9) A bus drove 48 mi at a speed of 20 mi/h.
How much time bus spend?

10) Of 16 children, 25% were boys. How many were girls?

Summary 1 Chapter 10 Lesson 1

Table 1
Addition and Subtraction of Whole Numbers

All numbers are made up of the digits: 0, 1, 2, 3, 4, 5, 6, 7, 8, and 9. Basic operations are: addition, subtraction, multiplication, and division. Natural (counting) numbers are numbers that use for calculations 1, 2, 3 ... Whole numbers are natural numbers and zero: 0, 1, 2, 3, 4, 5 ... Whole numbers can have represented with points on a number line

The first operation with numbers is addition, which combines two or more numbers together.

The first number in addition is called the addend	2
The second number in addition is called the addend	3
The result of addition is called the sum	5
Addend + Addend = Sum	2 + 3 = 5

The second operation with numbers is subtraction, which means to take something away.

The first number in subtraction is called the minuend	5
The second number in subtraction is called the subtrahend	2
The result of subtraction is called the difference	3
Minuend – Subtrahend = Difference	5 – 2 = 3

To check addition use subtraction

200 + 300 = 500	500 – 200 = 300	500 – 300 = 200

To check subtraction use addition or subtraction.

500 – 200 = 300	500 – 300 = 200	200 + 300 = 500

Remember

10 = 9 + 1 = 8 + 2 = 7 + 3 = 6 + 4 = 5 + 5	15 = 9 + 6 = 8 + 7
11 = 9 + 2 = 8 + 3 = 7 + 4 = 6 + 5	16 = 9 + 7 = 8 + 8
12 = 9 + 3 = 8 + 4 = 7 + 5 = 6 + 6	17 = 9 + 8
13 = 9 + 4 = 8 + 5 = 7 + 6	18 = 9 +9
14 = 9 + 5 = 8 + 6 = 7 + 7	19 = 10 + 9

Summary 1 Chapter 10 Lesson 1

Table 2
Multiplication and Division of Whole Numbers

Multiplication of Whole Numbers

The third operation with numbers is multiplication × or ·

Multiplication is the operation of repeated addition of the same numbers $2\times 3 = 2 + 2 + 2 = 6$ or $2 \cdot 3 = 2 + 2 + 2 = 6$

The first number in multiplication is called the factor 2

The second number in multiplication is called the factor 3

The result of multiplication is called the product 6

Factor × Factor = Product $2 \times 3 = 6$

Multiplying Numbers by 10, 100 …

To multiply numbers by 10, add zero to the end of a number:
$5\times10=50$

To multiply numbers by 100, add two zeros to end of a number:
$5\times100=500$

Division of Whole Numbers

The fourth operation with numbers is division ÷ or :

Division means to separated number by some equal groups

The first number in division is called the dividend 6

The second number in division is called the divisor 2

The result of division is called the quotient 3

Dividend ÷ Divisor = Quotient $6\div2 = 3$

Division by 10

To divide numbers by 10, delete one zero: $40 \div 10 = 4$

Division by 100

To divide numbers by 100, delete two zeros: $400 \div 100 = 4$

To check multiplication use division

$2\times3=6$ Check $6\div2=3$ or $6\div3=2$

To check division use multiplication or division

$6\div2=3$ Check $2\times3=6$ or $6\div3=2$

Even numbers 16, 48

Numbers where last digits are even 0, 2, 4, 6, 8 called even numbers

Odd numbers 31, 27

Number where last digits are odd 1, 3, 5, 7, 9, called odd numbers

Summary 1 Chapter 10 Lesson 1

Table 3
Multiplication and Division

$A + 0 = A$ $0 + A = A$ $A - 0 = A$ $A - A = 0$
$A \times 1 = A$ $1 \times A = A$ $A \div 1 = A$ $A \div A = 1$
$A \times 0 = 0$ $0 \times A = 0$ $0 \div A = 0$ $A \div 0 = \emptyset$
$A + B = B + A$ $A \times B = B \times A$ $\emptyset$ no solutions

Number 5 from the left column multiply by number 4 from first row to get answer 20. $5\times4 = 20$, $20\div5=4$, $20\div4=5$
Memorize Table

×	1	2	3	<u>4</u>	5	6	7	8	9	10
1	1	2	3	4	5	6	7	8	9	10
2	2	4	6	8	10	12	14	16	18	20
3	3	6	9	12	15	18	21	24	27	30
4	4	8	12	16	20	24	28	32	36	40
<u>5</u>	5	10	15	<u>20</u>	25	30	35	40	45	50
6	6	12	18	24	30	36	42	48	54	60
7	7	14	21	28	35	42	49	56	63	70
8	8	16	24	32	40	48	56	64	72	80
9	9	18	27	36	45	54	63	72	81	90
10	10	20	30	40	50	60	70	80	90	100

Summary 1 Chapter 10 Lesson 1

Table 4

Perfect Square Numbers

Take number 20 from the left column add number 5 from the first row to get number 25. Answer is in intersection row of 20 and column of 5

$25^2 = 25 \cdot 25 = 625$ Memorize Table of Perfect Square Numbers 3 rows and 1 column

a^2	0	1	2	3	4	5	6	7	8	9
	0	1	4	9	16	25	36	49	64	81
10	100	121	144	169	196	225	256	289	324	361
20	400	441	484	529	576	625	676	729	784	841
30	900	961	1024	1089	1156	1225	1296	1369	1444	1521
40	1600	1681	1764	1849	1936	2025	2116	2209	2304	2401
50	2500	2601	2704	2809	2916	3025	3136	3249	3364	3481
60	3600	3721	3844	3969	4096	4225	4356	4489	4624	4761
70	4900	5041	5184	5329	5476	5625	5776	5929	6084	6241
80	6400	6561	6724	6889	7056	7225	7396	7569	7744	7921
90	8100	8281	8464	8649	8836	9025	9216	9409	9604	9801

Square Roots from Perfect Square Numbers

The square root of three and four-digit perfect square numbers is always a two-digit number. Divide the number under the root by groups of two digits, from right to left. The last group can be one or two digits. To find the first digit, find the number whose square is smaller than or equal to the last group. To find the second digit, use the rule:

Last digit of given number	1	4	5	6	9
Last digit of answer	1 or 9	2 or 8	5	4 or 6	3 or 7

$31^2 = 31 \times 31 = 961$ $\quad$ $32^2 = 32 \times 32 = 1024$

$\sqrt{961} = 31$ $\quad$ $\sqrt{1024} = 32$

Summary 1 Chapter 10 Lesson 1

Table 5 Tests for division

Test for division is a way to know, without doing division, if some other number can divide a number evenly or with a remainder of zero. A number is divisible by another number when there is no remainder.

Test for 2

A number is divisible by 2 if its last digit is even digits (0, 2, 4, 6, 8)
If a number can divide by 2 this number called even. Example: 1,234

Test for 5

A number is divisible by 5 if its last digit is 0 or 5
Example: 100 can be divided by 5 $100 \div 5 = 20$

Test for 10

A number is divisible by 10 if its last digits have one or more zeros
Example: 240 can be divided by 10 $240 \div 10 = 24$

Test for 25

A number is divisible by 25 if the last two digits to the right are 00, 25, 50, 75. Example: 100, 250, 675, 925 $100 \div 25 = 4$

Test for 4

A number is divisible by 4 if the last two digits to the right of the number are 00 or make a new number that can be divided by 4.
Example: 1$\underline{00}$, 7$\underline{28}$, 5$\underline{40}$ $100 \div 4 = 25$ $28 \div 4 = 7$ $40 \div 4 = 10$

Test for 8

A number is divisible by 8 if the last three digits to the right of number are 000 or make a new number that can be divided by 8.
Example: 1$\underline{000}$, 2$\underline{160}$, 6$\underline{240}$, $1000 \div 8 = 125$, $160 \div 4 = 40$, $240 \div 4 = 60$

Test for 3

A number is divisible by 3 if the sum of its digits is divisible by 3
$342 = 300 + 40 + 2 = 3\times100 + 4\times10 + 2 = 3(99+1) + 4(9+1) + 2$
$3 \times 99+3+4\times9+4+2 = 3\times99+4\times9 + (3+4+2)$
Example: 342 $3+4+2 = 9$ $9\div3 = 3342\div3 = 114$

Test for 9

A number is divisible by 9 if the sum of its digits is divisible by 9.
$468= 400 +60 + 8 = 4\times100+6\times10+8 = 4(99+1) + 6(9+1) +8$
$4 \times 99+ 4 + 6\times 9 + 6 + 8 = 4\times99 + 6\times9 + (4 + 6 + 8)$
Example 468 $4+6+8 =18$ $18\div9 = 2$ $468 \div9 = 52$

Summary 1 Chapter 10 Lesson 1

Table 6 Tests for division

Test for 6 A number is divisible by 6 if it is divisible by both 2 and 3.

Example: 246 Last digit 6 2+4+6=12 12÷3 = 4 246÷6=41

Test for 15

A number is divisible by 15 if it is divisible by both 5 and 3.

Example: 465 Last digit 5 4 + 6 + 5 = 15 15÷3 = 5 465÷15 = 31

Test for Sum

If each addend can divide by that number the sum divisible by a number

(a + b) ÷ c = a ÷ c + b ÷ c (18 + 12) ÷ 3 = 18 ÷ 3 + 12 ÷ 3 = 6 + 4 = 10

Test for Difference

If minuend and subtrahend can divide by a number the difference divisible by a number (a – b) ÷ c = a ÷ c – b ÷ c

(18 – 12) ÷ 3 = 18 ÷ 3 – 12 ÷ 3 = 6–4 = 2

Test for Product

If one of factors divide by number the product divisible by a number

(A × B) ÷ C = (A ÷ C) × B (6 × 5) ÷ 3 = (6 ÷ 3) × 5 = 2 × 5 = 10

(A × B) ÷ C = A × (B ÷ C) (7 × 6) ÷ 3 = 7 × (6 ÷ 3) = 7 × 2 = 14

Mental Math (Do operations mentally)

If numbers can divide evenly, try to multiply or divide easily

To multiply a number by 5, divide this number by 2, and multiply that result by 10. 16 × 5 = 16 ÷ 2 × 10 = 8 × 10 = 80

24×5=24÷2×10=12×10=120

To multiply a number by 25, divide this number by 4, and multiply that result by 100. 16×25=16÷4×100=4×100=400

24×25=24÷4×100=6×100=600

To multiply a number by 125, divide this number by 8, and multiply result by 1000. 16 × 125 = 16 ÷ 8 × 1000 = 2 × 1000 = 2000

24 × 125 = 24 ÷ 8 × 1000 = 3 × 1000 = 3000

To divide a number by 5, divide this number by 10, and multiply that result by 2. 50 ÷ 5 = 50 ÷ 10 × 2 = 5 × 2 = 10

80 ÷ 5 = 80 ÷ 10 × 2 = 8 × 2 = 16

To divide a number by 25, divide this number by 100, and multiply that result by 4. 500 ÷ 25 = 500 ÷ 100 ×4 = 5 × 4 = 20

800÷25=800÷100×4=8×4=32

To divide a number by 125, divide this number by 1000, and multiply that result by 8. 5000 ÷ 125 = 5000 ÷ 1000 × 8 = 5 × 8 = 40

8000 ÷ 125 = 8000 ÷ 1000 × 8 = 8 × 8 = 6

Summary 1 Chapter 10 Lesson 2

Table 7 Divisors and Multiples

Greatest Common Factor (in short GCF)

The Greatest Common Factor is the greatest number by which all given numbers can be divided.

1) Find the GCF of numbers 12 and 18 $(\div 1, 2, 3, 4, \ldots)$
Divisors of number 12 are 1, 2, 3, 4, 6, 12
Divisors of number 18 are 1, 2, 3, 6, 9, 18
Common divisors are 1, 2, 3, 6 Greatest common divisors are 6
GCD (12, 18) = 6 $12 \div 6 = 2$ $18 \div 6 = 3$

2)If greatest of the numbers can be divided evenly by the smallest number, then the greatest common factor of numbers is equal to the smallest of given numbers:
Find the GCF of numbers 12 and 4
Factors of 12 are 1, 2, 3, 4, 6, 12 Factors of 4 are 1, 2, 4
Common factors are 1, 2, 4 Greatest common factors is 4
GCF (12, 4) = 4 $12 \div 4 = 3$

3) The greatest common factor of prime numbers equal one:
GCF (5,7) = 1

Least Common Multiples (in short LCM)

The least number is the smallest number which can be evenly divided by each of the given numbers

1) Find the LCM of numbers 12 and 18 $(\times 1, 2, 3, 4, 5\ldots)$
Multiples of 12 are 12, 24, 36, 48, 60, 72…
Multiples of 18 are 18, 36, 54, 72…
Common multiples are 36, 72…Least common multiples is 36
LCM (12, 18) = 36 $36 \div 12 = 3$ $36 \div 18 = 2$

2)If the greatest of given numbers is evenly divided by the smallest number, the LCM of the numbers is equal to the greatest of the given numbers:Find the LCM of numbers 4, 12
Multiples of 4 are 4, 8, 12, 16, 20, 24… Multiples of 12 are 12, 24…
Common multiples are 12, 24… LCM (4, 6, 12) = 12

3)The least common multiple of prime numbers is equal to the product of the numbers. Find the LCM of numbers 2 and 5
Multiples of 2 are 2, 4, 6, 8, 10… Multiples of 5 are 5, 10…

Summary 1 Chapter 10 Lesson 2

Table 8 Decimals

Properties of Decimal

$4 = 4. = 4.0 = 4.00$ $\quad 5.00 = 5$

Rounding of Decimals

$2.83 \approx 2.8$ $\quad 2.38 \approx 2.4$

Comparison of Decimals

$1.48 < 3.6$ $\quad 0.5 > 0.05$

Addition and Subtraction of Decimals

$1.25 + 0.80 = 2.05$ $\quad 1.25 - 0.80 = 0.45$

Multiplication of Decimals

$0.24 \times 10 = 2.4$ $\quad 0.24 \times 100 = 24$ $\quad 0.24 \times 1000 = 240$

$2.5 \times 4 = 10$ $\quad 1.25 \times 8 = 10$ $\quad 0.5 \times 0.2 = 0.1$

Powers of Decimals

$3.5^2 = 3.5 \times 3.5 = 12.25$ $\quad 0.5^3 = 0.5 \times 0.5 \times 0.5 = 0.25 \times 0.5 = 0.125$

Division of Decimals

$5 \div 10 = 0.5$ $\quad 5 \div 100 = 0.05$ $\quad 5 \div 1000 = 0.005$

$2 \div 5 = 0.4$ $\quad 0.2 \div 5 = 0.04$ $\quad 0.02 \div 5 = 0.004$

$2 \div 0.5 = 20 \div 5 = 4$ $\quad 0.2 \div 0.5 = 2 \div 5 = 0.4$ $\quad 0.02 \div 0.5 = 0.2 \div 5 = 0.04$

Order of Operations with the Same Level

$0.8 + 0.6 + 0.2 = 0.8 + 0.2 + 0.6 = 1.6$ $\quad 2.5 \times 6 \times 4 = 2.5 \times 4 \times 6 = 60$

$0.8 - 0.2 + 0.4 = 0.6 + 0.4 = 1$ $\quad 1.6 \div 4 \times 2 = 0.4 \times 2 = 0.8$

Order of Operations of Two Levels

$0.2 \times 4 - 0.2 = 0.8 - 0.2 = 0.6$ $\quad 1.8 \div 9 + 0.4 = 0.2 + 0.4 = 0.6$

$0.6 + 0.4 \times 2 = 0.6 + 0.8 = 1.4$ $\quad 0.8 - 1.5 \div 3 = 0.8 - 0.5 = 0.3$

Order of Operations in Expression with Parentheses

$(0.5 + 0.4) \times 2 = 0.9 \times 2 = 1.8$ $\quad 3 \times (0.9 - 0.7) = 3 \times 0.2 = 0.6$

$(1.8 - 0.4) \div 7 = 1.4 \div 7 = 0.2$ $\quad 2 \div (0.4 - 0.2) = 2 \div 0.2 = 10$

Groping Symbols: 1) Parentheses ( ) 2) Bracket [ ] 3) Brace { }

Order of Five Operations

Grouping Symbols (1) [2] {3} **Exponents (Power) a^n**

Multiplication or Division from left to right $\times$ or $\div$ $L \rightarrow R$

Addition or Subtraction from left to right $+$ or $-$ $L \rightarrow R$

Summary 1 Chapter 10 Lesson 2

Table 9

Order of Four Operations

An arithmetic expression contains numbers, sign of operations and sometimes parentheses. Result of all operations called value of expressions. Expressions with numbers can be in different form: as a sum 8 + 4, a difference 8 - 4, a product 8 × 4, or a quotient 8 ÷ 4.
Operations of the first level are addition (+) and subtraction (-)
Operations of the second level are multiplication (×) and division (÷)
Operations of the third level are power (a^n), root, and logarithm

Order of Operations of the Same Level (+, -) or (×, ÷)

An arithmetic expression without parenthesis, that has just addition or just multiplication, add or multiply in any order:
80 + 60 + 20 = 80 + 20 + 60 = 160 25 × 6 × 4 = 25 × 4 × 6 = 600
11 + 22 + 19 + 28 = (11 + 19) + (22 + 28) = 30 + 50 = 80
4 × 5 × 25 × 2 = (4 × 25) × (5 × 2) = 100 × 10 = 1000

An expression without parenthesis that has operations with the same level of operations (+, -) or (×, ÷), work in order, from left to the right:
80 – 20 + 40 = 60 + 40 = 100 160 ÷ 4 × 2 = 40 × 2 = 80
38 – 8 + 32 – 2 = 30 + 32 – 2 = 62 – 2 = 60
64 ÷ 8 × 4 ÷ 2 = 8 × 4 ÷ 2 = 32 ÷ 2 = 16

Order of Operations of Two Levels (+, -, ×, ÷)

To work an expression without parentheses that has operations of two levels of operations, at first work in order with the second level of operation (×, ÷) from left to right, then work in order with the first level of operations (+, -,) from left to right
60 + 40 × 2 = 60 + 80 = 140 180 ÷ 9 + 40 = 20 + 40 = 60
20 × 4 – 20 = 80 – 20 = 60 80 – 150 ÷ 3 = 80 – 50 = 30
12 + 18 × 2÷2 = 12 + 36 ÷ 2 = 12 + 18 = 30
18 – 24 ÷ 6 × 2 = 18 – 4 × 2 = 18 – 8 = 10
16 × 4 ÷ 2 – 6 = 64 ÷ 2 – 6 = 32 – 6 = 26
18 ÷ 9 × 2 + 4 = 2 × 2 + 4 = 4 + 4 = 8

Summary 1 Chapter 10 Lesson 2

Table 10

Order of Operations in an Expression with Parentheses

In an expression with parentheses: the first step is to work with the operations in parentheses, and the second step is to work before and after parentheses from left to right:

$(50 + 40) \times 2 = 90 \times 2 = 180$ $\quad$ $3 \times (90 - 70) = 3 \times 20 = 60$

$(180 - 40) \div 7 = 140 \div 7 = 20$ $\quad$ $200 \div (40 - 20) = 200 \div 20 = 10$

$15 \div 3 \times (5 + 4) \times 2 = 15 \div 3 \times 9 \times 2 = 5 \times 9 \times 2 = 45 \times 2 = 90$

$20 \div (4 - 2) \times 3 = 20 \div 2 \times 3 = 10 \times 3 = 30$

$6 \times (9 - 7) \div 3 = 6 \times 2 \div 3 = 12 \div 3 = 4$

$(18 - 4) \div 7 \times 5 = 14 \div 7 \times 5 = 2 \times 5 = 10$

Grouping Symbols

Expressions have three kinds of grouping symbols: first do the operations in parentheses, second do the operations - in brackets, and third do the operations - in braces:

Parentheses ( ) $\quad$ **Brackets [ ]** $\quad$ **Braces { }**

Order of Five Operations

Expressions have five operations: first do the operations of the third level (a^n); second, do the operations of the second level (× or ÷), and lastly, do the operations of the first level (+ or -), in order as they occur from left to right.

Remember (in short):

Grouping Symbols	**(1) [2] {3}**	
Exponents (Power)	a^n	
Multiplication or Division from left to right	**× or ÷**	**L → R**
Addition or Subtract from left to right	**+ or –**	**L → R**

Example 1

$6^2 \div 2^2 \times 3 - 3^2 \times 6 \div 3 = 36 \div 4 \times 3 - 9 \times 6 \div 3 = 9 \times 3 - 54 \div 3 = 27 - 18 = 9$

Example 2

$50 - \{200 \div [60 - (30 + 20)]\} = 50 - \{200 \div [60 - 50]\} = 50 - 200 \div 10 = 50 - 20 = 30$

Summary 1 Chapter 10 Lesson 2

Table 11 Fractions

Properties of Fraction

$\frac{a}{b}=\frac{a\cdot n}{b\cdot n}$

$\frac{2}{3}=\frac{2\times 6}{3\times 6}=\frac{12}{18}$

$\frac{a\div n}{b\div n}=\frac{a}{b}$

$\frac{60}{75}=\frac{60\div 3}{75\div 3}=\frac{20}{25}=\frac{20\div 5}{25\div 5}=\frac{4}{5}$

Comparison of Fractions

$\frac{6}{11}>\frac{7}{22}$ $\quad\frac{6}{11}=\frac{6\cdot 2}{11\cdot 2}=\frac{12}{22}$ $\quad\frac{12}{22}>\frac{7}{22}$

Addition of Fractions

$\frac{a}{c}+\frac{b}{c}=\frac{a+b}{c}$ $\quad\frac{2}{5}+\frac{3}{8}+\frac{7}{20}=\frac{2\cdot 8}{5\cdot 8}+\frac{3\cdot 5}{8\cdot 5}+\frac{7\cdot 2}{20\cdot 2}=\frac{16}{40}+\frac{15}{40}+\frac{14}{40}=\frac{45}{40}=\frac{9}{8}=1\frac{1}{8}$

Subtraction of Fractions

$\frac{a}{c}-\frac{b}{c}=\frac{a-b}{c}$ $\quad\frac{3}{10}-\frac{1}{4}=\frac{6}{20}-\frac{5}{20}=\frac{1}{20}$ $\quad\frac{1}{12}-\frac{1}{18}=\frac{3}{36}-\frac{2}{36}=\frac{3-2}{36}=\frac{1}{36}$

Multiplication of Fractions

$\frac{a}{b}\times\frac{c}{d}=\frac{a\times c}{b\times d}=\frac{ac}{bd}$

$\frac{a}{b}\cdot c=\frac{a}{b}\cdot\frac{c}{1}=\frac{a\cdot c}{b}$

$a\cdot\frac{b}{c}=\frac{a}{1}\cdot\frac{b}{c}=\frac{ab}{c}$

$\frac{4}{6}\times\frac{9}{8}=\frac{4\times 9}{6\times 8}=\frac{1\times 3}{2\times 2}=\frac{3}{4}$

$\frac{3}{8}\times 4=\frac{3}{8}\times\frac{4}{1}=\frac{3\times 1}{2\times 1}=\frac{3}{2}=1\frac{1}{2}$

$8\times\frac{5}{4}=\frac{8}{1}\times\frac{5}{8}=\frac{5}{1}=5$

Division of Fractions

$\frac{a}{b}\div\frac{c}{d}=\frac{a}{b}\cdot\frac{d}{c}=\frac{ad}{dc}$

$\frac{a}{b}\div c=\frac{a}{b}\cdot\frac{1}{c}=\frac{a}{bc}$

$a\div\frac{b}{c}=\frac{a}{1}\cdot\frac{c}{b}=\frac{ac}{b}$

$\frac{5}{6}\div\frac{10}{3}=\frac{5}{6}\times\frac{3}{10}=\frac{1\times 1}{2\times 2}=\frac{1}{4}$

$\frac{3}{4}\div 6=\frac{3}{4}\div\frac{6}{1}=\frac{3}{4}\times\frac{1}{6}=\frac{1}{8}$

$6\div\frac{3}{4}=\frac{6}{1}\times\frac{4}{3}=\frac{8}{1}=8$

Power of Fractions

$\left(\frac{a}{b}\right)^n=\frac{a^n}{b^n}$ $\quad\left(\frac{1}{3}\right)^2=\frac{1^2}{3^2}=\frac{1}{9}$

Summary 1 Chapter 10 Lesson 2

Table 12
Mixed Numbers

Mixed Number $2+\frac{1}{2}=2\frac{1}{2}$ $\frac{1}{4}+1=1\frac{1}{4}$ $1\frac{3}{4}=\frac{4\cdot 1+3}{4}=\frac{7}{4}$

Comparison of Mixed Numbers: $1\frac{1}{3}<1\frac{5}{6}$ $1\frac{2}{6}<1\frac{5}{6}$

Addition of Mixed Numbers

$1\frac{1}{8}+2\frac{3}{8}=3\frac{4}{8}=3\frac{1}{2}$ $1\frac{1}{3}+2\frac{3}{9}=1\frac{1}{3}+2\frac{1}{3}=3\frac{2}{3}$

$1\frac{2}{6}+2\frac{1}{9}=1\frac{1}{3}+2\frac{1}{9}=1\frac{3}{9}+2\frac{1}{9}=3\frac{4}{9}$ $1\frac{1}{2}+1\frac{1}{3}+2\frac{1}{6}=1\frac{3}{6}+1\frac{2}{6}+2\frac{1}{6}=4\frac{6}{6}=5$

Subtraction of Mixed Numbers

$6\frac{7}{8}-5\frac{3}{8}=1\frac{4}{8}=1\frac{1}{2}$ $4\frac{3}{4}-2\frac{1}{4}=2\frac{2}{4}=2\frac{1}{2}$ $3\frac{1}{2}-2=1\frac{1}{2}$

$1-\frac{2}{3}=\frac{3}{3}-\frac{2}{3}=\frac{1}{3}$ $4-\frac{5}{8}=3\frac{8}{8}-\frac{5}{8}=3\frac{3}{8}$ $8-3\frac{5}{6}=7\frac{6}{6}-3\frac{5}{6}=4\frac{1}{6}$

$6\frac{2}{7}-2\frac{5}{7}=5\frac{9}{7}-2\frac{5}{7}=3\frac{4}{7}$ $5\frac{1}{3}-2\frac{5}{6}=5\frac{2}{6}-2\frac{5}{6}=4\frac{8}{6}-2\frac{5}{6}=2\frac{3}{6}=2\frac{1}{2}$

Multiplication of Mixed Numbers

$3\frac{1}{3}\cdot 2\frac{2}{5}=\frac{10}{3}\cdot\frac{12}{5}=\frac{2\cdot 4}{1\cdot 1}=8$ $\frac{5}{8}\cdot 4\frac{4}{5}\cdot\frac{1}{6}=\frac{5}{8}\cdot\frac{24}{5}\cdot\frac{1}{6}=\frac{3}{6}=\frac{1}{2}$

Division of Mixed Numbers

$1\frac{1}{6}\div 4\frac{2}{3}=\frac{7}{6}\div\frac{14}{3}=\frac{7}{6}\times\frac{3}{14}=\frac{1}{4}$ $3\frac{1}{5}\div 2\frac{2}{5}=\frac{16}{5}\times\frac{5}{12}=\frac{4}{3}=1\frac{1}{3}$

$8\div 4\frac{4}{5}=\frac{8}{1}\div\frac{24}{5}=\frac{8}{1}\times\frac{5}{24}=\frac{5}{3}=1\frac{2}{3}$ $4\frac{1}{3}\div 13=\frac{13}{3}\div\frac{13}{1}=\frac{13}{3}\times\frac{1}{13}=\frac{1}{3}$

Power of Mixed Numbers

$\left(\frac{3}{4}\right)^3=\frac{3^3}{4^3}=\frac{27}{64}$ $\left(1\frac{1}{2}\right)^2=\left(\frac{3}{2}\right)^2=\frac{3^2}{2^2}=\frac{9}{4}=2\frac{1}{4}$

Summary 1 Chapter 10 Lesson 3

Table 13 Fractions and Decimals

Converting Fractions to Terminating Decimals

$10=2\times5 \qquad 100=4\times25 \qquad 1000 = 8\times125$

$\frac{1}{10}=0.1, \frac{1}{100}=0.01, \frac{1}{1000}=0.001, \frac{1}{2}=\frac{1\times5}{2\times5}=\frac{5}{10}=0.5 \quad \frac{1}{5}=\frac{1\times2}{5\times2}=\frac{2}{10}=0.2$

$\frac{1}{4}=\frac{1\times25}{4\times25}=\frac{25}{100}=0.25, \frac{15}{20}=\frac{3}{4}=\frac{3\times25}{4\times25}=\frac{75}{100}=0.75, 2\frac{1}{8}=2\frac{1\times125}{8\times125}=2\frac{125}{1000}=2.125$

Memorize

$\frac{1}{2}=0.5 \qquad \frac{1}{4}=0.25 \qquad \frac{1}{5}=0.2 \qquad \frac{1}{8}=0.125 \qquad \frac{3}{4}=0.75$

Converting Terminating Decimals to Mixed Numbers

$1.125=1\frac{125}{1000}=1\frac{125\div5}{1000\div5}=1\frac{25}{200}=1\frac{25\div5}{200\div5}=1\frac{5}{40}=1\frac{1}{8}$

Combine Operations with Fractions and Decimals

Easy: + and - in decimals, × and ÷ in fractions

$\frac{1}{2}-0.25=0.5-0.25=0.25 \qquad 0.25\div\frac{1}{8}=\frac{1}{4}\times\frac{8}{1}=2$

$\frac{2}{3}+0.6=\frac{2}{3}+\frac{6}{10}=\frac{2}{3}+\frac{3}{5}=\frac{10}{15}+\frac{9}{15}=\frac{19}{15}=1\frac{4}{15}$

$2.8-2\frac{2}{3}=2\frac{8}{10}-2\frac{2}{3}=2\frac{4}{5}-2\frac{2}{3}=2\frac{12}{15}-2\frac{10}{15}=\frac{2}{15}$

$\frac{2}{3}\times0.36=0.36\times\frac{2}{3}=0.36\div3\times2=0.12\times2=0.24$

$3.6\div4\frac{1}{2}=3.6\div4.5=\frac{3.6}{4.5}=\frac{36}{45}=\frac{4}{5}=0.8$

Converting Simple Repeating Decimals to Fractions

Yes $\frac{1}{2},\frac{1}{4},\frac{1}{5},\frac{1}{8},\frac{1}{10}$ No $\frac{1}{3},\frac{1}{6},\frac{1}{7},\frac{1}{9}$ a) $x=0.3333\ldots\ (\times10)\ 0.\overline{3}=\frac{3}{9}=\frac{1}{3}$

b) $x=0.1818\ldots\ 0.\overline{18}=\frac{18}{99}=\frac{2}{11}$ c) $0.\overline{6}=\frac{6}{9}=\frac{2}{3}$, d) $1.\overline{27}=1\frac{27}{99}=1\frac{3}{11}$

Converting Composite Repeating Decimals in the Fractions

$x=0.8333\ldots\ (\times10)$

a) $0.8\overline{3}=\frac{83-8}{90}=\frac{75}{90}=\frac{5}{6}=\frac{5}{6}$ b) $0.1\overline{3}=\frac{13-1}{90}=\frac{12}{90}=\frac{2}{15}$

c) $1.1\overline{6}=1\frac{16-1}{90}=1\frac{15}{90}=1\frac{1}{6}$ d) $1.8\overline{3}=1\frac{83-8}{90}=1\frac{75}{90}=1\frac{5}{6}$

Summary 1 Chapter 10 Lesson 3

Table 14 Rational Numbers

Addition and Subtraction of Rational Numbers

$(-1.25) + (-0.8) = -1.25 - 0.8 = -2.05$ $\quad -1.25 + 0.8 = -0.45$

$-1.25 - (-0.8) = -1.25 + 0.8 = -0.45$ $\quad 0.8 - 1.25 = -0.45$

Multiplication of Rational Numbers

$(+6)(+2) = +12$ $\quad (-6)(-2) = +12$ $\quad (+)(+) = +$ $\quad (-)(-) = +$

$(+6)(-2) = -12$ $\quad (-6)(+2) = -12$ $\quad (+)(-) = -$ $\quad (-)(+) = -$

$(-0.25)(10) = -2.5$ $\quad (0.25)(-100) = -25$ $\quad (-0.25)(-1000) = 250$

$(-4)(2.5) = -10$ $\quad 8(-12.5) = -100$ $\quad (1.25)(-0.8) = -1$

Division of Rational Numbers

$(+6)\div(+2) = +3$ $\quad (-6)\div(-2) = +3$ $\quad (+)\div(+) = +$ $\quad (-)\div(-) = +$

$(-6)\div(+2) = -3$ $\quad (+6)\div(-2) = -3$ $\quad (+)\div(-) = -$ $\quad (-)\div(+) = -$

$(-5)\div 10 = -0.5$ $\quad 5 \div (-100) = -0.05$ $\quad (-5)\div(-1000) = 0.005$

$(-2)\div 5 = -0.4$ $\quad 0.2 \div (-5) = -0.04$ $\quad (-2)\div(-0.5) = 20\div 5 = 4$

$0.2\div(-0.5) = (-2)\div 5 = -0.4$ $\quad (-1.2)\div 0.8 = (-12)\div 8 = -1.5$

Powers of Rational Numbers

$2^3 = 2\times 2\times 2 = 8$ $\quad (-3)^2 = (-3)(-3) = 9$

$(-2)^3 = -2(-2)(-2) = -8$ $\quad (-2)^4 = (-2)(-2)(-2)(-2) = 16$

$-2^3 = -2\times 2\times 2 = -8$ $\quad -3^2 = -3\times 3 = -9$

$(-0.1)^2 = (-0.1)(-0.1) = 0.01$ $\quad -0.3^2 = -0.09$

Order of Operations

$-25(-6)(-4) = -25\times 4\times 6 = -600$ $\quad -20(-4) + 20 = -80 + 20 = -60$

$-60 - 40(-2) = -60 + 80 = 20$ $\quad -16 \div (-4)(-20) = 4(-20) = -80$

$(-50 - 40)\times 2 = -90\times 2 = -180$ $\quad 3(-90 + 70) = 3(-20) = -60$

$-180 \div 9 - 40 = -20 - 40 = -60$ $\quad -80 + 150 \div 3 = -80 + 50 = -30$

$(-180 + 40)\div 7 = -140 \div 7 = -20$ $\quad 200 \div (-40 + 20) = 200\div(-20) = -10$

Examples

a) $5(-3^2 - 2^2\times 2^3) = 5(-9 - 4\times 8) = 5(-9 - 32) = 5(-41) = -205$

b) $\{-200 \div [-60 - (-30 - 20)]\} = \{-200 \div [-60 + 50]\} = -200 \div(-10) = 20$

Summary 1 Chapter 10 Lesson 3

Table 15 Laws for operations 1

A formula is a statement of math relations between variables.
Formulas include letters that can represent any numbers.
Equality is two expressions joined by the equal symbol.
Laws of Addition: First Law of Addition (commutative)
Changing the order of two addends does not change the sum:
$A + B = B + A$ $2 + 3 = 5$ $3 + 2 = 5$ $2 + 3 = 3 + 2$
Second Law of Addition (associative)
Changing the grouping of addends does not change the sum:
$A + (B + C) = (A + B) + C$ $8 + (2 + 5) = (8 + 2) + 5 = 10 + 5 = 15$
Use the commutative law of addition as the first step.
Use the associative law of addition as the second step.
Add in any order. Use the two laws together.
$2 + 7 + 3 + 8 = (2 + 8) + (3 + 7) = 10 + 10 = 20$

Laws of Multiplication

First Law of Multiplication (commutative)
Changing the order of two factors does not change the product:
$A B = B A$ $2 \times 3 = 6$ $3 \times 2 = 6$ $2 \times 3 = 3 \times 2$
Second Law of Multiplication (associative)
Changing the grouping of factors does not change the product:
$A (B C) = (AB) C$ $5 \times (2 \times 4) = (5\times2) \times 4 = 10 \times 4 = 40$
Use the commutative law of multiplication as the first step
Use the associative law of multiplication as the second step
Multiply in any order; use the two laws together:
$5 \times 8 \times 2 = 5 \times 2 \times 8 = 10 \times 8 = 80$

First Distributive Law of Multiplication with Addition

To multiply a number by a sum, multiply that number by each of the addends and add the results
$A (B + C) = AB + AC$ $2(3+5) = 2\times 3+ 2\times5= 6 +10=16$
$(A + B) C = AC + BC$ $14 \times 5= (10 + 4) \times 5 = 10 \times 5 + 4 \times 5 =50+20=70$

Second Distributive Law of Multiplication with Subtraction

To multiply a number by a difference, multiply that number by the minuend and subtrahend, and subtract the results:
$A (B – C) = AB – AC$ $2(10 – 4) = 2 \times 10 – 2 \times 4 = 20 – 8 = 12$
$(A - B) C = AC – B C$ $(20 - 2) \times 5 = 20 \times 5 – 2 \times 5 = 100 – 10 = 90$
To do mental calculations means to carry out operations by laws or properties.

Summary 1 Chapter 10 Lesson 3

Table 16 Laws for rational numbers 2

Distributive laws of multiplication

1) Whole Numbers (Put A before parentheses)

AB + AC = A (B + C)   AB – AC = A (B - C)

If two products have a common factor, put the common factor before parentheses. AB + AC   A – common factor

In parentheses, put results of division products by common factor

AB ÷ A = B   AC ÷ A = C

$16 \cdot 5 + 4 \cdot 5 = 5\ (16 + 4) = 5\ (20) = 100$

$22 \cdot 5 - 2 \cdot 5 = 5\ (22 - 2) = 5\ (20) = 100$

2) Decimals

a) (B + C) A = AB + AC $(0.4 + 0.1)5 = 0.4 \times 5 + 0.1 \times 5 = 2 + 0.5 = 2.5$

AB + AC = A (B + C) $0.4 \times 5 + 0.1 \times 5 = 5\ (0.4 + 0.1) = 5(0.5) = 2.5$

b) A (B - C) = AB – AC $5\ (0.4 - 0.1) = 5 \times 0.4 - 5 \times 0.1 = 2 - 0.5 = 1.5$

AB – AC = A (B – C) $5 \times 0.4 - 5 \times 0.1 = 5\ (0.4 - 0.1) = 5(0.3) = 1.5$

3) Negative Numbers

If first term negative, put (- A) before parentheses and change each sign inside parentheses: -AB – AC = - A (B + C) - AB + AC = - A (B - C)

a) $-16 \cdot 5 - 4 \cdot 5 = -5\ (16 + 4) = -5\ (20) = -100$

$-22 \cdot 5 + 2 \cdot 5 = -5\ (22 - 2) = -5\ (20) = -100$

b) $-7.5 \times 5 - 2.5 \times 5 = -5(7.5 + 2.5) = -5(10) = -50$

$-12.5\ (5) + 2.5(5) = -5(12.5 - 2.5) = -5(10) = -50$

4) To solve problems work by order of operations or change division for multiplication then put A before parentheses.

AB + AC = A (B + C)   AB – AC = A (B - C)

a) $2\frac{5}{8} \div \frac{1}{8} + 1\frac{3}{8} \div \frac{1}{8} = 2\frac{5}{8} \times 8 + 1\frac{3}{8} \times 8 = 8(2\frac{5}{8} + 1\frac{3}{8}) = 8(4) = 32$

b) $2\frac{5}{8} \div \frac{1}{8} - 1\frac{3}{8} \div \frac{1}{8} = 2\frac{5}{8} \times 8 - 1\frac{3}{8} \times 8 = 8(2\frac{5}{8} - 1\frac{3}{8}) = 8(1\frac{2}{8}) = 8(\frac{10}{8}) = 10$

5) To solve problems work by order of operations or change division for multiplication then put (- A) before parentheses and change each sign inside parentheses

- AB - AC = - A (B + C)   - AB + A C = - A (B - C)

a) $-2\frac{5}{8} \div \frac{1}{8} - 1\frac{3}{8} \div \frac{1}{8} = -2\frac{5}{8} \times 8 - 1\frac{3}{8} \times 8 = -8\ (2\frac{5}{8} + 1\frac{3}{8}) = -8\ (4) = -32$

b) $-2\frac{5}{8} \div \frac{1}{8} + 1\frac{3}{8} \div \frac{1}{8} = -2\frac{5}{8} \times 8 + 1\frac{3}{8} \times 8 = -8\ (2\frac{5}{8} - 1\frac{3}{8}) = -8\ (1\frac{2}{8}) = -8\ (\frac{10}{8}) = -10$

Summary 1 Chapter 10 Lesson 3

Table 17

Common Factor for Whole Numbers 1

$AB + AC = A(B + C)$ $AB - AC = A(B - C)$

If two products have a common factor, put the common factor before parentheses. AB + AC, A – common factor, Put A before parentheses

In parentheses, put results of division products by common factor

$AB \div A = B$ $AC \div A = C$

$16 \cdot 5 + 4 \cdot 5 = 5(16 + 4) = 5(20) = 100$

$22 \cdot 5 - 2 \cdot 5 = 5(22 - 2) = 5(20) = 100$

$75\times5 + 25\times5 = 5(75 + 25) = 5(100) = 500$

$125(5) - 25(5) = 5(125 - 25) = 5(100) = 500$

$36 \cdot 5 = (30 + 6) \cdot 5 = 30 \cdot 5 + 6 \cdot 5 = 150 + 30 = 180$

$8 \cdot 9 + 2 \cdot 9 = 9 \cdot (8 + 2) = 9 \cdot 10 = 90$

Memorize Formulas

1) $A + (B + C) = (A + B) + C$
2) $A \times (B \times C) = (A \times B) \times C$
3) $A \cdot (B + C) = A{\cdot}B + A{\cdot}C$
4) $A \cdot (B - C) = A{\cdot}B - A{\cdot}C$
5) $A{\cdot}B + A{\cdot}C = A \cdot (B + C)$
6) $A{\cdot}B - A{\cdot}C = A \cdot (B - C)$
7) $A \div C + B \div C = (A + B) \div C$
8) $A \div C - B \div C = (A - B) \div C$

Examples

a) $A(B + C) = AB + AC$

   $5 \times (40 + 10) = 5 \times 40 + 5 \times 10 = 200 + 50 = 250$

b) $A(B - C) = AB - AC$

   $5 \times (40 - 10) = 5 \times 40 - 5 \times 10 = 200 - 50 = 150$

c) $AB + AC = A(B + C)$

   $40 \times 5 + 10 \times 5 = 5(40 + 10) = 5(50) = 250$

d) $AB - AC = A(B - C)$

   $5 \times 40 - 5 \times 10 = 5(40 - 10) = 5(30) = 150$

Summary 1 Chapter 10 Lesson 3

Table 18

Common Factor for Fractions 2

If first term negative, put (- A) before parentheses and change the sign before each term: -AB – AC = - A (B + C)   - AB + AC = - A (B - C)

- 16 · 5 - 4 · 5 = - 5 (16 + 4) = - 5 (20) = - 100

- 22 · 5 + 2 · 5 = - 5 (22 - 2) = - 5 (20) = - 100

Examples: Evaluate

1) $-1\frac{3}{5}(2\frac{1}{8})-1\frac{3}{5}(2\frac{7}{8})=-1\frac{3}{5}(2\frac{1}{8}+2\frac{7}{8})=-\frac{8}{5}(5)=-8$

2) $-\frac{3}{4}(4\frac{2}{3})+\frac{3}{4}(3\frac{1}{3})=-\frac{3}{4}(4\frac{2}{3}-3\frac{1}{3})=-\frac{3}{4}(1\frac{1}{3})=-\frac{3}{4}\times\frac{4}{3}=-1$

3) $\frac{5}{6}(-4\frac{3}{5})+\frac{5}{6}\times1\frac{3}{5}=-\frac{5}{6}(4\frac{3}{5}-1\frac{3}{5})=-\frac{5}{6}(3)=-\frac{5}{2}=-2\frac{1}{2}=-2.5$

4) $-4\frac{1}{2}\times\frac{1}{6}-\frac{1}{6}\times1\frac{1}{2}=-\frac{1}{6}(4\frac{1}{2}+1\frac{1}{2})=-\frac{1}{6}(6)=-1$

5) $-3\frac{3}{4}\times1\frac{1}{6}+2\frac{1}{4}\times1\frac{1}{6}=-1\frac{1}{6}(3\frac{3}{4}-2\frac{1}{4})=-1\frac{1}{6}(1\frac{2}{4})=-\frac{7}{6}\times\frac{3}{2}=-\frac{7}{4}=-1.75$

6) $(-\frac{1}{3}+\frac{1}{5})\div\frac{1}{15}=(-\frac{1}{3}+\frac{1}{5})\times15=-\frac{1}{3}\times\frac{15}{1}+\frac{1}{5}\times\frac{15}{1}=-5+3=-2$

7) $(-1\frac{3}{4}+1\frac{1}{2})\div\frac{1}{8}=(-\frac{7}{4}+\frac{3}{2})\times8=-\frac{7}{4}\times\frac{8}{1}+\frac{3}{2}\times\frac{8}{1}=-14+12=-2$

8) $(-1\frac{1}{2}+1\frac{1}{6})\div\frac{1}{6}=(-\frac{3}{2}+\frac{7}{6})\times6=-\frac{3}{2}\times\frac{6}{1}+\frac{7}{6}\times\frac{6}{1}=-9+7=-2$

9) $(-2\frac{1}{2}-1\frac{1}{3})\div\frac{1}{6}=(-\frac{5}{2}-\frac{4}{3})\times6=-\frac{5}{2}\times\frac{6}{1}-\frac{4}{3}\times\frac{6}{1}=-15-8=-23$

10) $(-1\frac{1}{3}-1\frac{3}{4})\div\frac{1}{12}=(-\frac{4}{3}-\frac{7}{4})\times12=-\frac{4}{3}\times\frac{12}{1}-\frac{7}{4}\times\frac{12}{1}=-16-21=-37$

Summary 1 Chapter 10 Lesson 4

Table 19 Conditional Equations

Six Properties of True Equality: if a = b then

$a + c = b + c \quad a - c = b - c \quad ac = bc \quad a \div c = b \div c \quad c \neq 0$

$a + c = b + c \quad a = b \quad 3y + 3 = 6 + 3 \quad 3y = 6 \quad y = 2$

$a - b = c \quad a = c + b \quad 3y + 2 = 8 \quad 3y = 8 - 2 \quad 3y = 6 \quad y=2$

$a + c = b + c \quad a = b \quad 3y - 9 = 0 \quad 3y = 9 \quad y = 3$

$a + b = c \quad a = c - b \quad 3y + 2 = 8 \quad 3y = 8 - 2 \quad 3y = 6 \quad y = 2$

Conditional equation has one solution $2y = 8 \quad y = 4$

$2x = 0x = 0 \quad 0x = 2$ ∅ no solutions

Rule for Solving Equations

1) Open parentheses $a(b + c) = ab + ac$
2) Combine like terms 3) Eliminate number on left side
4) Eliminate letter on right side 5) Multiplication property

Equation with Whole Numbers

$-3(y - 1) = -2(y + 2) + 6 \quad -3y + 3 = -2y - 4 + 6 \quad -3y + 3 = -2y + 2$

$-3y + 2y = -3 - 2 \quad -y = -5 \quad y = 5$

Equations with Decimals

$3(y - 5.8) = 6.5(y + 1.2) - 12.25 \quad 3y - 17.4 = 6.5y + 7.8 - 12.25$

$3y - 17.4 = 6.5y - 4.45 \quad 3y = 6.5y + 12.95$

$-3.5y = 12.95 \quad y = -3.7$

Equations with Fraction (× LCD)

$\frac{1}{6} - \frac{y}{4} = \frac{2-y}{3} \quad \frac{12}{1} \times \frac{1}{6} - \frac{12}{1} \times \frac{y}{4} = \frac{12}{1} \times \frac{2-y}{3} \quad 2 - 3y = 4(2 - y)$

$2 - 3y = 8 - 4y \quad -3y + 4y = 8 - 2 \quad y = 6$

Proportions

$12 \div 4 = 18 \div 6 \quad a \div b = c \div d \quad \frac{12}{4} = \frac{18}{6} \quad 3 = 3 \quad \frac{a}{b} = \frac{c}{d}$

$12 \times 6 = 72 \quad 4 \times 18 = 72 \quad 12 \times 6 = 4 \times 18 \quad ad = bc$

$12 \div 18 = 4 \div 6 \quad 6 \div 4 = 18 \div 12 \quad 6 \div 18 = 4 \div 12$

$\frac{12}{4} = \frac{18}{y} \quad 12\,y = 4 \cdot 18 \quad 12y = 72 \quad y = 72 \div 12 \quad y = 6$

$\frac{12}{y} = \frac{18}{6} \quad 18\,y = 12 \cdot 6 \quad 18y = 72 \quad y = 72 \div 18 \quad y = 4$

Summary 1 Chapter 10 Lesson 4

Table 20 Plane Geometry

Right triangles

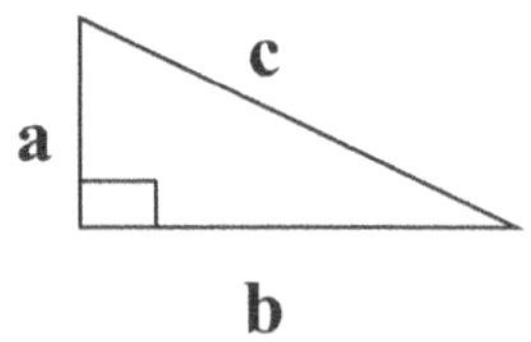

1) Pythagorean Theorem 1) $c^2 = a^2 + b^2$

2) Area 2) $A_\Delta = 0.5ab$

Triangles

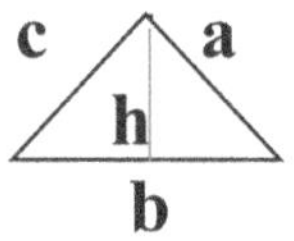

3) Perimeter 3) $P = a + b + c$

4) Area 4) $A_\Delta = 0.5bh$

Parallelogram

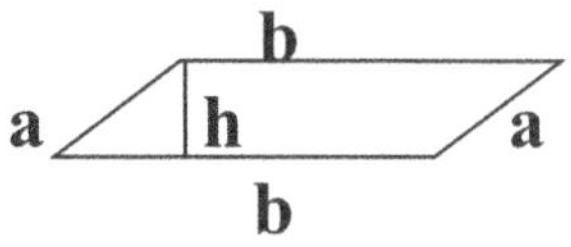

5) Perimeter 5) $P_{par} = 2a + 2b$

6) Area 6) $A_{par} = b\ h$

Rhombus

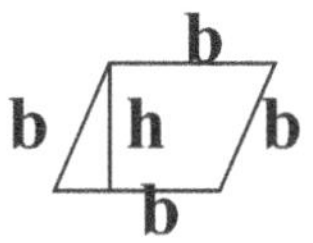

7) Perimeter 7) $P_{rh} = 4b$

8) Area 8) $A_{rh} = b \cdot h$

Rectangle

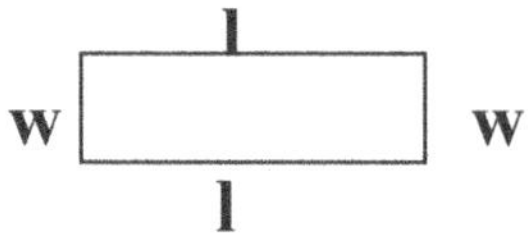

9) Perimeter 9) $P_{rec} = 2l + 2w$

10) Area 10) $A_{rec} = l \cdot w$

Square

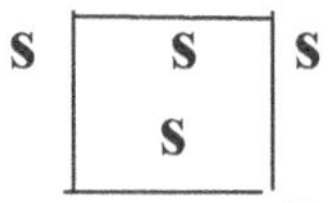

11) Perimeter 11) $P_\blacksquare = 4s$

12) Area 12) $A_\blacksquare = s^2$

Trapezoid

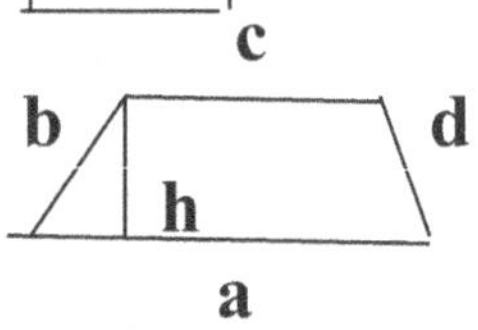

13) Perimeter 13) $P_{tr} = a + b + c + d$

14) Area 14) $A_{tr} = 0.5\ h\ (a + c)$

Circle

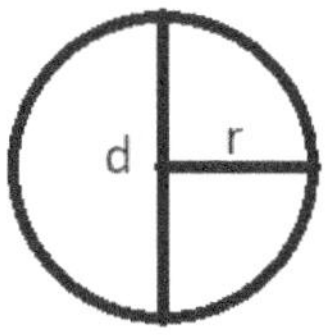

15) Number π in fraction $\approx \frac{22}{7}$, 16) Number π in decimal ≈ 3.14

17) Diameter $d = 2r$, 18) Circumference 1 (d) $C = \pi d$

19) Circumference 2 (r) $C = 2\pi r$ 20) Area $A_c = \pi r^2$

Solid Geometry

Cube 1)$V = s^3$, 2)$SA = 6s^2$, 3) $d = s\sqrt{3}$, Sphere 4)$V = \frac{4}{3}\pi r^3$ 5) $SA = 4\pi r^2$

Rectangular Solid:6)$V = l \cdot w \cdot h$, 7)$SA = 2lw + 2lh + 2wh$, 8)$d^2 = l^2 + w^2 + h^2$

Regular Prism:9)$V = Base \cdot h$, Regular Pyramid: 10)$V = \frac{1}{3} Base \cdot h$

Right Cylinder: 11)$V = \pi r^2 h$, Right Cone 12) $V = \frac{1}{3}\pi r^2 h$

Memorize Formulas of Geometry

Summary 1 Chapter 10 Lesson 4

Table 21

Plane Geometry

Right triangles

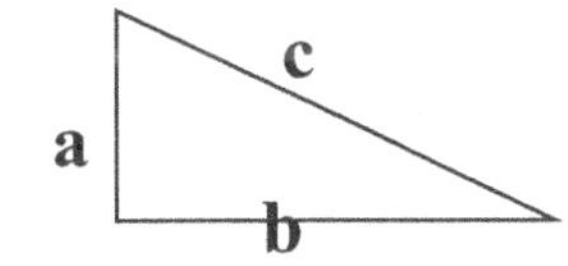

1) Pythagorean Theorem	1) $c^2 =$
2) Area	2) $A_\Delta =$

Triangles

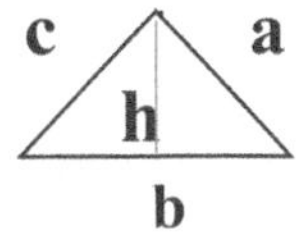

3) Perimeter	3) $P =$
4) Area	4) $A_\Delta =$

Parallelogram

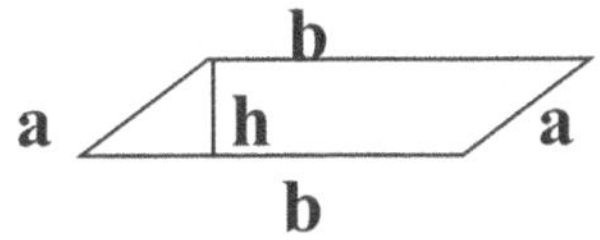

5) Perimeter	5) $P_{par} =$
6) Area	6) $A_{par} =$

Rhombus

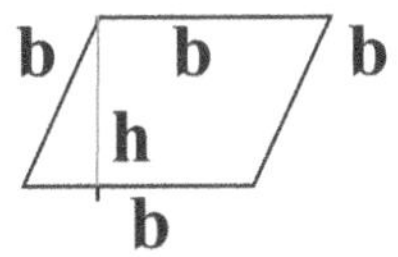

7) Perimeter	7) $P_{rh} =$
8) Area	8) $A_{rh} =$

Rectangle

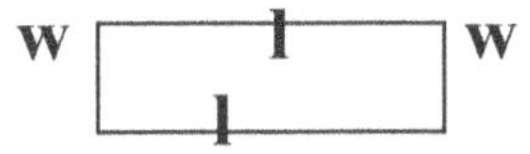

9) Perimeter	9) $P_{rec} =$
10) Area	10) $A_{rec} =$

Square

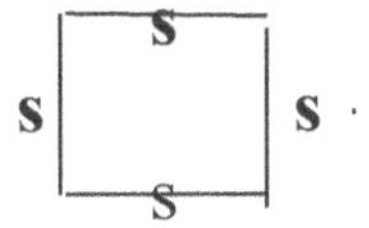

11) Perimeter	11) $P_{\blacksquare} =$
12) Area	12) $A_{\blacksquare} =$

Trapezoid

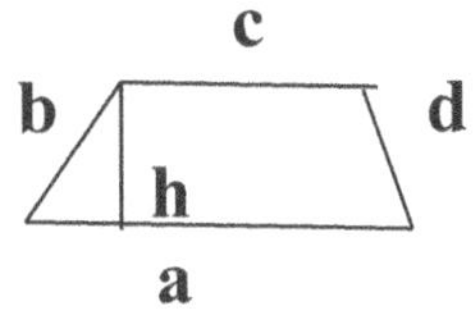

13) Perimeter	13) $P_{tr} =$
14) Area	14) $A_{tr} =$

Circle

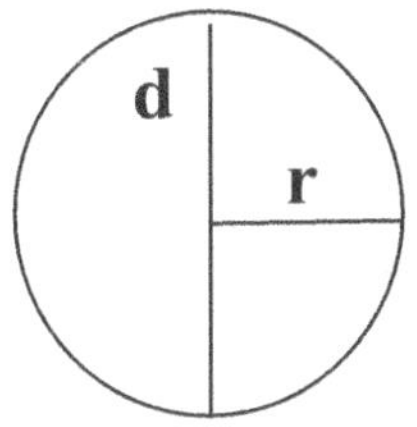

15) Number π in fraction	15) $\pi \approx$
16) Number π in decimal	16) $\pi \approx$
17) Diameter	17) $d =$
18) Circumference 1 (d)	18) $C =$
19) Circumference 2 (r)	19) $C =$
20) Area	20) $A_c =$

Memorize Formulas of Geometry

Summary 1 Chapter 10 Lesson 4

Table 22 Solid Geometry 1

Unit for volume

1 cm = 10 mm — **$1\ cm^3 = 1000\ mm^3$**

1 yd = 3 ft — **$1\ yd^3 = 27\ ft^3$** — **s is a side**

SA is a surface area — **V is a volume** — **d is a diagonal**

Cube

The volume of a cube is equal cube of the its side: $V = s^3$

Surface Area of Cube $SA = 6s^2$

Diagonal of Cube is $B_1D = d$ $d = s\sqrt{3}$

Example 1
Find the volume, the surface of area, and the diagonal of cube, if s = 4 ft.

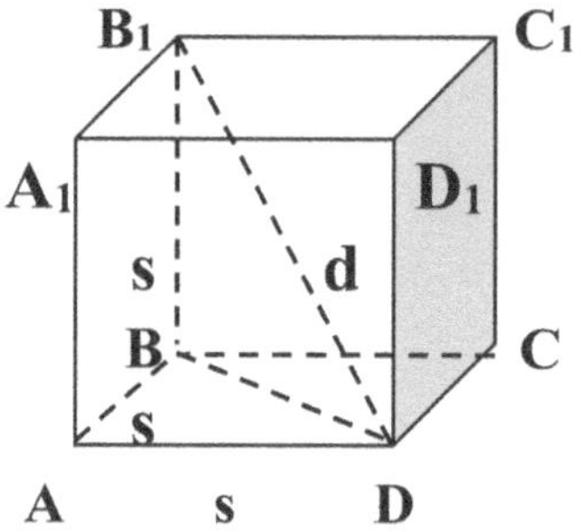

Figure 1

$V = s^3 = 4^3 = 64\ ft^3$

$SA = 6 \cdot 4^2 = 6 \cdot 16 = 96\ ft^2$

$d = s\sqrt{3} = 4\sqrt{3}\ ft$

Rectangular Solid

The volume of a rectangular solid is equal to the product length, width, and height. A rectangular solid has length l, width w, and height h.

Volume of Rectangular Solid $V = lwh$

Surface Area of Rectangular Solid $SA = 2lw + 2lh + 2wh$

Diagonal of Rectangular Solid $d^2 = l^2 + w^2 + h^2$

Example 2 Find the volume, the surface of area, and the diagonal of rectangular solid, if l = 10 in, w = 6 in, h = 8 in.

Figure 2

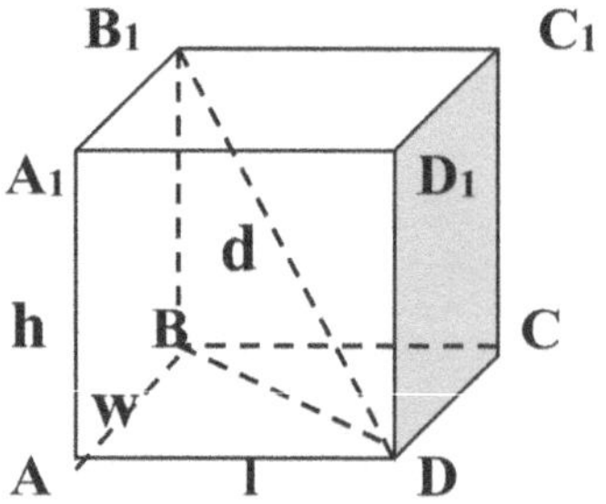

$V = lwh = 10 \cdot 6 \cdot 8 = 60 \cdot 8 = 480\ in^3$

$SA = 2lw + 2lh + 2wh = 2 \cdot 10 \cdot 6 + 2 \cdot 10 \cdot 8 + 2 \cdot 6 \cdot 8 =$

$20 \cdot 6 + 20 \cdot 8 + 12 \cdot 8 = 120 + 160 + 96 = 376\ in^2$

$d^2 = l^2 + w^2 + h^2 = 10^2 + 6^2 + 8^2 = 100 + 36 + 64 = 100 + 100 = 200$

$d = \sqrt{200} = \sqrt{100 \cdot 2} = 10\sqrt{2}\ in$

Summary 1 Chapter 10 Lesson 4

Table 23 Solid Geometry 2

Sphere: The volume of a sphere is equal to the product four over three of the number π by cube of radius. An example of a sphere is any balls.

Volume of sphere is $V = \frac{4}{3}\pi r^3$ where $\pi \approx \frac{22}{7}$ or $\pi \approx 3.14$

Surface Area of Sphere $SA = 4\pi r^2$ OB = r

Example 3 Figure 3

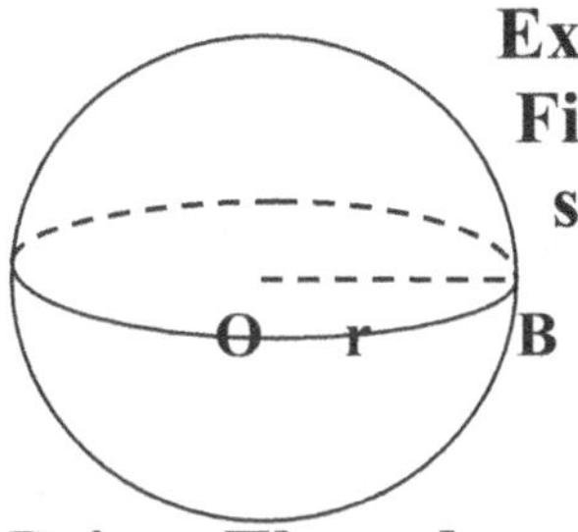

Find exact volume and exact surface area of a sphere if the radius is 3 m.

$$V = \frac{4}{3}\pi r^3 = \frac{4}{3}\pi \times 3^3 = \frac{4}{3} \times \frac{27}{1}\pi = 4 \times 9\pi = 36\pi \text{ m}^3$$

$$SA = 4\pi r^2 = 4\pi \times 3^2 = 4 \times 9\pi = 36\pi \text{ m}^2$$

Regular Prism: The volume of a regular prism is equal to the product of the base and height: V = Base · h Base means area of ABC or $A_1B_1C_1$

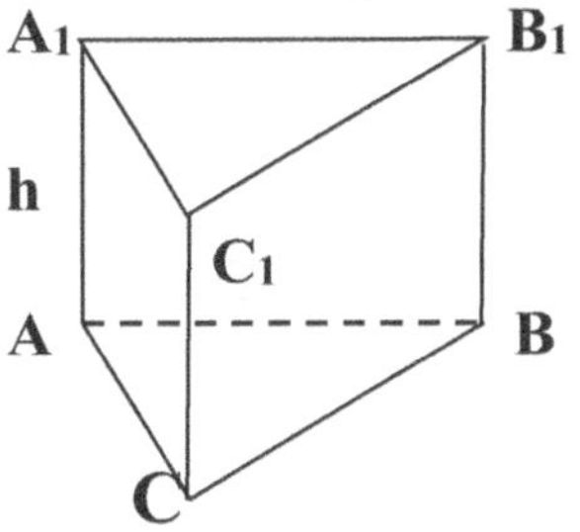

Figure 4

Example 4 Find the volume of a regular prism if the base is 25 m^2 and the height is 10 m. Base=Area of triangle ABC = 25 m^2 h = 10 m

$V = B_{base} \cdot h = 25 \cdot 10 = 250 \text{ m}^3$

Right Circular Cylinder: The volume of a right circular cylinder is equal to the product of the number π, and the square of radius by height:

$V = \pi r^2 h$

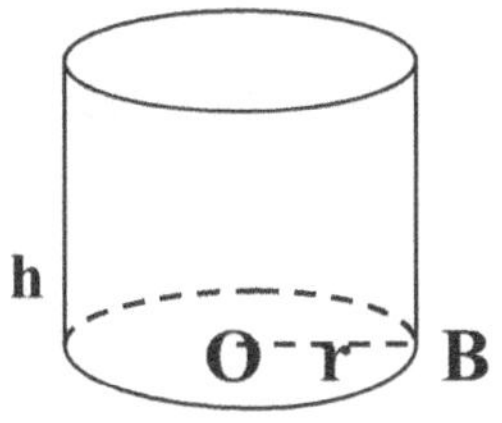

Figure 5

Example 5 Find volume of a right circular cylinder if r = 7 in h = 10 in and $\pi \approx \frac{22}{7}$, $V = \pi r^2 h = \frac{22}{7} \cdot 7^2 \cdot 10 = \frac{22}{7} \cdot \frac{49}{1} \cdot 10 = 22 \cdot 7 \cdot 10 = 1540 \text{ in}^3$

Summary 1 Chapter 10 Lesson 4

Table 24

Solid Geometry 3

Regular Pyramid

The volume of a regular pyramid is equal to the product of one-third of base by height: $V = \frac{1}{3}$ Base · h   Base means area of triangles ABC

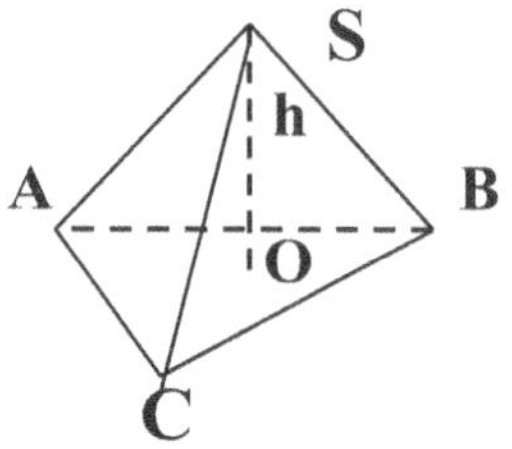

Figure 6

Example 6 Find the volume of a regular pyramid if the base is 120 m^2 and height is 5 m. The base is the area of the triangle ABC = 120 m^2

h = 5 m $V = \frac{1}{3}$ Base · h = $\frac{1}{3} \cdot 120 \cdot 5 = 40 \cdot 5 = 200\ m^3$

Right Circular Cone

The volume of a right circular cone is equal to the product of one-third of square radius by height: $V = \frac{1}{3}\pi r^2 h$ $\pi \approx 3.14$ or $\pi \approx \frac{22}{7}$

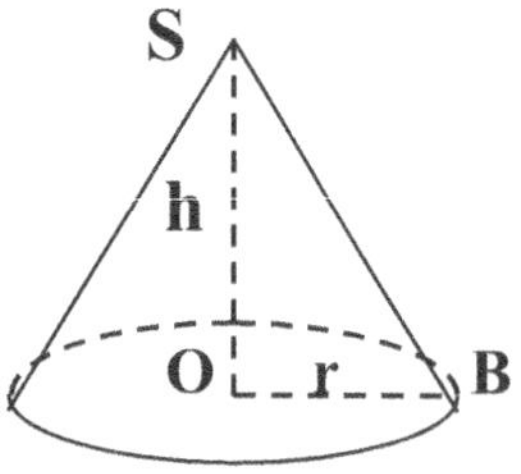

Figure 7

Example 7 Find the volume of a right circular cone if its height is 10 ft and radius is 3 ft: h = 10 ft   r = 3 ft

$V = \frac{1}{3}\pi r^2 h = \frac{1}{3}\pi \cdot 3^2 \cdot 10 = \frac{1}{3} \cdot 9 \cdot 10\pi = 3 \cdot 10\pi = 30\pi = 30\,(3.14) = 94.2\ ft^3$

Memorize Formulas of Geometry

Cube 1) V =   2) SA =   3) d =   Sphere 4) V =   5) SA =
Rectangular Solid: 6) V =   7) SA =   8) d^2 =
Regular Prism 9) V =   Regular Pyramid 10) V =

Right Cylinder 11) V =   Right Cone 12) V =

Summary 1 Chapter 10 Lesson 5

Table 25 Word Problems

1) Decimals

a) To find the amount by known base, multiply the base by a decimal: Find the amount 0.6 of 40 $40 \times 0.6 = 24$

b) To find the base by known amount, divide the amount by a decimal: Finding the base:0.6 is equivalent \$ 24, $24 \div 0.6 = 40$

c) To find the ratio between two numbers, divide the numbers.
Find the ratio 24 from 40 $24 \div 40 = 0.6$

2) Fractions

a) To find the amount by known base, multiply the base by a fraction. Finding the amount: $\frac{3}{5}$ of 40 $40 \times \frac{3}{5} = \$24$

b) To find the base by known amount, divide amount by a fraction. Finding the base: $\frac{3}{5}$ equivalent 24 $24 \div \frac{3}{5} = \$40$

c) To find the ratio between two numbers, divide the numbers.
Find the ratio: 24 from \$40 $24 \div 40 = \frac{24}{40} = \frac{3}{5}$

3) Uniform Motion: Distance – d, Time – t, Rate – r (Speed)

a) To find distance, multiply rate by time: $d = rt$
What distance did a man go for 2 hours if his rate was 5 mi/hr.? $d = 5 \cdot 2 = 10$ mi

b) To find time, divide distance by speed. $t = d \div r$
How much time does a man needs if he runs 8 mi with a rate 4 mi/hr.? $t = 8 \div 4 = 2$ hrs.

c) To find speed, divide distance by time. $r = d \div t$
What is the rate a girl has, if she runs 6 mi for 2 hours?
$r = 6 \div 2 = 3$ mi / hr.

4) Directly proportional quantities

If 4 apples sell for \$1.32. How much do 11 apples cost?
$132 \div 4 = 33$ c $33 \times 11 = 363$ c = \$3.63

Summary 1 Chapter 10 Lesson 5

Table 26 Percent Problems

Solve Percent Problems by Equations ("is" and "of")

1) Find A: What numbers is 25% of 40? $A = 0.25 \cdot 40$ $\underline{A = 10}$

2) Find B: 10 is 25% of what number?

$10 = \frac{25}{100}B$ $\frac{10}{0.25} = \frac{0.25B}{0.25}$ $\frac{1000}{25} = B$ $\underline{B = 40}$

3) Find R: What percent of 40 is 10?

$\frac{R}{100} \times 40 = 10$ $100 \times \frac{R}{100} \times 40 = 10 \times 100$ $40R = 1000$ $\underline{R = 25\%}$

Solve Percent Problems by Proportion

1)Find A: Michael solves 80% of 20 problems in test. How many problems he solved? $\frac{A}{B} = \frac{R}{100}$ $\frac{A}{20} = \frac{80}{100}$ $\frac{A}{20} = \frac{8}{10}$ $10A = 160$, $A = \underline{16}$ **<u>problems</u>**

2) Find B:David solved 16 problems in test, it is equal 80% of all test. How many problems were in test?

$\frac{A}{B} = \frac{R}{100}$ $\frac{16}{B} = \frac{80}{100}$ $\frac{16}{B} = \frac{8}{10}$ $8B = 160$ $\underline{B = 20 \text{ problems}}$

3) Find R: What percent of 20 is 16?

$\frac{A}{B} = \frac{R}{100}$ $\frac{16}{20} = \frac{R}{100}$ $\frac{4}{5} = \frac{R}{100}$ $5R = 400$ $\underline{R = 80\%}$

Solve Percent Problems by Formulas

1) Find A: Olivia had $ 30, she spent 70% of all her money. How much money she spent? To find the amount, multiply the base by the percent, which converted to decimal.

Formula 1 $A = B \cdot \frac{R}{100}$ $A = 30 \cdot \frac{70}{100}$ $A = 30 \cdot 0.7$ $A = \underline{\$\ 21}$

2) Find B: Anna spent $ 21, it is 70% of all her money. How much money she had? To find the base, divide the amount by the percent, which converted to decimal.

Formula 2 $B = A \div \frac{R}{100}$ $B = 21 \div \frac{70}{100}$ $B = 21 \div 0.7 = 210 \div 7 = \underline{\$\ 30}$

3) Find R: Vicki had $ 30, she spent $ 21.What percent of money she spent? To find a percent, divide the numbers and multiply the result by 100%.

Formula 3 $R = \frac{A}{B} \cdot 100\%$ $R = \frac{21}{30} \cdot 100\%$ $R = \frac{7}{10} \times 100\%$ $R = \underline{70\%}$ **To solve the percent problems, choose one of three ways which is easy for you.**

Summary 1 Chapter 10 Lesson 5

Table 27

Uniform Motion 1

The motion of a moving object whose speed and direction do not change is called uniform motion.
Word problem at moving content three quantities
Distance - d Time - t Rate – r (Speed)
Rate – Distance passed for time unit (for 1 hour, 1 minute)
Movement units
Distance (d) measure mi, m, km
Time (t) measure in min, hrs.
Rate (r) measure in mi/hrs., m/min, km/hrs.

Distance d
To find distance, multiply rate by time: $d = rt$
What distance did a man runs in 2 hours, if his rate was 5 mi/hr.?
$d = 5 \times 2 = 10$ mi
If men drive at the speed of 70mph for 2 hours, how far will he travel?
$d = 70 \times 2 = 140$ mi

Time t
To find time, divide distance by speed: $t = d \div r$
How much time needed if a man went 8 mi with a rate of 4 mi/hrs.?
$t = 8 \div 4 = 2$ hrs.
If a man drives at a speed of 52 mph for 91 mi, how long will he travel?
$t = d \div r = 91$ mi $\div$ 52 mph $= 7 \div 4 = 1.75$ hours

Rate (Speed) r
To find the speed, divide distance by time: $r = d \div t$
How much rate does a girl have, if she runs 6 mi in 2 hours.
$r = 6 \div 2 = 3$ mi / hr.

Woman drove a distance of 252 mi in $4\frac{2}{3}$ hours. What was her speed?

$r = d \div t = 252 \div 4\frac{2}{3} = \frac{252}{1} \times \frac{3}{14} = 18 \times 3 = 54$ mph

Summary 1 Chapter 10 Lesson 5

Table 28
Uniform Motion 2
Find the Unknowns of Uniform Motion

#		Distance	Time	Speed	Answers
1	Pedestrian	X mi	4 hrs.	5 mi/hr.	20 mi
2	Tractor	X mi	2 hrs.	30 mi/hr.	60 mi
3	Train	X mi	3 hrs.	120 mi/hr.	360 mi
4	Plane	X mi	2 hrs.	1000 mi/hr.	2000 mi
5	Skier	26 mi	X hrs.	13 mi/hr.	2 hrs.
6	Train	240 mi	X hrs.	60 mi/hr.	4 hrs.
7	Car	160 mi	X hrs.	80 mi/hr.	2 hrs.
8	Plane	5000 mi	X hrs.	2500 mi/hr.	2 hrs.
9	Bicycle	28 mi	2 hrs.	X mi/hr.	14 mi/hr.
10	Car	210 mi	3 hrs.	X mi/hr.	70 mi/hr.
11	Plane	15000 mi	3 hrs.	X mi/hr.	5000 mi/hr.
12	Eagle	X mi	1 hr.	120 mi/hr.	120 mi
13	Starling	X mi	2 hrs.	80 mi/hr.	160 mi
14	Swallow	X mi	3 hrs.	70 mi/hr.	210 mi
15	Deer	X mi	3 hrs.	10 mi/hr.	30 mi
16	Shark	72 mi	X hrs.	36 mi/hr.	2 hrs.
17	Ship	280 mi	4 hrs.	X mi/hr.	70 mi/hr.
18	Sparrow	120 mi	2 hrs.	X mi/hr.	60 mi/hr.
19	Zebra	45 mi	X hrs.	15 mi/hr.	3 hrs.
20	Boat	X mi	4 hrs.	24 mi/hr.	96 mi
21	Car	The same	2 hrs.	60 mi/hr.	120 mi
22	Car	The same	3 hrs.	X mi/hr.	40 mi/hr.
23		X mi	2 hrs.	60 mi/hr.	120 mi
24		120 mi	X hrs.	60 mi/hr.	2 hrs.
25		120 mi	2 hrs.	X mi/hr.	60 mi/hr.
26		X mi	2 hrs.	40 mi/hr.	80 mi
27		24 mi	X hrs.	6 mi/hr.	4 hrs.
28		300 mi	5 hrs.	X mi/hr.	60 mi/hr.
29		70 m	7 secs.	The same	10 m/sec.
30		90 m	X secs.	The same	9 secs.

Summary 1 Chapter 10 Lesson 5
Table 29 Application of Percent 1

1. **Sales Tax Formula**

Amount of sales tax = Cost of merchandise × Tax rate
Example: A graphing calculator cost \$ 100. The sales tax rate is 5.5%.
Compute the total cost. 5.5% = 0.55, 100×0.055 =\$5.5, 100+5.5=\$105.5

2. **Percent of commission**

Amount of commission = Dollar sales × Percent of commission
Example: Diana earned 5% on a house she sold for \$ 100,000
Find her dollar commission. 100000× 0.05 = \$ 5000

3. **Percent of decrease**

Amount of decrease = original value - new value
Amount of decrease = original value × percent of decrease
Percent of decrease = amount of decrease ÷ original value ×100%
Example: A paper shredder was decreased price from \$ 80 to \$ 60
What is the percent decreased in price?
80 – 60 = 20 20 ÷ 80 × 100% = 0.25 × 100% = 25%

4. **Percent of discount**

Amount of decrease = original value - new value
Amount of decrease = original value × percent of decrease
Percent of decrease = amount of decrease ÷ original value ×100%
Example: The Company put suits that usually sell for \$ 250 on sale for \$ 170. What the percent discount?
250 – 170 = \$ 80 80 ÷ 250 × 100% = 0.32 × 100% = 32%

5. **Percent of increase**

Amount of increase = new value – original value
Amount of increase = original value × percent of increase
Percent of increase = amount of increase ÷ original value × 100%
Example: The sales in one-year Honda Civic increased from 400,000 to 408,000. Compute the percent increase.
408000 – 400000=8000 8000÷400000 ×100 = 0.2×100=20%

6. **Percent markup**

Amount of markup = new value – original value
Amount of markup = original value × percent of markup (rate)
Percent of markup = amount of markup ÷ original value × 100%
Example: A college bookstore marks up of math book is 30%.
What is mark up for math book that has a price of \$ 70.
\$70 × 0.3 = \$ 21 70 + 21 = \$ 91

Summary 1 Chapter 10 Lesson 5

Table 30 Application of Percent 2

7. **Percent of increase**

Amount of increase = new value – original value
Amount of increase = original value × percent of increase
Percent of increase = amount of increase ÷ original value × 100%
Example: The sales in one-year Honda Civic increased from 400,000 to 408,000. Compute the percent increase.
408000 – 400000=8000 $\qquad$ 8000÷400000 ×100 = 0.2×100=20%

8. **Percent markup**

Amount of markup = new value – original value
Amount of markup = original value × percent of markup (rate)
Percent of markup = amount of markup ÷ original value × 100%
Example: A college bookstore marks up of math book is 30%. What is mark up for math book that has a price of $ 70.
$70 × 0.3 = $ 21 $\qquad$ 70 + 21 = $ 91

9. **Percent of Profit**

Amount profit = Total Amount income - Amount Expenses
Amount profit = Total Amount income × Percent Profit
Percent of profit = Amount expenses ÷ Total Amount income ×100%
Example: The company had expenses of 148,000 and a profit of $ 12,000. Find the percent profit.
Total amount income = $ 148,000 + $ 12, 000= $ 160, 000

$$\frac{12000}{160000}=\frac{n}{100} \quad \frac{12}{160}=\frac{n}{100} \quad 160n=1200 \quad 16n=120 \quad 4n=30 \quad n=7.5\%$$

10)Amount of expenses = total amount income × percent of expenses
Example: The company had a total income of $ 150,000 and 7,500 profit last month. Amount expenses $ 150000 – 7,500 = 142,500

$$\frac{142500}{150000}=\frac{n}{100} \quad \frac{1425}{1500}=\frac{n}{100} \quad 1500n=142500 \quad 15n=1425 \quad n=95\%$$

Summary 1 Chapter 10 Lesson 6
Mental Math 1
Add

1) 9 + 1
2) 8 + 2
3) 7 + 3
4) 6 + 4
5) 5 + 5
6) 9 + 2
7) 8 + 3
8) 7 + 4
9) 6 + 5
10) 9 + 3
11) 8 + 4
12) 7 + 5
13) 6 + 6
14) 9 + 4
15) 8 + 5
16) 7 + 6
17) 9 + 5
18) 8 + 6
19) 7 + 7
20) 9 + 6
21) 8 + 7
22) 9 + 7
23) 8 + 8
24) 9 + 8
25) 9 + 9

Summary 1 Chapter 10 Lesson 6
Mental Math 2
Subtract

1) 10 – 1
2) 10 – 2
3) 10 – 3
4) 10 – 4
5) 10 – 5
6) 11 - 2
7) 11 – 3
8) 11 – 4
9) 11 – 5
10) 12 – 3
11) 12 – 4
12) 12 – 5
13) 12 – 6
14) 13 – 4
15) 13 – 5
16) 13 – 6
17) 14 – 5
18) 14 – 6
19) 14 – 7
20) 15 – 6
21) 15 – 7
22) 16 – 7
23) 16 – 8
24) 17 – 8
25) 18 - 9

Summary 1 Chapter 10 Lesson 6
Mental Math 3
Multiply

1) 5×6
2) 6×6
3) 7×6
4) 8×6
5) 9×6
6) 5×7
7) 6×7
8) 7×7
9) 8×7
10) 9×7
11) 5×8
12) 6×8
13) 7×8
14) 8×8
15) 9×8
16) 5×9
17) 6×9
18) 7×9
19) 8×9
20) 9×9
21) 5×10
22) 6×10
23) 7×10
24) 8×10
25) 9×10

Summary 1 Chapter 10 Lesson 6
Mental Math 4
Divide

1) $30 \div 6$
2) $36 \div 6$
3) $42 \div 6$
4) $48 \div 6$
5) $54 \div 6$
6) $35 \div 7$
7) $42 \div 7$
8) $49 \div 7$
9) $56 \div 7$
10) $63 \div 7$
11) $40 \div 8$
12) $48 \div 8$
13) $56 \div 8$
14) $64 \div 8$
15) $72 \div 8$
16) $45 \div 9$
17) $54 \div 9$
18) $63 \div 9$
19) $72 \div 9$
20) $81 \div 9$
21) $50 \div 10$
22) $60 \div 10$
23) $70 \div 10$
24) $80 \div 10$
25) $90 \div 10$

Summary 1 Chapter 10 Lesson 6
Mental Math 5 Exponents 1

1)	0^2	11)	10^2
2)	1^2	12)	20^2
3)	2^2	13)	30^2
4)	3^2	14)	40^2
5)	4^2	15)	50^2
6)	5^2	16)	60^2
7)	6^2	17)	70^2
8)	7^2	18)	80^2
9)	8^2	19)	90^2
10)	9^2	20)	100^2

Summary 1 Chapter 10 Lesson 6
Mental Math 6
Laws for Whole Numbers 1

1) $8\times3 + 2\times3$
2) $7\times5 - 2\times5$
3) $7\times4 + 3\times4$
4) $6\times2 - 4\times2$
5) $4\times5 + 6\times5$
6) $9\times4 - 8\times4$
7) $5\times2 + 5\times8$
8) $3\times7 - 3\times6$
9) $3\times4 + 3\times6$
10) $5\times9 - 5\times8$

11) $8\times75 + 8\times25$
12) $25\times6 - 15\times6$
13) $2\times17 + 2\times13$
14) $75\times4 - 35\times4$
15) $3\times27 + 3\times23$
16) $93\times 2 - 2\times53$
17) $14\times4 + 4\times16$
18) $6\times28 - 6\times18$
19) $29\times3 + 21\times3$
20) $67\times2 - 2\times47$

Summary 1 Chapter 10 Lesson 7
Power of Whole Numbers.

The four arithmetic operations with numbers are: Addition, Subtraction, Multiplication, Division $3 + 2 = 5$ $5 - 2 = 3$ $3 \times 2 = 6$ $6 \div 2 = 3$
The fifth operation with numbers is power of numbers.
A power of a number indicates repeated multiplication.
Base is a number that has to be multiplied 3
Exponent is a number that shows how many times to multiply the base 2
Power is result of multiplication of the same numbers 9

$3^2 = 3 \times 3 = 9$ **$\text{Base}^{\text{Exponent}} = \text{Power}$**
Exponent 2 called squared **$4^2 = 4 \times 4 = 16$**
Exponent 3 called cubed **$2^3 = 2\cdot2\cdot2 = 4\cdot2 = 8$**
Special powers $0^{10} = 0$ $0^n = 0$ $1^{10} = 1$ $1^n = 1$ $a^1 = a$
Memorize perfect square numbers: 1, 4, 9, 16, 25, 36, 49, 64, 81, 100
$10^2 + 11^2 + 12^2 = 13^2 + 14^2$ $100 + 121 + 144 = 169 + 196$ $365 = 365$
$3^4 = 3\times3\times3\times3 = 81$ $3^5 = 81 \times 3 = 243$ $3^6 = 243 \times 3 = 729$
$4^4 = 4\times4\times4\times4 = 256$ $4^5 = 256 \times 4 = 1024$ $5^4 = 125 \times 5 = 625$

Mental Math 7 Exponents 2

1) 1^3
2) 2^3
3) 3^3
4) 4^3
5) 5^3
6) 6^3
7) 7^3
8) 8^3
9) 9^3
10) 10^3
11) 2^1
12) 2^2
13) 2^3
14) 2^4
15) 2^5
16) 2^6
17) 2^7
18) 2^8
19) 2^9
20) 2^{10}

Summary 1 Chapter 10 Lesson 7
Prime Factorization

Factor × Factor = Product $3 \times 2 = 6$

A factor of a number n is a nonzero whole number that divides evenly into n. All divisors of 12 are 1, 2, 3, 4, 6, 12, $1\times12 = 2\times6 = 3\times4 = 12$

All divisors of 24 are 1, 2, 3, 4, 6, 8, 12, 24, $1\times24=2\times12=3\times8=4\times6=24$

Prime Numbers: A natural number called a prime number if it has only two different divisors: one and the number itself 2, 3, 5, 7, 11, 13, 17, 19 …

Composite Numbers: A natural number that has more than two divisors called a composite number 4, 6, 8, 9, 10, 12, 14, 15, 16…

The number 1 has only one divisor: which is why 1 is not a prime or a composite. Prime numbers cannot factorize.

Remember the first 10 prime numbers: 2, 3, 5, 7, 11, 13, 17, 19, 23, 29

To factorize composite numbers to prime factors means to represent composite numbers in the product of prime factors:

$8 = 4 \times 2 = 2 \times 2 \times 2 = 2^3$

Prime Factorization

Example 1: Find prime factorization for 100

Try to divide by prime numbers: by 2, by 3, by 5, by 7…

$100 \div 2$ $50 \div 2$ $25 \div 5$ $5 \div 5$ $100 = 2\times2\times5\times5 = 2^2\cdot5^2$

Example 2: Represent your numbers like the product of just prime numbers $50 = 2 \times 25 = 2 \times 5 \times 5 = 2\times 5^2$

Mental Math 8
Find Prime Factorization

1) 4	6) 49	11) 10	16) 20
2) 9	7) 64	12) 12	17) 24
3) 16	8) 81	13) 14	18) 27
4) 25	9) 6	14) 15	19) 30
5) 36	10) 8	15) 18	20) 32

Summary 1 Chapter 10 Lesson 7
Mental Math 9
Decimals

1) $0.2 + 0$
2) $0 + 1.5$
3) $0.2 + 0.5$
4) $0.5 + 0.2$
5) $1.8 + 0.2$
6) $0.6 - 0$
7) $1.7 - 1.7$
8) $0.5 - 0.2$
9) 0.6×0
10) 0×1.7
11) 1×1.9
12) 0.8×1
13) 0.2×5
14) 0.4×5
15) 5×0.6
16) 0.5×8
17) 1.2×5
18) 5×1.4
19) $0 \div 0.6$
20) $0.7 \div 0$
21) $0.8 \div 1$
22) $1.9 \div 1.9$
23) 0.2^2
24) 0.3^2
25) 0.4^2
26) 0.5^2
27) 0.6^2
28) 0.7^2
29) 0.8^2
30) 0.9^2

Summary 1 Chapter 10 Lesson 7
Mental Math 10
Laws for Decimals 2

1) $8\times7.5+8\times2.5$
2) $3.7\times4+4\times1.3$
3) $2.5\times4-1.5\times4$
4) $5.4\times5-3.4\times5$
5) $7\times3.7+6.3\times7$
6) $4.9\times2+2\times5.1$
7) $7.5\times6-3.5\times6$
8) $3.9\times8-8\times1.9$
9) $3\times2.7+3\times2.3$
10) $2.4\times3+1.6\times3$

11) $9.3 \times 2 - 2 \times 5.3$
12) $5.8 \times 3 + 4.2 \times 3$
13) $4.3 \times 4 + 4 \times 1.7$
14) $9.6 \times 2 - 2.6 \times 2$
15) $5 \times 2.8 - 5 \times 1.8$
16) $3.8 \times 4 - 1.8 \times 4$
17) $2.9 \times 3 + 3.1 \times 3$
18) $0.3 \times 6 + 0.7 \times 6$
19) $6.7 \times 4 - 4 \times 4.7$
20) $8 \times 5.7 - 8 \times 4.7$

Summary 1 Chapter 10 Lesson 7
Mental Math 11
Fractions

1) $\frac{1}{2}\times\frac{2}{5}$

2) $\frac{1}{3}\times\frac{3}{4}$

3) $\frac{5}{8}\times 8$

4) $4\times\frac{1}{4}$

5) $\frac{1}{2}\div\frac{1}{2}$

6) $\frac{1}{3}\div\frac{1}{6}$

7) $\frac{1}{4}\div\frac{1}{12}$

8) $2\div\frac{1}{2}$

9) $(\frac{1}{2})^2$

10) $(\frac{1}{5})^2$

11) $(\frac{2}{5})^2$

12) $(\frac{3}{5})^2$

13) $\frac{1}{4}+\frac{3}{4}$

14) $\frac{1}{2}+\frac{3}{2}$

15) $\frac{7}{8}-\frac{5}{8}$

16) $\frac{7}{8}-\frac{3}{8}$

17) $\frac{1}{2}+\frac{1}{4}$

18) $\frac{3}{4}+\frac{1}{2}$

19) $\frac{1}{2}-\frac{1}{4}$

20) $\frac{3}{4}-\frac{1}{2}$

Summary 1 Chapter 10 Lesson 7
Mental Math 12
Mixed numbers

1) $1+\frac{1}{2}$

2) $\frac{1}{4}+2$

3) $1-\frac{1}{2}$

4) $2\frac{1}{2}-1$

5) $2\frac{3}{8}+1\frac{1}{8}$

6) $3\frac{3}{4}+2\frac{1}{4}$

7) $4\frac{3}{4}-2\frac{1}{4}$

8) $3\frac{5}{8}-2\frac{1}{8}$

9) $3\frac{1}{2}+2\frac{1}{2}$

10) $4\frac{1}{2}+2\frac{1}{2}$

11) $2\frac{3}{4}-1\frac{1}{4}$

12) $4\frac{5}{6}-1\frac{2}{6}$

13) $2\frac{3}{8}-1\frac{1}{8}$

14) $3\frac{3}{8}-2\frac{1}{8}$

15) $4\frac{7}{8}-3\frac{3}{8}$

16) $5\frac{5}{8}-4\frac{1}{8}$

17) $2-1\frac{1}{2}$

18) $2-1\frac{3}{4}$

19) $3-1\frac{1}{4}$

20) $3-1\frac{1}{5}$

Summary 1 Chapter 10 Lesson 8
Mental Math 13
Laws for Fractions 3

1) $\frac{1}{2}\times 8+\frac{1}{2}\times 2$

2) $\frac{4}{3}\times 6-\frac{1}{3}\times 6$

3) $\frac{1}{4}\times 8+\frac{3}{4}\times 8$

4) $\frac{6}{5}\times 5-\frac{1}{5}\times 5$

5) $\frac{1}{3}\times 9+\frac{2}{3}\times 9$

6) $\frac{3}{2}\times 8-\frac{1}{2}\times 8$

7) $\frac{1}{3}\times 6+\frac{2}{3}\times 6$

8) $\frac{5}{4}\times 4-\frac{1}{4}\times 4$

9) $\frac{1}{5}\times 5+\frac{4}{5}\times 5$

10) $\frac{7}{6}\times 6-\frac{1}{6}\times 6$

11) $\frac{1}{6}\times 6+\frac{5}{6}\times 6$

12) $\frac{5}{2}\times 4-\frac{1}{2}\times 4$

13) $\frac{3}{7}\times 7+\frac{4}{7}\times 7$

14) $\frac{7}{3}\times 3-\frac{1}{3}\times 3$

15) $\frac{3}{8}\times 8+\frac{5}{8}\times 8$

16) $\frac{5}{4}\times 8-\frac{1}{4}\times 8$

17) $\frac{4}{9}\times 9+\frac{5}{9}\times 9$

18) $\frac{7}{5}\times 5-\frac{2}{5}\times 5$

19) $\frac{3}{5}\times 5+\frac{2}{5}\times 5$

20) $\frac{7}{6}\times 6-\frac{1}{6}\times 6$

Summary 1 Chapter 10 Lesson 8
Mental Math 14
Write the Fractions as Decimals

1) $\frac{3}{10}$

2) $\frac{9}{100}$

3) $\frac{7}{1000}$

4) $\frac{1}{2}$

5) $\frac{12}{24}$

6) $\frac{1}{5}$

7) $\frac{2}{5}$

8) $\frac{3}{5}$

9) $\frac{4}{5}$

10) $\frac{12}{15}$

11) $\frac{1}{4}$

12) $\frac{2}{4}$

13) $\frac{3}{4}$

14) $\frac{3}{12}$

15) $\frac{15}{20}$

16) $\frac{7}{25}$

17) $\frac{7}{50}$

18) $\frac{11}{50}$

19) $\frac{3}{20}$

20) $\frac{9}{20}$

Summary 1 Chapter 10 Lesson 8
Mental Math 15
Negative Fractions

1) $12(-\frac{5}{6})$

2) $-\frac{5}{9}\times 18$

3) $24(-\frac{3}{4})$

4) $-\frac{5}{8}\times 32$

5) $\frac{1}{2}\div(-\frac{1}{2})$

6) $\frac{1}{3}\div(-\frac{1}{6})$

7) $-\frac{1}{4}\div\frac{1}{12}$

8) $-2\div\frac{1}{2}$

9) $(-\frac{1}{2})^2$

10) $(-\frac{1}{5})^2$

11) $(-\frac{2}{5})^2$

12) $-(-\frac{3}{5})^2$

13) $-\frac{1}{4}-\frac{3}{4}$

14) $-\frac{1}{2}-\frac{3}{2}$

15) $-\frac{7}{8}+\frac{5}{8}$

16) $-\frac{7}{8}+\frac{3}{8}$

17) $-\frac{1}{2}-\frac{1}{4}$

18) $-\frac{3}{4}-\frac{1}{2}$

19) $-\frac{1}{2}+\frac{1}{4}$

20) $-\frac{3}{4}+\frac{1}{2}$

Summary 1 Chapter 10 Lesson 8
Mental Math 16
Negative Mixed Numbers

1) $-1-\frac{1}{2}$

2) $-\frac{1}{4}-2$

3) $-1+\frac{1}{2}$

4) $-2\frac{1}{2}+1$

5) $-2\frac{3}{8}-1\frac{1}{8}$

6) $-3\frac{3}{4}-2\frac{1}{4}$

7) $-4\frac{3}{4}+2\frac{1}{4}$

8) $-3\frac{5}{8}+2\frac{1}{8}$

9) $-3\frac{1}{2}-2\frac{1}{2}$

10) $-4\frac{1}{2}-2\frac{1}{2}$

11) $-2\frac{3}{4}+1\frac{1}{4}$

12) $-4\frac{5}{6}+1\frac{2}{6}$

13) $-2\frac{3}{8}+1\frac{1}{8}$

14) $-3\frac{3}{8}+2\frac{1}{8}$

15) $-4\frac{7}{8}+3\frac{3}{8}$

16) $-5\frac{5}{8}+4\frac{1}{8}$

17) $-2+1\frac{1}{2}$

18) $-2+1\frac{3}{4}$

19) $-3+1\frac{1}{4}$

20) $-3+1\frac{1}{5}$

Summary 1 Chapter 10 Lesson 9
Mental Math 17
Exponents 3

1) 11^2
2) 12^2

3) 13^2
4) 14^2

5) 15^2
6) 16^2

7) 17^2
8) 18^2

9) 19^2
10) 20^2

11) 21^2
12) 22^2

13) 23^2
14) 24^2

15) 25^2
16) 26^2

17) 27^2
18) 28^2

19) 29^2
20) 30^2

Summary 1 Chapter 10 Lesson 9
Mental Math 18
Square Roots

1) $\sqrt{64}$
2) $\sqrt{81}$
3) $\sqrt{100}$
4) $\sqrt{121}$
5) $\sqrt{144}$
6) $\sqrt{169}$
7) $\sqrt{196}$
8) $\sqrt{225}$
9) $\sqrt{256}$
10) $\sqrt{289}$
11) $\sqrt{324}$
12) $\sqrt{361}$
13) $\sqrt{400}$
14) $\sqrt{441}$
15) $\sqrt{484}$
16) $\sqrt{529}$
17) $\sqrt{576}$
18) $\sqrt{625}$
19) $\sqrt{676}$
20) $\sqrt{729}$
21) $\sqrt{784}$
22) $\sqrt{841}$
23) $\sqrt{900}$
24) $\sqrt{961}$
25) $\sqrt{1600}$
26) $\sqrt{2500}$
27) $\sqrt{3600}$
28) $\sqrt{4900}$
29) $\sqrt{6400}$
30) $\sqrt{8100}$

Summary 1 Chapter 10 Lesson 9
Mental Math 19
Solve Equations

1) $x + 6 = 10$
2) $7 + x = 10$

3) $x - 9 = 5$
4) $x - 8 = 9$

5) $x - 7 = 8$
6) $12 - y = 3$

7) $13 - y = 8$
8) $14 - y = 5$

9) $15 - y = 7$
10) $16 - y = 9$

11) $8 \cdot x = 72$
12) $x \cdot 9 = 63$

13) $x \div 8 = 7$
14) $x \div 7 = 6$

15) $x \div 6 = 9$
16) $72 \div z = 9$

17) $72 \div z = 8$
18) $63 \div z = 9$

19) $56 \div z = 7$
20) $54 \div z = 9$

Summary 1 Chapter 10 Lesson 9
Mental Math 20
Proportions

1) $\frac{1}{3}=\frac{x}{9}$ 2) $\frac{4}{3}=\frac{20}{x}$

3) $\frac{9}{x}=\frac{3}{2}$ 4) $\frac{3}{2}=\frac{15}{x}$

5) $\frac{3}{2}=\frac{x}{6}$ 6) $\frac{12}{x}=\frac{3}{4}$

7) $\frac{x}{5}=\frac{18}{6}$ 8) $\frac{x}{4}=\frac{6}{2}$

9) $\frac{9}{3}=\frac{6}{x}$ 10) $\frac{44}{x}=\frac{8}{2}$

11) $y \div 6 = 4 \div 3$ 12) $z \div 8 = 2 \div 4$

13) $y \div 3 = 6 \div 2$ 14) $2 \div 8 = z \div 4$

15) $3 \div 6 = 2 \div y$ 16) $8 \div 2 = z \div 4$

17) $y \div 3 = 12 \div 2$ 18) $z \div 3 = 8 \div 4$

19) $y \div 4 = 4 \div 2$ 20) $z \div 3 = 4 \div 2$

Summary 1 Chapter 10 Lesson 10
Reviews for Final Exam
Review 10 A

Date | **Name**
Read directions carefully | **Answers must be in the circle.**
Show all steps for full credits. | **Do only 1 operation per 1 step.**
Each problem is worth 5 points. | **Solve 1 problem 5 minutes.**

Put - C before parentheses and change each sign inside parentheses
$-AC - BC = -C(A + B)$
$-7.5 \times 5 - 2.5\times5 = -5(7.5 + 2.5) = -5(10) = -50$

Simplify by laws for Negative Numbers

1) $-3.7 \times 4 - 1.3 \times 4$

2) $-4.9 \times 2 - 5.1 \times 2$

3) $-2.4 \times 3 - 1.6 \times 3$

4) $-9.6 \times 2 - 0.4 \times 2$

5) $-0.3 \times 6 - 0.7 \times 6$

6) $-0.7 \times 2 - 0.3\times 2$

7) $-0.3 \times 8 - 0.3\times 2$

8) $-3.6 \times 3 - 3.6 \times 7$

9) $-4.7\times 9 - 5.3 \times 9$

10) $-2.4 \times 7 - 7.6\times 7$

Summary 1 Chapter 10 Lesson 10
Review 10 B

Date **Name**

Read directions carefully **Answers must be in the circle.**

Show all steps for full credits. **Do only 1 operation per 1 step.**

Each problem is worth 5 points. **Solve 1 problem 5 minutes.**

Put - C before parentheses and change each sign inside parentheses

$-AC + BC = -C(A - B)$

$-12.5(5) + 2.5(5) = -5(12.5 - 2.5) = -5(10) = -50$

Simplify by Laws for Negative Numbers

1) $-5.4 \times 5 + 3.4 \times 5$

2) $-3.9 \times 6 + 1.9 \times 6$

3) $-5.8 \times 3 + 1.8 \times 3$

4) $-3.8 \times 4 + 1.8 \times 4$

5) $-5.7 \times 8 + 4.7 \times 8$

6) $-2.4 \times 6 + 2.4 \times 5$

7) $-4.8 \times 3 + 4.8 \times 2$

8) $-6.6 \times 8 + 1.6 \times 8$

9) $-6.4 \times 3 + 1.4 \times 3$

10) $-6.2 \times 4 + 5.2 \times 4$

Summary 1 Chapter 10 Lesson 10
Review 10 C

Date | Name
Read directions carefully | Answers must be in the circle.
Show all steps for full credits. | Do only 1 operation per 1 step.
Each problem is worth 5 points. | Solve 1 problem 5 minutes.

To solve problems work by order of operations or change division for multiplication then put C before parentheses.

AC + BC = C (A + B) **AC – B C = C (A - B)**

a) $2\frac{5}{8} \div \frac{1}{8} + 1\frac{3}{8} \div \frac{1}{8} = 2\frac{5}{8} \times 8 + 1\frac{3}{8} \times 8 = 8(2\frac{5}{8} + 1\frac{3}{8}) = 8(4) = 32$

b) $2\frac{5}{8} \div \frac{1}{8} - 1\frac{3}{8} \div \frac{1}{8} = 2\frac{5}{8} \times 8 - 1\frac{3}{8} \times 8 = 8(2\frac{5}{8} - 1\frac{3}{8}) = 8(1\frac{2}{8}) = 8(\frac{10}{8}) = 10$

Simplify by Laws for Fractions

1) $1\frac{3}{4} \div \frac{1}{4} + 1\frac{1}{4} \div \frac{1}{4}$

2) $4\frac{3}{5} \div \frac{1}{5} + 1\frac{2}{5} \div \frac{1}{5}$

3) $2\frac{1}{2} \div \frac{1}{2} + 1\frac{1}{2} \div \frac{1}{2}$

4) $1\frac{5}{6} \div \frac{1}{6} + 1\frac{1}{6} \div \frac{1}{6}$

5) $1\frac{2}{3} \div \frac{1}{3} + 1\frac{1}{3} \div \frac{1}{3}$

6) $3\frac{1}{2} \div \frac{1}{2} - 1\frac{1}{2} \div \frac{1}{2}$

7) $1\frac{3}{4} \div \frac{1}{4} - 1\frac{1}{4} \div \frac{1}{4}$

8) $3\frac{2}{3} \div \frac{1}{3} - 1\frac{2}{3} \div \frac{1}{3}$

9) $4\frac{3}{5} \div \frac{1}{5} - 1\frac{3}{5} \div \frac{1}{5}$

10) $4\frac{5}{6} \div \frac{1}{6} - 2\frac{1}{6} \div \frac{1}{6}$

Summary 1 Chapter 10 Lesson 10
Review 10 D

Date **Name**

Read directions carefully **Answers must be in the circle.**

Show all steps for full credits. **Do only 1 operation per 1 step.**

Each problem is worth 5 points. **Solve 1 problem 5 minutes.**

To solve problems work by order of operations or change division for multiplication. Put - C before parentheses and change each sign inside parentheses. $-AC - BC = -C(A + B)$ $-AC + BC = -C(A - B)$

a) $-2\frac{5}{8} \div \frac{1}{8} - 1\frac{3}{8} \div \frac{1}{8} = -2\frac{5}{8} \times 8 - 1\frac{3}{8} \times 8 = -8(2\frac{5}{8} + 1\frac{3}{8}) = -8(4) = -32$

b) $-2\frac{5}{8} \div \frac{1}{8} + 1\frac{3}{8} \div \frac{1}{8} = -2\frac{5}{8} \times 8 + 1\frac{3}{8} \times 8 = -8(2\frac{5}{8} - 1\frac{3}{8}) = -8(1\frac{2}{8}) = -8(\frac{10}{8}) = -10$

Simplify by Laws for Fractions

1) $-1\frac{3}{4} \div \frac{1}{4} - 1\frac{1}{4} \div \frac{1}{4}$

2) $-4\frac{3}{5} \div \frac{1}{5} - 1\frac{2}{5} \div \frac{1}{5}$

3) $-2\frac{1}{2} \div \frac{1}{2} - 1\frac{1}{2} \div \frac{1}{2}$

4) $-1\frac{5}{6} \div \frac{1}{6} - 1\frac{1}{6} \div \frac{1}{6}$

5) $-1\frac{2}{3} \div \frac{1}{3} - 1\frac{1}{3} \div \frac{1}{3}$

6) $-3\frac{1}{2} \div \frac{1}{2} + 1\frac{1}{2} \div \frac{1}{2}$

7) $-1\frac{3}{4} \div \frac{1}{4} + 1\frac{1}{4} \div \frac{1}{4}$

8) $-3\frac{2}{3} \div \frac{1}{3} + 1\frac{2}{3} \div \frac{1}{3}$

9) $-4\frac{3}{5} \div \frac{1}{5} + 1\frac{3}{5} \div \frac{1}{5}$

10) $-4\frac{5}{6} \div \frac{1}{6} + 2\frac{1}{6} \div \frac{1}{6}$

Summary 1 Chapter 10 Lesson 11
Laws for operations

Date | **Name**

Read directions carefully. **Answers must be in the circle.**
Show all steps for full credits. **Each problem is worth 5 points.**
Solve 1 problem 5 minutes. **Do only 1 operation per 1 step.**

Distributive laws of multiplication

(A + B) C = AC + BC **(A - B) C = AC – B C**

Example a) $-\frac{5}{6}\times 3=-\frac{5}{6}\times\frac{3}{1}=-\frac{5}{2}=-2\frac{1}{2}=-2.5$

Example b) $\frac{5}{3}\times(-6)=-6\times\frac{5}{3}=-6\div 3\times 5=-2\times 5=-10$

Example c) $\left(-1\frac{1}{3}+\frac{5}{6}\right)\times 3=-\frac{4}{3}\times\frac{3}{1}+\frac{5}{6}\times\frac{3}{1}=-4+\frac{5}{2}=-4+2.5=-1.5$

Worksheet 7 A
Evaluate by using laws

1) $9(-1+\frac{4}{9})$

2) $(-\frac{2}{3}+\frac{3}{5})\times 15$

3) $9(-3+\frac{5}{9})$

4) $(-1-\frac{5}{6})\times 6$

5) $7(-2-\frac{3}{7})$

6) $(-1+\frac{1}{6})\times 6$

7) $6(-1\frac{1}{2}+\frac{2}{3})$

8) $(-\frac{5}{7}-\frac{4}{7})\times 7$

9) $7(-\frac{5}{7}+\frac{4}{7})$

10) $(-3-\frac{2}{3})\times 3$

11) $3(-2-\frac{1}{6})$

12) $(-2\frac{1}{4}+\frac{5}{8})\times 4$

13) $3(-1\frac{5}{6}+1\frac{1}{3})$

14) $(-3\frac{1}{2}+\frac{1}{4})\times 4$

15) $3(-1\frac{2}{3}+\frac{5}{6})$

16) $(-1\frac{2}{7}+\frac{1}{3})\times 21$

17) $6(-2\frac{1}{2}-1\frac{1}{3})$

18) $(-4-\frac{5}{8})\times 2$

19) $3(-1\frac{1}{3}+\frac{5}{12})$

20) $(-1\frac{1}{4}+\frac{5}{8})\times 4$

Summary 1 Chapter 10 Lesson 11

Date | Name
Read directions carefully. | Answers must be in the circle.
Show all steps for full credits. | Each problem is worth 5 points.
Solve 1 problem 5 minutes. | Do only 1 operation per 1 step.

$- AC - BC = - C (A + B)$ $\qquad - AC + BC = - C (A - B)$

Example $\quad -1\frac{1}{4}\times\frac{3}{4}-\frac{3}{4}\times2\frac{3}{4}=-\frac{3}{4}(1\frac{1}{4}+2\frac{3}{4})=-\frac{3}{4}\times\frac{4}{1}=-3$

Worksheet 7 B
Evaluate by using Common Factor

1) $-8\times1\frac{3}{4}+8\times\frac{5}{8}$
2) $-3\frac{2}{9}\times\frac{3}{5}-\frac{3}{5}\times6\frac{7}{9}$
3) $-7\times1\frac{1}{7}+7\times\frac{5}{14}$
4) $-4\times2\frac{1}{4}+4\times1\frac{7}{8}$
5) $-2\frac{6}{7}\times\frac{7}{10}-2\frac{1}{7}\times\frac{7}{10}$
6) $-5\frac{5}{6}\times\frac{3}{4}+1\frac{5}{6}\times\frac{3}{4}$
7) $-4\frac{4}{5}\times\frac{5}{8}-\frac{5}{8}\times3\frac{1}{5}$
8) $-3\frac{2}{3}\times1\frac{1}{2}+1\frac{2}{3}\times1\frac{1}{2}$
9) $-\frac{5}{6}\times5\frac{1}{5}+\frac{5}{6}\times2\frac{1}{5}$
10) $-3\frac{1}{3}\times1\frac{1}{5}-1\frac{1}{5}\times1\frac{2}{3}$
11) $-1\frac{3}{5}\times2\frac{1}{2}-1\frac{3}{5}\times7\frac{1}{2}$
12) $-4\frac{2}{3}\times\frac{3}{4}+3\frac{1}{3}\times\frac{3}{4}$
13) $-4\frac{3}{5}\times\frac{5}{6}+\frac{5}{6}\times1\frac{3}{5}$
14) $-4\frac{1}{2}\times\frac{1}{6}-\frac{1}{6}\times1\frac{1}{2}$
15) $-3\frac{3}{4}\times1\frac{1}{6}+2\frac{1}{4}\times1\frac{1}{6}$
16) $-\frac{1}{3}\times15+\frac{1}{5}\times15$
17) $-1\frac{3}{4}\times8+1\frac{1}{2}\times8$
18) $-1\frac{1}{2}\times6+1\frac{1}{6}\times6$
19) $-2\frac{1}{2}\times6+1\frac{1}{2}\times6$
20) $-1\frac{3}{4}\times12+1\frac{1}{4}\times12$
21) $(-1\frac{3}{5}-1\frac{2}{5})\div\frac{1}{5}$
22) $(-1\frac{3}{5}+1\frac{2}{5})\div\frac{2}{5}$
23) $(-4\frac{3}{5}-1\frac{2}{5})\div\frac{1}{5}$
24) $(-3\frac{2}{3}+1\frac{2}{3})\div\frac{2}{3}$
25) $(-1\frac{3}{4}+1\frac{1}{2})\div\frac{1}{8}$
26) $(-2\frac{1}{2}-1\frac{1}{2})\div\frac{1}{6}$
27) $(-1\frac{4}{5}-1\frac{1}{5})\div\frac{1}{5}$
28) $(-4\frac{3}{5}+1\frac{3}{5})\div\frac{1}{5}$
29) $(-1\frac{1}{3}-1\frac{2}{3})\div\frac{1}{3}$
30) $(-4\frac{1}{5}+2\frac{1}{5})\div\frac{1}{5}$

Summary 1 Chapter 10 Lesson 12
Proportions

Date **Name**

Read directions carefully. **Answers must be in the circle.**

Show all steps for full credits. **Each problem is worth 5 points.**

Solve 1 problem 5 minutes. **Do only 1 operation per 1 step.**

Example: Solve the equation

$3\frac{1}{3} \div 2\frac{1}{2} = 2\frac{2}{3} \div x$ $\quad \frac{3\frac{1}{3}}{2\frac{1}{2}} = \frac{2\frac{2}{3}}{x}$ $\quad 3\frac{1}{3}x = 2\frac{1}{2} \cdot 2\frac{2}{3}$ $\quad x = 2\frac{1}{2} \cdot 2\frac{2}{3} \div 3\frac{1}{3}$

$x = \frac{5}{2} \cdot \frac{8}{3} \div \frac{10}{3}$ $\quad x = \frac{5}{2} \cdot \frac{8}{3} \cdot \frac{3}{10}$ $\quad x = \frac{1 \cdot 4}{1 \cdot 2}$ $\quad x = 2$

Worksheet 8 A
Solve the proportions

1) $\frac{24}{72} = \frac{x}{9}$ 2) $\frac{60}{45} = \frac{24}{x}$

3) $\frac{42}{x} = \frac{51}{34}$ 4) $\frac{18}{12} = \frac{72}{x}$

5) $\frac{69}{46} = \frac{x}{54}$ 6) $\frac{4.2}{x} = \frac{5.7}{3.8}$

7) $\frac{x}{5.1} = \frac{2.7}{0.9}$ 8) $\frac{x}{4.7} = \frac{27.3}{9.1}$

9) $\frac{7.8}{2.6} = \frac{4.5}{x}$ 10) $\frac{4.4}{x} = \frac{8.8}{2.2}$

11) $\frac{y}{5\frac{2}{3}} = \frac{1\frac{7}{9}}{1\frac{7}{27}}$ 12) $\frac{z}{\frac{4}{5}} = \frac{3\frac{1}{8}}{1\frac{1}{4}}$

13) $y \div \frac{3}{5} = 3\frac{1}{8} \div 3\frac{3}{4}$ 14) $1\frac{7}{8} \div 8\frac{1}{3} = z \div \frac{4}{9}$

15) $3\frac{3}{4} \div 1\frac{1}{8} = 2\frac{2}{3} \div y$ 16) $\frac{3}{8} \div \frac{3}{11} = z \div \frac{4}{11}$

17) $y \div \frac{1}{3} = \frac{3}{4} \div \frac{1}{2}$ 18) $z \div \frac{3}{14} = 3\frac{1}{9} \div \frac{4}{9}$

19) $y \div 4\frac{1}{6} = \frac{4}{7} \div \frac{20}{21}$ 20) $z \div 3\frac{1}{5} = 4\frac{1}{2} \div 2\frac{1}{4}$

Summary 1 Chapter 10 Lesson 12
Proportions

Date **Name**

Read directions carefully. **Answers must be in the circle.**

Show all steps for full credits. **Each problem is worth 5 points.**

Solve 1 problem 5 minutes. **Do only 1 operation per 1 step.**

Example: Solve the equation

$3\frac{1}{3} \div 2\frac{1}{2} = 2\frac{2}{3} \div x$ $\qquad \frac{3\frac{1}{3}}{2\frac{1}{2}} = \frac{2\frac{2}{3}}{x}$ $\qquad 3\frac{1}{3}x = 2\frac{1}{2} \cdot 2\frac{2}{3}$ $\qquad x = 2\frac{1}{2} \cdot 2\frac{2}{3} \div 3\frac{1}{3}$

$x = \frac{5}{2} \cdot \frac{8}{3} \div \frac{10}{3}$ $\qquad x = \frac{5}{2} \cdot \frac{8}{3} \cdot \frac{3}{10}$ $\qquad x = \frac{1 \cdot 4}{1 \cdot 2}$ $\qquad x = 2$

Worksheet 8 B
Solve the proportions

21) $\frac{3}{14} \div 3\frac{1}{7} = y \div 3\frac{2}{3}$

22) $1\frac{1}{2} \div y = 4\frac{3}{4} \div 2\frac{3}{8}$

23) $y \div \frac{4}{25} = \frac{3}{4} \div \frac{4}{5}$

24) $y \div 2\frac{1}{3} = \frac{4}{7} \div 4\frac{1}{6}$

25) $7\frac{1}{3} \div 2\frac{1}{2} = 3\frac{2}{3} \div y$

26) $6\frac{1}{2} \div 3\frac{1}{4} = 3\frac{1}{2} \div k$

27) $m \div 2\frac{1}{2} = 3\frac{1}{4} \div 6\frac{1}{2}$

28) $n \div 2.7 = 8\frac{1}{3} \div 0.5$

29) $2.8 \div p = 1\frac{2}{3} \div 2\frac{6}{7}$

30) $d \div 6.4 = 1\frac{7}{8} \div 7\frac{1}{2}$

31) $c \div 3\frac{1}{3} = 0.9 \div 0.4$

32) $3\frac{1}{3} \div d = 3\frac{3}{5} \div 2.7$

33) $1\frac{7}{9} \div g = 9\frac{1}{3} \div 2\frac{5}{8}$

34) $h \div \frac{7}{12} = \frac{2}{3} \div \frac{7}{9}$

35) $3\frac{3}{5} \div k = 1\frac{4}{5} \div 1\frac{1}{4}$

36) $2\frac{1}{2} \div 1\frac{1}{2} = 6\frac{2}{3} \div m$

37) $3\frac{3}{5} \div \frac{4}{15} = n \div \frac{5}{9}$

38) $2 \div p = 1\frac{4}{7} \div 3\frac{1}{7}$

39) $5\frac{1}{4} \div 3\frac{1}{2} = 4\frac{1}{2} \div u$

40) $3\frac{1}{3} \div 2\frac{1}{2} = 2\frac{2}{3} \div w$

Summary 1 Chapter 10 Lesson 13
College Placement Test A (50 minutes)

Date	**Name**
Read directions carefully	**Answers must be in the circle.**
Show all steps for full credits.	**Do only 1 operation per 1 step.**
Each problem is worth 5 points.	**Solve 1 problem 5 minutes.**

Evaluate

1) $39600 \div (85^2 - 25^2)$

2) $4.9 \times 7 - 1.8 \times 7 + 6.9 \times 7$

3) $2\frac{1}{2} \times 16 - 1\frac{1}{4} \times 16$

4) $-2\frac{6}{7} \div 1\frac{3}{7} + 2\frac{1}{7} \div 1\frac{3}{7}$

5) Simplify $4(\frac{1}{2}c - \frac{1}{4}d) - \frac{2}{3}(6c + 3d)$

6) Solve the equation $0.3y + 0.2(2y - 3) = 0.8 - 0.7(y - 2)$

Solve the word problems

7) Find the perimeter of right triangle if hypotenuse is $5\frac{1}{10}$ ft and leg is $2\frac{2}{5}$ ft. Draw pictire.

8) Find the area of trapezoid, if height is $4\frac{2}{3}$ m, bases are $8\frac{1}{2}$ m and $6\frac{1}{2}$ m. Draw Picture.

9) A bus drove 75 mi at a speed of 30 mi/h.
How much time it spent?

10) Of 32 children, 62.5% were boys. How many were girls?

Summary 1 Chapter 10 Lesson 13
College Placement Test B (50 minutes)

Date **Name**
Read directions carefully **Answers must be in the circle.**
Show all steps for full credits. **Do only 1 operation per 1 step.**
Each problem is worth 5 points. **Solve 1 problem 5 minutes.**

Perform the indicated operations.

1) $70200 \div (95^2 - 35^2)$

2) $3.4 \times 5 - 1.7 \times 5 + 8.3 \times 5$

3) $1\frac{3}{4} \times 12 - 1\frac{1}{4} \times 12$

4) $-4\frac{4}{5} \div 1\frac{3}{5} - 3\frac{1}{5} \div 1\frac{3}{5}$

5) Simplify $16(\frac{3}{4}x + \frac{1}{4}y) - 15(\frac{2}{3}x + \frac{1}{3}y)$

6) Solve the equation $-5(4.2y + 1) + 4(1.4y - 2) = -20.7$

7) Find the perimeter of right triangle if hypotenuse is $7\frac{4}{5}$ ft and leg is 3 ft. Draw Picture.

8) Find the area of trapezoid, if it height is $2\frac{2}{3}$ m, bases are $7\frac{1}{2}$ m and $4\frac{1}{2}$ m. Draw Picture.

9) A bus drove 36 mi at a speed of 15 mi/h.
How much time it spent?

10) Of 24 children, 37.5% were boys. How many were girls?

Summary 1 Chapter 10 Lesson 14
Final Exam A

Date **Name**

Read directions carefully **Answers must be in the indicated place.**

Show all steps for full credits. **Do only 1 operation per 1 step.**

Each problem is worth 5 points. **Solve 1 problem 5 minutes.**

Always try to find the easiest way to calculate, fewer mistakes will made.

Perform the indicated operations

1) $3^2 - 0.9^2 \div 0.3^2$

1) ______

2) $\left(\frac{1}{4}-\frac{1}{8}\right)^2 \div \frac{1}{32}$

2) ______

3) $2\frac{3}{4} \div \frac{1}{16} - 1\frac{1}{4} \div \frac{1}{16}$

3) ______

4) $(-\frac{2}{5})^2 - (-\frac{3}{5})^2$

4) ______

5) $-\frac{5}{6} \times 5\frac{1}{5} + \frac{5}{6} \times 2\frac{1}{5}$

5) ______

Summary 1 Chapter 10 Lesson 14
Final Exam A

Date **Name**

Read directions carefully **Answers must be in the indicated place.**
Show all steps for full credits. **Do only 1 operation per 1 step.**
Each problem is worth 5 points. **Solve 1 problem 5 minutes.**

Simplify expression

6) $-6(\frac{1}{6}-\frac{1}{2}x)-2(1-\frac{1}{2}x)$

6)_____

Solve the equation

7) $0.3x + 0.2\,(2x - 3) = 0.8 - 0.7\,(x - 2)$

7) _____

Solve the word problems

8)Find area of trapezoid, if height is $5\frac{1}{3}$ ft, bases are $6\frac{1}{2}$ ft and $5\frac{1}{2}$ ft. Draw Picture.

8) _____

9) What speed does a boy have, if he drove 57 mi in 4 hours?

9) _____

10) A factory has 5 new machines and 80% were old machine.
How many machines are in the factory?

10)_____

Summary 1 Chapter 10 Lesson 14
Final Exam B

Date **Name**

Read directions carefully **Answers must be in the indicated place.**

Show all steps for full credits. **Do only 1 operation per 1 step.**

Each problem is worth 5 points. **Solve 1 problem 5 minutes.**

Always try to find the easiest way to calculate, fewer mistakes will made.

Perform the indicated operations

1) $2^2 - 0.4^2 \div 0.2^2$ 1) _____

2) $\left(\frac{1}{5} - \frac{1}{10}\right)^2 \div \frac{1}{10}$ 2) _____

3) $1\frac{3}{4} \div \frac{1}{12} - 1\frac{1}{4} \div \frac{1}{12}$ 3) _____

4) $-(-\frac{4}{5})^2 + (-\frac{1}{5})^2$ 4) _____

5) $-3\frac{1}{3} \times 1\frac{1}{5} - 1\frac{1}{5} \times 1\frac{2}{3}$ 5) _____

Summary 1 Chapter 10 Lesson 14
Final Exam B

Date **Name**

Read directions carefully **Answers must be in the indicated place.**

Show all steps for full credits. **Do only 1 operation per 1 step.**

Each problem is worth 5 points. **Solve 1 problem 5 minutes.**

Always try to find the easiest way to calculate, fewer mistakes will made.

Simplify expression

6) $-8(\frac{1}{2}-\frac{1}{4}y)-3(\frac{2}{3}-\frac{1}{3}y)$

6)_____

Solve the equation

7) $-2(3.1x-1)+3(1.2x+1)=-14.5$

7) _____

Solve the word problems

8) Find area of trapezoid, if height is $2\frac{2}{3}$ ft, bases are $7\frac{1}{2}$ ft and $4\frac{1}{2}$ ft.

Draw the picture.

8)_____

9) What speed does a boy have, if he drove 73 mi in 4 hours? 9) _____

10) A factory has 10 new machines and 60% were old machine. 10) _____
How many machines are in the factory?

My Grading Scale without minuses

A=91–100% B+=86–90% B=81–85% C+=76–80%

C=71–75% D+=66–70% D=61–65 % F=0 - 60%

Lesson 14 Final Exam B

1) 0 2) 0.1 3) 6 4) -0.6 5) -6

6) 3y - 6 7) 7. 5 8) 16 cm^2 9) 18.25 mi/hr. 10) 25 machines

Summary 1 Answers Chapter 10

Lesson 1 College Placement Test

1) 5, 2) 0 3) 21 4) - 2.8 5) - 4x – 4y

6) 7.5, 7) 18.2 ft 8) 60 in^2 9) 2.4 hr. 10) 12 girl

Lesson 7 Mental Math 7

1)1, 2)8 , 3)27, 4)64, 5)125, 6) 216, 7)343, 8)512, 9)729, 10)1000, 11)2, 12)4, 13)8, 14)16, 15) 32, 16) 64, 17)128, 18)256, 19)512, 20) 1024

Lesson 7 Mental Math 8

1)2^2, 2)3^2, 3)2^4, 4)5^2, 5)$2^2 \cdot 3^2$,6) 7^2,7) 2^6, 8)3^4 , 9)$2 \cdot 3$, 10)2^3, 11)$2 \cdot 5$,12)$2^2 \cdot 3$, 13)$2 \cdot 7$, 14)$3 \cdot 5$, 15)$2 \cdot 3^2$, 16)$2^2 \cdot 5$, 17)$2^3 \cdot 3$, 18)3^3, 19) $2 \cdot 3 \cdot 5$, 20) 2^5

Lesson 7 Mental Math 9

1) 0.2, 2) 1.5, 3) 0.7, 4) 0.7, 5) 2, 6) 0.6 , 7) 0 , 8) 0.3, 9) 0, 10) 0, 11) 1.9 , 12) 0.8, 13) 1, 14) 2 , 15) 3 , 16) 4 , 17) 6, 18) 7, 19) 0; 20) Ø , 21) 0.8, 22) 1, 23) 0.04 , 24) 0.09 , 25) 0.16, 26) 0.25, 27) 0.36, 28) 0.49, 29) 0.64, 30) 0.81,

Lesson 7 Mental Math 10

1) 80, 2) 20, 3) 4, 4) 10 , 5) 70, 6) 20, 7) 24, 8) 16, 9) 15, 10) 12, 11) 8, 12) 30, 13) 24, 14) 14, 15) 5, 16) 8, 17) 18, 18) 6 , 19) 8, 20) 8,

Lesson 7 Mental Math 11

1) $\frac{1}{5}$, 2) $\frac{1}{4}$, 3) 5 , 4)1, 5)1, 6)2, 7) 3, 8)4 , 9)$\frac{1}{4}$, 10) $\frac{1}{25}$, 11) $\frac{4}{25}$, 12) $\frac{9}{25}$,

13) 1, 14) 2, 15) $\frac{1}{4}$, 16) $\frac{1}{2}$, 17) $\frac{3}{4}$, 18) $1\frac{1}{4}$, 19) $\frac{1}{4}$, 20) $\frac{1}{4}$

Lesson 7 Mental Math 12

1) $1\frac{1}{2}$, 2) $2\frac{1}{4}$, 3) $\frac{1}{2}$, 4)$1\frac{1}{2}$, 5)$3\frac{1}{2}$, 6)6 , 7)$2\frac{1}{2}$, 8)$1\frac{1}{2}$, 9)6 ,10)7,

11)$1\frac{1}{2}$, 12)$3\frac{1}{2}$, 13)$1\frac{1}{4}$, 14) $1\frac{1}{4}$, 15)$1\frac{1}{2}$, 16)$1\frac{1}{2}$, 17)$\frac{1}{2}$, 18)$\frac{1}{4}$, 19)$1\frac{3}{4}$, 20)$1\frac{4}{5}$

Lesson 8 Mental Math 13

1) 5 , 2) 6, 3) 8 , 4) 5, 5) 9, 6) 8, 7) 6, 8) 4, 9) 5, 10) 6, 11) 6, 12) 8, 13) 7, 14) 6 , 15) 8, 16) 8, 17) 9, 18) 5, 19) 5, 20) 6

Lesson 8 Mental Math 14

1)0.3, 2)0.09, 3)0.007, 4)0.5, 5)0.5, 6)0.2, 7)0.4, 8)0.6 , 9)0.8 ,10)0.8,11)0.25, 12)0.5, 13)0.75, 14)0.25, 15)0.75, 16)0.28, 17)0.14, 18)0.22, 19)0.15, 20)0.45

Summary 1 Answers Chapter 10

Lesson 8 Mental Math 15

1)– 10, 2)-10,3)– 18,4)– 20,5)– 1,6) -2,7)– 3,8) - 4 ,9)$\frac{1}{4}$, 10)$\frac{1}{25}$,11)$\frac{4}{25}$,

12) - $\frac{9}{25}$,13) – 1,14) -2,15) - $\frac{1}{4}$,16) - $\frac{1}{2}$,17) - $\frac{3}{4}$,18) -1$\frac{1}{4}$,19) -$\frac{1}{4}$,20) -$\frac{1}{4}$.

Lesson 8 Mental Math 16

1)- 1$\frac{1}{2}$, 2) - 2$\frac{1}{4}$, 3) -$\frac{1}{2}$, 4) -1$\frac{1}{2}$, 5)- 3$\frac{1}{2}$, 6) – 6, 7) - 2$\frac{1}{2}$,8) - 1$\frac{1}{2}$, 9) – 6, 10) – 7,

11)-1$\frac{1}{2}$, 12)-3$\frac{1}{2}$,13)–1$\frac{1}{4}$,14)-1$\frac{1}{4}$, 15)-1$\frac{1}{2}$, 16)-1$\frac{1}{2}$, 17)-$\frac{1}{2}$, 18)-$\frac{1}{4}$,19)-1$\frac{3}{4}$, 20)- 1$\frac{4}{5}$

Lesson 9 Mental Math 17

1)121, 2)144, 3)169, 4)196, 5)225, 6)256, 7)289, 8)324, 9)361, 10)400, 11)441 12)484, 13)529, 14)576,15)625, 16)676,17)729, 18)784, 19) 841,20) 900.

Lesson 9 Mental Math 18

1) 8, 2) 9, 3) 10, 4) 11, 5) 12, 6) 13, 7) 14, 8) 15, 9) 16, 10) 17 , 11) 18, 12) 19, 13) 20, 14) 21, 15) 22, 16) 23, 17) 24, 18) 25, 19) 26, 20) 27, 21) 28, 22) 29, 23) 30, 24) 31, 25) 40, 26) 50, 27) 60, 28) 70, 29) 80, 30) 90.

Lesson 9 Mental Math 19

1) 4, 2) 3, 3) 14, 4) 17, 5) 15 , 6) 9, 7) 5, 8) 9, 9) 8, 10) 7, 11) 9, 12) 7, 13) 56, 14) 42, 15) 54, 16) 8, 17) 9, 18) 7, 19) 8, 20) 6.

Lesson 9 Mental Math 20

1) 3, 2) 15, 3) 6, 4) 10, 5) 9, 6) 16, 7) 15, 8) 12, 9) 2, 10) 11, 11) 8, 12) 4 , 13) 9, 14) 1, 15) 4, 16) 16, 17) 18, 18) 6 , 19) 8, 20) 6.

Lesson 10 Reviews for Final Exam Review 10 A

1) - 20, 2) - 20, 3) - 12, 4)- 20, 5) - 6,
6) - 2, 7) - 3, 8) - 36, 9)- 90, 10) - 70.

Review 10 B

1) - 10, 2) - 12, 3) - 12 , 4) - 8, 5) - 8 ,
6) - 2.4, 7) - 4.8, 8) - 40 , 9) - 15 , 10) - 4.

Review 10 C

1) 12, 2) 30, 3) 8 , 4) 18 , 5) 9 ,
6) 4, 7) 2, 8) 6 , 9) 15 , 10) 16.

Review 10 D

1) - 12, 2) - 30 , 3) - 8, 4) - 18, 5) - 9,
6) - 4 , 7) - 2, 8) - 6, 9) - 15, 10) - 16.

Summary 1 Answers Chapter 10

Lesson 11 Worksheet 7 A

1) -5, 2) -1, 3) -22 , 4) -11, 5) -17,
6) -5, 7) -5, 8) -9, 9) -1, 10) -11,
11) -6.5, 12) -6.5, 13) -1.5, 14) - 13, 15) -2.5,
16) -20, 17) -23, 18) -9.25, 19) -2.75, 20) -2.5.

Lesson 11 Worksheet 7 B

1) -9, 2) -6, 3) -5.5, 4) -1.5, 5) -3.5,
6) - 3, 7) -5, 8) -3, 9) -2.5, 10) -6,
11) - 16, 12) - 1, 13) - 2.5, 14) - 1 , 15) -1.75,
16) - 2, 17) - 2, 18) - 2, 19) - 6 , 20) - 6 ,
21) - 15, 22) - 0.5, 23) - 30, 24) - 3, 25) - 2 ,
26) - 24, 27) - 15, 28) - 15, 29) - 9, 30) - 10 ,

Lesson 12 Worksheet 8 A

1) 3, 2) 18, 3) 28, 4) 50, 5) 81,
6) 2.8, 7) 15.3, 8) 14.1, 9) 1.5, 10) 1.1,
11) 8, 12) 2, 13) 0.5, 14) 0.1, 15) 0.8,
16) 0.5, 17) 0.5, 18) 1.5, 19) 2.5, 20) 6.4,

Lesson 12 Worksheet 8 B

21) 0.25 , 22) 0.75, 23) 0.15, 24) 0.32, 25) 1.25,
26) 1.75 , 27) 1.25 , 28) 45, 29) 4.8, 30) 1.6,
31) 7.5 , 32) 2.5, 33) 0.5 , 34) 0.5, 35) 2.5,
36) 4 , 37) 7.5, 38) 4, 39) 3, 40) 2.

Lesson 13 College Placement Test A

1)6, 2)70,3)20,4)– 0.5,5) -2c – 3d,6) 2,7)12 ft,8)35 in^2,9)2.5 hr. ,10)12 girls.

Lesson 13 College Placement Test B

1) 9,2) 50,3) 6,4) – 1,5)2x – y,6) 0.5,7)18 ft,8)16m^2,9)2.4 hr.,10)15 girls.

Lesson 14 Final Exam A

1) 0, 2) 0.5 , 3) 24, 4) -0.2, 5) -2.5,
6) 4x – 3, 7) - 5 , 8) 32 ft^2 , 9) 14.25 mi/hr. , 10) 25 machines

Lesson 14 Final Exam B

1) 0, 2) 0.1, 3) 6 , 4) -0.6, 5) -6 ,
6) 3y – 6, 7) 7. 5, 8) 16 cm^2 , 9) 18.25 mi/hr. , 10) 25 machines.

Memorize Exponents 4

1)121,2)144,3)169,4)196,5)225,6)256,7)289,8)324,9)361,10)400,
11)441,12)484,13) 529,14)576,15)625,16)676,17)729,18)784,
19)841,20)900,21)100,22)400,23)900,24)1600,25)2500,26)3600,
27)4900,28)6400, 29)8100,30) 10000 .

Summary 1
Memorize Exponents 4
(for good memory)

1) 11^2	11) 21^2	21) 10^2
2) 12^2	12) 22^2	22) 20^2
3) 13^2	13) 23^2	23) 30^2
4) 14^2	14) 24^2	24) 40^2
5) 15^2	15) 25^2	25) 50^2
6) 16^2	16) 26^2	26) 60^2
7) 17^2	17) 27^2	27) 70^2
8) 18^2	18) 28^2	28) 80^2
9) 19^2	19) 29^2	29) 90^2
10) 20^2	20) 30^2	30) 100^2

Summary 2

Chapter 11

Extra

Reviews

Students!

Students, you are smarter than a calculator!
Students, you are smarter than a computer!
Students, you need to believe in yourself.
Students, be patient and do your best.

Do not be afraid to make mistakes.
Tell yourself positive words.
I am young! I am smart!
I am strong! I can do anything!

Smart students use their brain!
Your brain is always with you!
A calculator is not always with you.
Mathematics is work for the brain.

Be open to new ideas in math!
Work hard and try your best!
Check answers and ask questions!
Practice! Practice! Practice!

Summary 2 Chapter 11 Contents
Extra Reviews
Lessons 1 - 10

Summary 2 Chapter 11 Lesson 1
College Placement Test (50 minutes)

Date **Name**

Evaluate

1) $18000 \div (75^2 - 45^2)$

2) $5.6 \times 4 + 2.7 \times 4 - 8.3 \times 4$

3) $3\frac{3}{5} \times 15 - 2\frac{1}{5} \times 15$

4) $-3\frac{2}{3} \div \frac{5}{6} + 1\frac{1}{3} \div \frac{5}{6}$

5) Simplify $\frac{3}{7}(14x - 21y) - 15(\frac{2}{3}x - \frac{1}{3}y)$

6) Solve the equation $-2(3.1x - 1) + 3(1.2x + 1) = -14.5$

7)Find the perimeter of right triangle if it hypotenuse is $8\frac{1}{2}$ ft and another leg is 1.3 ft. Draw Picture.

8) Find the area of trapezoid, if it height is $6\frac{2}{3}$ in, bases are $8\frac{1}{2}$ in and $9\frac{1}{2}$ in. Draw Picture.

9) A bus drove 48 mi at a speed of 20 mi/h.
How much time bus spend?

10) Of 16 children, 25% were boys. How many were girls?

How do you feel about Math?
a) Love b) Like c) Nothing d) Do not Like e) Difficult

Summary 2 Chapter 11 Lesson 1
Review 1 A Whole Numbers

Date **Name**

Read directions carefully. **Answers must be in the circle.**

Show all steps for full credits. **Each problem is worth 5 points.**

Solve 1 problem 5 minutes. **Do only 1 operation per 1 step.**

Evaluate

1) $450607 + 506078$

2) $900000 - 887766$

3) 908×605

4) 4506×809

5) $686588 \div 98$

6) $641048 \div 908$

7) $363600 \div 80 \times 9$

8) $8100 - 7263 \div 9$

9) $500 - 320 \div 8 \times 5$

10) $150 \div 5 \times 3 - 120 \div 4 \times 3$

Summary 2 Chapter 11 Lesson 1
Review 2 A Decimals

Date

Name

Read directions carefully.

Answers must be in the circle.

Show all steps for full credits.

Each problem is worth 5 points.

Solve 1 problem 5 minutes.

Do only 1 operation per 1 step.

Evaluate

1) $37.37 + 13.73$

2) $61.1 - 46.46$

3) 0.375×4.8

4) $36.45 \div 0.9$

5) $9.5\times5.8 - 9.5\times4.8$

6) $33 \div 6.6 \times 5$

7) $96.96 - 68.68 \div 4$

8) $6^2 - 0.6^2 \div 0.3^2$

9) $0.4^2 \div (0.6 - 0.4)^2$

10) $12 \div [8.4 - (2.6^2 + 1.24)]$

Summary 2 Chapter 11 Lesson 1
Review 3 A Fractions

Date **Name**

Read directions carefully. **Answers must be in the circle.**

Show all steps for full credits. **Each problem is worth 5 points.**

Solve 1 problem 5 minutes. **Do only 1 operation per 1 step.**

1) Reduce the fraction $\frac{60}{100}$.

Evaluate

2) $2\frac{5}{12}+1\frac{5}{6}$

3) $4-2\frac{3}{4}$

4) $2\frac{2}{5}\times 5\frac{5}{6}$

5) $4\frac{5}{8}\div 6\frac{1}{6}$

6) $\frac{2}{3}\div 1\frac{1}{3}\div\frac{5}{8}$

7) $\left(1\frac{1}{2}\right)^3\div\left(1\frac{1}{2}\right)$

8) $\left(\frac{1}{2}-\frac{1}{3}\right)^2\div\frac{1}{18}$

9) $6\frac{1}{2}\div\left(6\frac{1}{3}-4\frac{1}{6}\right)$

10) $1\frac{3}{4}\times 8-1\frac{1}{2}\times 8$

Summary 2 Chapter 11 Lesson 1
Review 4 A Decimals and Fractions

Date **Name**

Read directions carefully. **Answers must be in the circle.**

Show all steps for full credits. **Each problem is worth 5 points.**

Solve 1 problem 5 minutes. **Do only 1 operation per 1 step.**

1) **Change terminating decimal 3.625 to fraction.**

2) **Change fraction $2\frac{3}{8}$ to terminating decimal.**

Evaluate

3) $0.875 + \frac{3}{6}$

4) $1\frac{1}{4} - 0.45$

5) $1.6 \times \frac{3}{8}$

6) $0.15 \div \frac{3}{4}$

7) $3.75 \div 3\frac{1}{8}$

8) $3^3 \div \left(1\frac{1}{2}\right)^2$

9) $(0.3 - \frac{3}{20}) \div \frac{1}{20}$

10) $\frac{5}{6} \div (1\frac{2}{3} - 1.25)$

Summary 2 Chapter 11 Lesson 2
Review 5 A Rational Numbers

Date | **Name**

Read directions carefully.
Show all steps for full credits.
Solve 1 problem 5 minutes.

Answers must be in the circle.
Each problem is worth 5 points.
Do only 1 operation per 1 step.
Evaluate

1) $\frac{1}{5}+(-2)$

2) $-(+1\frac{3}{10})-(-2\frac{7}{10})-(-\frac{1}{5})$

3) $(-1\frac{2}{5})(0.35)$

4) $(-4.5)\div(-1\frac{1}{2})$

5) $-(-\frac{1}{5})^2-(-\frac{2}{5})^2$

6) $-4\frac{4}{5}\times\frac{5}{8}-\frac{5}{8}\times3\frac{1}{5}$

7) $3\frac{1}{3}\div(-4\frac{5}{6}+3\frac{1}{2})$

8) $(\frac{1}{5}-0.3)\div0.2$

9) $0.15\div(-\frac{7}{10}+0.4)$

10) $\frac{3}{20}\div(-2.8+2\frac{1}{2})$

Summary 2 Chapter 11 Lesson 2
Review 6 A Equations

Date **Name**
Read directions carefully. **Answers must be in the circle.**
Show all steps for full credits. **Each problem is worth 5 points.**
Solve 1 problem 5 minutes. **Do only 1 operation per 1 step.**

Simplify the expressions

1) $100(0.01y - 0.03) - 1000(0.001y - 0.05)$

2) $4(\frac{1}{2}x - \frac{1}{4}y) - \frac{2}{3}(6x + 3y)$

Solve the equations

3) $\frac{1}{10}x + 10 = 50$

4) $81072 \div z = 9$

5) $28x - 987 - 45x = 9247$

6) $0.1y + 0.2y + 0.3y = 1.08$

7) $\frac{3}{10}y - \frac{1}{5}y - \frac{3}{10} = 1\frac{1}{5}$

8) $-5(\frac{2}{5}x + \frac{4}{5}) - 4(\frac{3}{4}x - \frac{1}{4}) = 0$

Solve the proportions

9) $\frac{7.8}{2.6} = \frac{4.5}{x}$

10) $5\frac{1}{4} \div 3\frac{1}{2} = 4\frac{1}{2} \div x$

Summary 2 Chapter 11 Lesson 2
Review 7 A Geometry

Date | **Name**
Read directions carefully. | **Answers must be in the circle.**
Show all steps for full credits. | **Each problem is worth 5 points.**
Solve 1 problem 5 minutes. | **Do only 1 operation per 1 step.**
Draw Picture. | **Solve the problems**

1)Find the perimeter of an isosceles triangle if its base is $3\frac{1}{2}$ in and leg $2\frac{1}{4}$ in.

2) Find one of leg of a right triangle if its other leg is 4.8 in and hypotenuse is 6 in.

3) Find the area of a square, if its side is $\frac{1}{2}$ in.

4) Find the area of a rectangle, if its length is $6\frac{2}{3}$ cm and the width is $4\frac{1}{2}$ cm.

5)Find the area of a trapezoid, if its height is $6\frac{2}{3}$ ft and bases are $9\frac{1}{2}$ ft and $8\frac{1}{2}$ ft.

6) Find the area of a triangle if its base is $6\frac{1}{4}$ m and height is $4\frac{4}{5}$ m.

7) Find the area of a right triangle if its legs are $4\frac{4}{5}$ ft and $5\frac{5}{6}$ ft.

8) Find the area of a triangle if its sides are 5 m, 12 m, and 13 m.

Right Triangle $a^2 + b^2 = c^2$ $A = 0.5ab$

9) Find the circumference of a circle, if its diameter is 20 yd and $\pi \approx 3.14$

10) Find the volume of a cube if its side is $1\frac{1}{2}$ in.

Summary 2 Chapter 11 Lesson 2
Review 8 A Word Problems

Date
Name
Read directions carefully.
Answers must be in the circle.
Show all steps for full credits.
Each problem is worth 5 points.
Solve 1 problem 5 minutes.
Do only 1 operation per 1 step.

Solve the word problems

1) A book has 80 pages. David read 0.35 of the whole book. How many pages more does he have left to read?

2) In the class, 7 students can dance. This equal $\frac{1}{4}$ of the whole class. How many students are in the class? How many students cannot dance?

3)Kristy read 30 pages. What part of the book did she read, if the book has 50 pages? Write the ratio as a reduced fraction. How many pages more does she have left to read?

4) Find the average for the numbers: $1\frac{1}{2}$, $1\frac{1}{5}$, $2\frac{1}{5}$, $2\frac{1}{10}$.

5) A girl paid $24 for 4 books. How much would 6 of the same books cost?

6) A bus drove 65 mi at a speed of 25 mi/h. How much time it spent?

7) What speed does a boy have, if he drove 53 mi in 4 hours?

8) Of 32 children, 37.5% were boys. How many were girls?

9) A factory has 4 new machines and 80% were old machine. How many machines are in the factory?

10) In 50 minutes of lesson, students solve an equation in 25 minutes. What percent of lesson time did students use to solve the equation?

Summary 2 Chapter 11 Lesson 3
Review 1 B Whole Numbers

Date | Name
Read directions carefully. | Answers must be in the circle.
Show all steps for full credits. | Each problem is worth 5 points.
Solve 1 problem 5 minutes. | Do only 1 operation per 1 step.

Evaluate

1) $(100 - 20) \times 25$

2) $14528 \div 16 - 9680 \div 16$

3) $58800 \div (95^2 - 25^2)$

4) GCF (18, 12)

5) LCM (18, 12)

6) $6^3 - 6^2 \div 2^2$

7) $8^2 \div 4^2 - 2^2$

8) $3^4 \div (6 - 3)^2$

9) $6^3 - 2^3(10 - 2^3 \div 2^2)$

10) $\{3 \cdot 10^2 - 2^2[2^2 + (2 + 3)^2]\} + 2^3$

Summary 2 Chapter 11 Lesson 3
Review 2 B Decimals

Date **Name**

Read directions carefully. **Answers must be in the circle.**
Show all steps for full credits. **Each problem is worth 5 points.**
Solve 1 problem 5 minutes. **Do only 1 operation per 1 step.**

Evaluate

1) $46.46 + 14.64$

2) $51.1 - 37.37$

3) 0.375×2.4

4) $45.54 \div 0.9$

5) $8.6\times4.7 - 8.6\times3.7$

6) $33 \div 5.5 \times 6$

7) $48.48 - 34.34 \div 2$

8) $4^2 - 0.4^2 \div 0.2^2$

9) $0.6^2 \div (0.5 - 0.3)^2$

10) $9.6 \div [11.4 - (2.8^2 + 1.16)]$

Summary 2 Chapter 11 Lesson 3
Review 3 B Fractions

Date **Name**

Read directions carefully. **Answers must be in the circle.**

Show all steps for full credits. **Each problem is worth 5 points.**

Solve 1 problem 5 minutes. **Do only 1 operation per 1 step.**

Evaluate

1) Reduce the fraction $\frac{40}{100}$.

2) $2\frac{1}{12}+1\frac{1}{6}$

3) $4-1\frac{1}{5}$

4) $1\frac{5}{6}\times 2\frac{2}{11}$

5) $3\frac{3}{8}\div 5\frac{5}{8}$

6) $\frac{1}{6}\div 1\frac{1}{3}\div\frac{5}{8}$

7) $(1\frac{1}{2})^2\div\left(1\frac{1}{4}\right)$

8) $\left(\frac{1}{3}-\frac{1}{6}\right)^2\div\frac{1}{9}$

9) $4\frac{1}{2}\div(6\frac{1}{2}-4\frac{1}{4})$

10) $1\frac{1}{2}\times 6-1\frac{1}{6}\times 6$

Summary 2 Chapter 11 Lesson 3
Review 4 B Decimals and Fractions

Date | **Name**
Read directions carefully. | **Answers must be in the circle.**
Show all steps for full credits. | **Each problem is worth 5 points.**
Solve 1 problem 5 minutes. | **Do only 1 operation per 1 step.**

Evaluate

1) **Change terminating decimal 2.375 to fraction.**

2) **Change fraction to terminating decimal** $3\frac{5}{8}$.

Evaluate

3) $1\frac{3}{6}+1.125$

4) $2\frac{1}{4}-1.75$

5) $3.2\times\frac{5}{8}$

6) $0.45\div\frac{3}{4}$

7) $4\frac{1}{8}\div 2.75$

8) $(1\frac{1}{2})^2\div 3^2$

9) $(2.5-2\frac{1}{3})\div 1\frac{1}{3}$

10) $6.5\div(4\frac{1}{3}-2\frac{1}{6})$

Summary 2 Chapter 11 Lesson 4
Review 5 B Rational Numbers

Date | **Name**

Read directions carefully. **Answers must be in the circle.**
Show all steps for full credits. **Each problem is worth 5 points.**
Solve 1 problem 5 minutes. **Do only 1 operation per 1 step.**

Evaluate

1) $-4-(-3\frac{2}{5})$

2) $-\frac{1}{5}-(-\frac{3}{10})-(-2\frac{1}{10})$

3) $-1\frac{4}{5}(0.45)$

4) $1.2\div(-1\frac{1}{2})$

5) $-(-\frac{3}{5})^2-(-\frac{4}{5})^2$

6) $-3\frac{2}{3}\times1\frac{1}{2}+1\frac{2}{3}\times1\frac{1}{2}$

7) $1\frac{11}{16}\div(-4\frac{1}{4}+3\frac{1}{8})$

8) $(-3\frac{1}{5}+1.1)\div4.2$

9) $\frac{8}{25}\div(-0.8+\frac{2}{5})$

10) $\frac{7}{20}\div(-0.3+\frac{1}{20})$

Summary 2 Chapter 11 Lesson 4
Review 6 B Equations

Date **Name**

Read directions carefully. **Answers must be in the circle.**

Show all steps for full credits. **Each problem is worth 5 points.**

Solve 1 problem 5 minutes. **Do only 1 operation per 1 step.**

Simplify the expressions

1) $10(0.1x-4) - 100(0.01x - 5)$

2) $\frac{4}{9}(18x - 9y) - 12(\frac{5}{6}x - \frac{1}{6}y)$

Solve equations

3) $\frac{1}{11}x - 5 = 55$

4) $72056 \div z = 8$

5) $125x - 876 - 201x = 37580$

6) $0.1z + 0.2z + 0.3z = 1.5$

7) $\frac{4}{5}z - 0.3z - 3 = 1.5$

8) $1.6(\frac{3}{4}y - \frac{5}{8}) - 1.8(\frac{2}{9}y - \frac{7}{9}) = 0$

Solve the proportions

9) $\frac{4.4}{z} = \frac{8.8}{2.2}$

10) $3\frac{1}{3} \div 2\frac{1}{2} = 2\frac{2}{3} \div x$

Summary 2 Chapter 11 Lesson 4
Review 7 B Geometry

Date | **Name**

Read directions carefully.
Show all steps for full credits.
Solve 1 problem 5 minutes.
Draw Picture.
Answers must be in the circle.
Each problem is worth 5 points.
Do only 1 operation per 1 step.
Solve the problems

1) Find the perimeter of an isosceles triangle if its base is $4\frac{1}{2}$ in and leg is $3\frac{1}{4}$ in.

2) Find the leg of right triangle if its hypotenuse is 9.5 m and other leg is 5.7 m.

3) Find the area of a square if its side is $3\frac{1}{2}$ in.

4) Find the area of a parallelogram if its base is $6\frac{1}{4}$ cm and height is $4\frac{4}{5}$ cm.

5) Find the area of a trapezoid, if its height is $4\frac{2}{3}$ ft and bases are $8\frac{1}{2}$ ft and $6\frac{1}{2}$ ft.

6) Find the area of a triangle if its base is $8\frac{1}{3}$ in and height is $7\frac{1}{5}$ in.

7) Find the area of a right triangle if its legs are $4\frac{1}{5}$ ft and $6\frac{2}{3}$ ft.

8) Find the area of a triangle if its sides are 15 m, 20 m, and 25 m.

Right Triangle $a^2 + b^2 = c^2$ $A = 0.5ab$

9) Find the circumference of a circle, if its diameter is 10 yd and $\pi \approx 3.14$

10) Find the volume of a cube if its side is $2\frac{1}{2}$ yd.

Summary 2 Chapter 11 Lesson 4
Review 8 B Word Problems

Date **Name**

Read directions carefully. **Answers must be in the circle.**
Show all steps for full credits. **Each problem is worth 5 points.**
Solve 1 problem 5 minutes. **Do only 1 operation per 1 step.**

Solve the word problems

1) A book has 60 pages. David read 0.75 of the whole book. How many pages more does he have left to read?

2) In the class, 12 students can dance; this is $\frac{2}{5}$ of the whole class. How many students are in the class? Write the ratio as a reduced fraction. How many students cannot dance?

3)Kristy read 24 pages. What part of the book did she read, if the book has 40 pages? Write the ratio as a reduced fraction. How many pages more does she have left to read?

4) Find the average for the numbers: $1\frac{1}{2}$, $1\frac{1}{5}$, $4\frac{1}{5}$, $4\frac{1}{10}$.

5)A girl paid $12 for 2 books. How much would 3 of the same books cost?

6) A bus drove 55 mi at a speed of 25 mi/h. How much time it spent?

7) What speed does a boy have, if he drove 69 mi in 4 hours?

8) Of 16 children, 37.5% were boys. How many were girls?

9) A factory has 8 new machines and 60% were old machine. How many machines are in the factory?

10) In 75 minutes of lesson, students solve an equation in 45 minutes. What percent of lesson time did students use to solve the equation?

Summary 2 Chapter 11 Lesson 5
Review 1 C Whole Numbers

Date
Name

Read directions carefully.
Show all steps for full credits.
Solve 1 problem 5 minutes.

Answers must be in the circle.
Each problem is worth 5 points.
Do only 1 operation per 1 step.

Evaluate

1) $\begin{array}{r} 340506 \\ +\ 405067 \\ \hline \end{array}$

2) $\begin{array}{r} 800000 \\ -\ 776655 \\ \hline \end{array}$

3) $\begin{array}{r} 807 \\ \times 504 \\ \hline \end{array}$

4) $\begin{array}{r} 3405 \\ \times\ \ 708 \\ \hline \end{array}$

5) $522435 \div 87$

6) $488235 \div 807$

7) $226240 \div 70 \times 8$

8) $6400 - 5648 \div 8$

9) $300 - 240 \div 6 \times 5$

10) $90 \div 3 \times 2 - 60 \div 4 \times 3$

Summary 2 Chapter 11 Lesson 5
Review 2 C Decimals

Date **Name**

Read directions carefully. **Answers must be in the circle.**

Show all steps for full credits. **Each problem is worth 5 points.**

Solve 1 problem 5 minutes. **Do only 1 operation per 1 step.**

Evaluate

1) $19.19 + 11.91$

2) $41.1 - 28.28$

3) 0.625×1.6

4) $18.36 \div 0.6$

5) $7.8 \times 6.2 - 7.8 \times 5.2$

6) $22 \div 5.5 \times 4$

7) $48.48 - 24.24 \div 2$

8) $3^2 - 0.9^2 \div 0.3^2$

9) $0.2^2 \div (0.4 - 0.2)^2$

10) $44 \div [\ 6.2 - (1.6^2 + 1.44)]$

Summary 2 Chapter 11 Lesson 5
Review 3 C Fractions

Date **Name**

Read directions carefully. **Answers must be in the circle.**

Show all steps for full credits. **Each problem is worth 5 points.**

Solve 1 problem 5 minutes. **Do only 1 operation per 1 step.**

1) Reduce the fraction $\frac{50}{100}$.

Evaluate

2) $1\frac{6}{15}+2\frac{1}{5}$

3) $5-1\frac{1}{5}$

4) $3\frac{3}{4}\times 5\frac{1}{3}$

5) $2\frac{7}{8}\div 3\frac{5}{6}$

6) $1\frac{1}{4}\div\frac{3}{4}\div 1\frac{2}{3}$

7) $\left(1\frac{1}{2}\right)^4\div\left(1\frac{1}{2}\right)^2$

8) $\left(\frac{1}{4}-\frac{1}{8}\right)^2\div\frac{1}{32}$

9) $2\frac{1}{2}\div\left(3\frac{1}{2}-2\frac{1}{4}\right)$

10) $2\frac{1}{2}\times 16-1\frac{1}{4}\times 16$

Summary 2 Chapter 11 Lesson 5
Review 4 C Decimals and Fractions

Date | **Name**
Read directions carefully. | **Answers must be in the circle.**
Show all steps for full credits. | **Each problem is worth 5 points.**
Solve 1 problem 5 minutes. | **Do only 1 operation per 1 step.**

1) Change decimal 1.625 to fraction.

2) Change fraction to terminating decimal $1\frac{7}{8}$.

Evaluate

3) $0.25 + \frac{4}{8}$

4) $1\frac{1}{4} - 0.75$

5) $2.4 \times \frac{3}{8}$

6) $0.36 \div \frac{3}{4}$

7) $4.25 \div 2\frac{1}{8}$

8) $3^2 \div \left(1\frac{1}{2}\right)^2$

9) $(0.6 - \frac{3}{20}) \div \frac{3}{20}$

10) $\frac{5}{6} \div (2\frac{2}{3} - 2.25)$

Summary 2 Chapter 11 Lesson 6
Review 5 C Rational Numbers

Date **Name**

Read directions carefully. **Answers must be in the circle.**

Show all steps for full credits. **Each problem is worth 5 points.**

Solve 1 problem 5 minutes. **Do only 1 operation per 1 step.**

Evaluate

1) $\frac{2}{5}+(-3)$

2) $-(+2\frac{7}{10})-(-1\frac{3}{10})-(-\frac{3}{5})$

3) $(-1\frac{3}{5})(0.75)$

4) $(4.5)\div(-1\frac{1}{4})$

5) $(-\frac{2}{5})^2-(-\frac{3}{5})^2$

6) $-\frac{5}{6}\times4\frac{1}{5}+\frac{5}{6}\times1\frac{1}{5}$

7) $4\frac{1}{2}\div(-4\frac{5}{6}+3\frac{1}{3})$

8) $(\frac{1}{5}-0.4)\div0.5$

9) $0.25\div(-\frac{8}{10}+0.3)$

10) $5\frac{1}{5}\div(-2.6+2\frac{1}{2})$

Summary 2 Chapter 11 Lesson 6
Review 6 C Equations

Date — **Name**

Read directions carefully. — **Answers must be in the circle.**
Show all steps for full credits. — **Each problem is worth 5 points.**
Solve 1 problem 5 minutes. — **Do only 1 operation per 1 step.**

Simplify the expressions

1) $0.3z + 0.2(2z - 3) - 0.8 - 0.7(z - 2)$

2) $16(\frac{3}{4}x + \frac{1}{4}y) - \frac{2}{3}(9x + 27y)$

Solve the equations

3) $\frac{1}{8}x - 10 = 40$

4) $49056 \div z = 7$

5) $-78x - 678 + 9x = 20298$

6) $0.8z - 0.3z - 3 = 1.5$

7) $\frac{1}{10}y + 5 + \frac{1}{10}y = 10$

8) $-6(\frac{1}{6} - \frac{1}{2}x) - 2(1 - \frac{1}{2}x) = 0$

Solve the proportions

9) $\frac{y}{5.1} = \frac{2.7}{0.9}$

10) $z \div \frac{7}{12} = \frac{2}{3} \div \frac{7}{9}$

Summary 2 Chapter 11 Lesson 6
Review 7 C Geometry

Date	**Name**
Read directions carefully.	**Answers must be in circle.**
Show all steps for full credits.	**Each problem is worth 5 points.**
Solve 1 problem 5 minutes.	**Do only 1 operation per 1 step.**
Draw Picture.	**Solve the problems**

1)Find the perimeter of a isosceles triangle if its base $6\frac{1}{2}$ in and leg $4\frac{1}{4}$ in.

2) Find the hypotenuse of a right triangle if its legs are 5.2 m and 3.9 m.

3) Find the area of a square, if its side is 2.5 in.

4)Find the area of a rectangle, if its length is $10\frac{1}{2}$ cm and width is $5\frac{1}{3}$ cm.

5)Find the area of a trapezoid, if its height is $5\frac{1}{3}$ ft, bases are $6\frac{1}{2}$ ft and $5\frac{1}{2}$ ft.

6) Find the area of a triangle if its base is $9\frac{3}{4}$ in and height is $5\frac{1}{3}$ in.

7) Find the area of a right triangle if its legs are $4\frac{4}{5}$ ft and $6\frac{1}{4}$ ft.

8) Find the area of a triangle if its sides are 6 m, 8 m, 10 m.

Right Triangle $a^2 + b^2 = c^2$ $A = 0.5ab$

9) Find the circumference of a circle, if its radius is 10 ft, $\pi \approx 3.14$

10) Find exact answer of volume of a sphere with radius is 3 yd.

Summary 2 Chapter 11 Lesson 6
Review 8 C Word Problems

Date **Name**

Read directions carefully. **Answers must be in the circle.**
Show all steps for full credits. **Each problem is worth 5 points.**
Solve 1 problem 5 minutes. **Do only 1 operation per 1 step.**

Solve the word problems

1) A book has 80 pages. David read 0.45 of the whole book. How many pages more does he have left to read?

2) In the class, 8 students can dance; this is $\frac{1}{5}$ of the whole class. How many students are in the class? How many students cannot dance?

3) Kristy read 40 pages. What part of the book did she read, if the book has 50 pages? Write the ratio as a reduced fraction. How many pages more does she have left to read?

4) Find the average for the numbers: $2\frac{1}{2}$, $2\frac{1}{5}$, $1\frac{1}{5}$, $1\frac{1}{10}$.

5) A girl paid $15 for 6 books. How much would 8 of the same books cost?

6) A bus drove 75 mi at a speed of 30 mi/h. How much time it spent?

7) What speed does a boy have, if he drove 57 mi in 4 hours?

8) Of 32 children, 62.5% were boys. How many were girls?

9) A factory has 5 new machines and 80% were old machine. How many machines are in the factory?

10) In 50 minutes of lesson, students solve an equation in 30 minutes. What percent of lesson time did students use to solve the equation?

Summary 2 Chapter 11 Lesson 7
Review 1 D Whole Numbers

Date
Read directions carefully.
Show all steps for full credits.
Solve 1 problem 5 minutes.

Name
Answers must be in the circle.
Each problem is worth 5 points.
Do only 1 operation per 1 step.

Evaluate

1) $(200 - 140) \times 25$

2) $12912 \div 16 - 11296 \div 16$

3) $39600 \div (85^2 - 25^2)$

4) GCF (16, 24)

5) LCM (16, 24)

6) $4^3 - 4^2 \div 2^2$

7) $6^3 \div 2^3 - 3^3$

8) $3^3 \div (6 - 3)^2$

9) $7^3 - 2^2(3^2 \times 2^3 - 2^2)$

10) $\{3 \cdot 8^2 - 2^2[2^3 + (1 + 3)^2]\} + 2^3$

Summary 2 Chapter 11 Lesson 7
Review 2 D Decimals

Date **Name**

Read directions carefully. **Answers must be in the circle.**

Show all steps for full credits. **Each problem is worth 5 points.**

Solve 1 problem 5 minutes. **Do only 1 operation per 1 step.**

Evaluate

1) $28.28 + 12.82$

2) $31.1 - 19.19$

3) 0.625×3.2

4) $24.36 \div 0.6$

5) $6.8\times3.9 - 6.8\times2.9$

6) $22 \div 4.4 \times 5$

7) $52.52 - 36.36 \div 4$

8) $2^2 - 0.4^2 \div 0.2^2$

9) $0.3^2 \div (0.6 - 0.3)^2$

10) $48 \div [8.4 - (1.8^2 + 2.76)]$

Summary 2 Chapter 11 Lesson 7
Review 3 D Fractions

Date | **Name**

Read directions carefully.
Show all steps for full credits.
Solve 1 problem 5 minutes.

Answers must be in the circle.
Each problem is worth 5 points.
Do only 1 operation per 1 step.

1) Reduce the fraction $\frac{20}{100}$

Evaluate

2) $1\frac{6}{15}+1\frac{2}{5}$

3) $3-1\frac{1}{4}$

4) $4\frac{2}{5}\times 1\frac{4}{11}$

5) $3\frac{3}{4}\div 6\frac{1}{4}$

6) $\frac{3}{8}\div\frac{3}{4}\div 1\frac{1}{4}$

7) $(1\frac{4}{5})\div\left(1\frac{1}{2}\right)^2$

8) $\left(\frac{1}{5}-\frac{1}{10}\right)^2\div\frac{1}{10}$

9) $1\frac{1}{2}\div(5\frac{1}{2}-4\frac{1}{4})$

10) $1\frac{3}{4}\times 12-1\frac{1}{4}\times 12$

Summary 2 Chapter 11 Lesson 7
Review 4 D Decimals and Fractions

Date | **Name**

Read directions carefully. | **Answers must be in the circle.**

Show all steps for full credits. | **Each problem is worth 5 points.**

Solve 1 problem 5 minutes. | **Do only 1 operation per 1 step.**

1) Change terminating decimal 2.875 to fraction

2) Change fraction $2\frac{5}{8}$ to terminating decimal.

Evaluate

3) $1\frac{4}{8}+1.25$

4) $2\frac{1}{4}-1.75$

5) $4.8\times\frac{5}{8}$

6) $0.57\div\frac{3}{4}$

7) $6\frac{7}{8}\div 2.75$

8) $3^4\div(1\frac{1}{2})^2$

9) $(3.5-3\frac{1}{3})\div 1\frac{1}{3}$

10) $6.5\div(3\frac{1}{3}-1\frac{1}{6})$

Summary 2 Chapter 11 Lesson 8
Review 5 D Rational Numbers

Date
Name

Read directions carefully.
Show all steps for full credits.
Solve 1 problem 5 minutes.

Answers must be in the circle.
Each problem is worth 5 points.
Do only 1 operation per 1 step.

Evaluate

1) $-5-(-4\frac{2}{5})$

2) $-\frac{2}{5}-(-\frac{7}{10})-(-2\frac{3}{10})$

3) $-1\frac{1}{5}(0.25)$

4) $1.5\div(-1\frac{1}{4})$

5) $-(-\frac{4}{5})^2+(-\frac{1}{5})^2$

6) $-3\frac{1}{3}\times1\frac{1}{5}-1\frac{1}{5}\times1\frac{2}{3}$

7) $1\frac{1}{4}\div(-4\frac{5}{8}+3\frac{1}{16})$

8) $(-2\frac{1}{5}+0.6)\div3.2$

9) $-\frac{4}{25}\div(-0.6+\frac{4}{5})$

10) $-\frac{9}{20}\div(-0.3+\frac{7}{20})$

Summary 2 Chapter 11 Lesson 8
Review 6 D Equations

Date **Name**

Read directions carefully.
Show all steps for full credits.
Solve 1 problem 5 minutes.
Answers must be in the circle.
Each problem is worth 5 points.
Do only 1 operation per 1 step.

Simplify the expressions

1) $1.6(y - 4) - 0.6 - 4(0.4y - 7)$

2) $\frac{3}{7}(14x - 21y) - 15(\frac{2}{3}x - \frac{1}{3}y)$

Solve the equations

3) $\frac{1}{9}x + 15 = 45$

4) $64072 \div z = 8$

5) $9y + 789 - 28y = 8541$

6) $0.6x - 0.3x - 0.2 = 2.5$

7) $0.1z - \frac{2}{5}z + \frac{1}{2}z = 1.5$

8) $-8(\frac{1}{2} - \frac{1}{4}y) - 3(\frac{2}{3} - \frac{1}{3}y) = 0$

Solve the proportions

9) $\frac{y}{4.7} = \frac{27.3}{9.1}$

10) $2 \div x = 1\frac{4}{7} \div 3\frac{1}{7}$

Summary 2 Chapter 11 Lesson 8
Review 7 D Geometry

Date | **Name**

Read directions carefully. | **Answers must be in the circle.**

Show all steps for full credits. | **Each problem is worth 5 points.**

Solve 1 problem 5 minutes. | **Do only 1 operation per 1 step.**

Draw Picture. | **Solve the problems**

1)Find the perimeter of an isosceles triangle if its base $7\frac{1}{2}$ in and leg is $5\frac{1}{4}$ in.

2) Find the hypotenuse of a right triangle if its legs are 4.2 m and 5.6 m.

3) Find the area of a square, if its side is 1.5 in.

4) Find the area of a rectangle, if its length is $7\frac{1}{2}$ cm, width is $4\frac{2}{3}$ cm.

5)Find the area of trapezoid, if its height is $2\frac{2}{3}$ ft, bases are $7\frac{1}{2}$ ft and $4\frac{1}{2}$ ft.

6) Find the area of a triangle if its base is $8\frac{3}{4}$ m and height is $6\frac{2}{5}$ m.

7) Find the area of a right triangle if its legs are $5\frac{2}{5}$ ft and $6\frac{2}{3}$ ft.

8) Find the area of a triangle if its sides are 9 in, 12 in, and 15 in.

Right Triangle $a^2 + b^2 = c^2$ $A = 0.5ab$

9) Find the circumference of a circle, if its radius is 20 ft, $\pi \approx 3.14$

10) Find exact answer of volume of a sphere with radius is 6 yd.

Summary 2 Chapter 11 Lesson 8
Review 8 D Word Problems

Date **Name**

Read directions carefully. **Answers must be in the circle.**

Show all steps for full credits. **Each problem is worth 5 points.**

Solve 1 problem 5 minutes. **Do only 1 operation per 1 step.**

Solve the word problems

1) A book has 60 pages. David read 0.85 of the whole book. How many pages more does he have left to read?

2) In the class, 15 students can dance; this is $\frac{3}{5}$ of the whole class. How many students are in the class? How many students cannot dance?

3)Kristy read 32 pages. What part of the book did she read, if the book has 40 pages? Write the ratio as a reduced fraction. How many pages more does she have left to read?

4) Find the average for the numbers: $4\frac{1}{2}$, $4\frac{1}{5}$, $1\frac{1}{5}$, $1\frac{1}{10}$.

5)A girl paid \$12 for 3 books. How much would 4 of the same books cost?

6) A bus drove 63 mi at a speed of 30 mi/h. How much time it spent?

7) What speed does a boy have, if he drove 73 mi in 4 hours?

8) Of 16 children, 62.5% were boys. How many were girls?

9) A factory has 10 new machines and 60% were old machine. How many machines are in the factory?

10) In 75 minutes of lesson, students solve an equation in 60 minutes. What percent of lesson time did students use to solve the equation?

Summary 2 Chapter 11 Lesson 9
Worksheet 9

Date **Name**

Show all steps for full credits not just answers
Each problem is worth 5 points
Evaluate

1) $\begin{array}{r} 700000 \\ \underline{-\ 234567} \end{array}$

2) $\begin{array}{r} 765 \\ \underline{\times 604} \end{array}$

3) $\begin{array}{r} 706 \\ \underline{\times 605} \end{array}$

4) $193884 \div 321$

5) $243714 \div 302$

6) $240240 \div 80 \times 3$

7) $7490 - 830 \times 3$

8) $56049 - 49042 \div 7$

9) $(30 \times 560) \div 8$

10) $289 \div (93 - 76)$

Summary 2 Chapter 11 Lesson 10
Worksheet 10

Date **Name**

Show all steps for full credits not just answers
Each problem is worth 5 points
Evaluate

1)
$$\begin{array}{r} 600000 \\ \underline{-\ 123456} \end{array}$$

2)
$$\begin{array}{r} 654 \\ \underline{\times 503} \end{array}$$

3)
$$\begin{array}{r} 605 \\ \underline{\times 403} \end{array}$$

4) $105630 \div 210$

5) $182508 \div 201$

6) $180180 \div 90 \times 2$

7) $6840 - 920 \times 2$

8) $42036 - 36030 \div 6$

9) $(20 \times 720) \div 9$

10) $256 \div (81 - 65)$

Summary 2 Answers Chapter 11

Lesson 1 College Placement Test

1) 5, 2) 0, 3) 21 , 4) - 2.8, 5) - 4x – 4y,
6) 7.5, 7) 18.2 ft , 8) 60 in^2 , 9) 2.4 hr. , 10) 12 girl .

Lesson 1 Review 1 A

1) 956,685, 2) 12,234, 3) 549,340, 4) 3,645,354, 5) 7006,
6) 706, 7) 40,905 , 8) 7,293, 9) 300, 10) 0.

Lesson 1 Review 2 A

1) 51.1, 2) 14.64, 3) 1.8, 4) 40.5, 5) 9.5,
6) 25, 7) 79.79, 8) 32, 9) 4, 10) 30.

Lesson 1 Review 3 A

1) $\frac{3}{5}$, 2) 3.25, 3) 1.25, 4) 14, 5) 0.75 ,
6) 0.8, 7) 2.25, 8) 0.5, 9) 3 , 10) 2.

Lesson 1 Review 4 A

1) $3\frac{5}{8}$, 2) 2.375, 3) 1.375, 4) 0.8 , 5) 0.6 ,
6) 0.2 , 7) 1.2 , 8) 12, 9) 3, 10) 2 .

Lesson 2 Review 5 A

1) -1.8 , 2) 1.6, 3) -0.49, 4) -3 , 5) - 0.2,
6) - 5 , 7) - 2.5, 8) - 0.5, 9) - 0.5, 10) - 0.5,

Lesson 2 Review 6 A

1) 47 , 2) - 2x – 3y , 3) 400, 4) 9008, 5) - 602,
6) 1.8, 7) 15, 8) - 0.6 , 9) 1.5, 10) 3.

Lesson 2 Review 7 A

1) 8 in, 2) 3.6 in, 3) 0.25in^2, 4) 30 cm^2, 5) 60 ft^2,
6) 15 m^2, 7) 14 ft^2 8) 30 m^2, 9) 62.8 yd, 10) 3.375 in^3.

Lesson 2 Review 8 A

1) 52 pages, 2) 28 students, 21 students, 3) $\frac{3}{5}$, 20 pages , 4) 1.75,
5) $36, 6) 2.6 hr, 7) 13.25 mi/hr. , 8) 20 girls ,
9) 20 machines, 10) 50%,

Summary 2 Answers Chapter 11

Lesson 3 Review 1 B

1) 2000, 2) 303, 3) 7, 4) 6, 5) 36,
6) 207, 7) 0,, 8) 9, 9) 152, 10) 192.

Lesson 3 Review 2 B

1) 61.1, 2) 13.73, 3) 0.9, 4) 50.6, 5) 8.6,
6) 36, 7) 31.31, 8) 12, 9) 9, 10) 4.

Lesson 3 Review 3 B

1) $\frac{2}{5}$, 2) 3.25, 3) 2.8, 4) 4, 5) 0.6,
6) 0.2, 7) 1.8, 8) 0.25, 9) 2, 10) 2,

Lesson 3 Review 4 B

1) $2\frac{3}{8}$, 2) 3.625, 3) 2.625, 4) 0.5, 5) 2,
6) 0.6, 7) 1.5, 8) 0.25, 9) 0.125, 10) 3.

Lesson 4 Review 5 B

1) -0.6, 2) 2.2, 3) - 0.81, 4) - 0.8, 5) - 1,
6) - 3, 7) - 1.5, 8) - 0.5, 9) - 0.8, 10) - 1.4,

Lesson 4 Review 6 B

1) 460, 2) $-2x-2y$, 3) 660, 4) 9007, 5) - 506,
6) 2.5, 7) 9, 8) - 0.5, 9) 1.1, 10) 2.

Lesson 4 Review 7 B

1) 11 in, 2) 7.6 m, 3) 12.25 in^2, 4) 30 cm^2, 5) 35 ft^2,
6) 30 in^2, 7) 14 ft^2, 8) 150 m^2, 9) 31.4 yd, 10) 15.625 yd^3.

Lesson 4 Review 8 B

1) 15 pages, 2) 30 students, 18 students, 3) $\frac{3}{5}$, 16 pages, 4) 2.75,
5) $ 18, 6) 2.2 hr, 7) 17.25 mi/hr, 8) 10 girls,
9) 20 machines, 10) 60%.

Summary 2 Answers Chapter 11

Lesson 5 Review 1 C

1) 745573, 2) 23345, 3) 406728, 4) 2410740, 5) 6005,
6) 605, 7) 25856, 8) 5694, 9) 100, 10) 15.

Lesson 5 Review 2 C

1) 31.1, 2) 12.82, 3) 1, 4) 30.6, 5) 7.8,
6) 16, 7) 36.36, 8) 0, 9) 1, 10) 20.

Lesson 5 Review 3 C

1) $\frac{1}{2}$, 2) 3.6, 3) 3.8, 4) 20, 5) 0.75,
6) 1, 7) 2.25, 8) 0.5, 9) 2, 10) 20.

Lesson 5 Review 4 C

1) $1\frac{5}{8}$, 2) 1.875, 3) 0.75, 4) 0.5, 5) 0.9,
6) 0.48, 7) 2, 8) 4, 9) 3, 10) 2.

Lesson 6 Review 5 C

1) -2.6, 2) -0.8, 3) - 1.2, 4) - 3.6, 5) -0.2,
6) -2.5, 7) - 3, 8) -0.4, 9) -0.5, 10) -52.

Lesson 6 Review 6 C

1) 0, 2) 6x – 14y, 3) 400, 4) 7008, 5) - 304,
6) 9, 7) 25, 8) 0.75, 9) 15.3, 10) 0.5.

Lesson 6 Review 7 C

1) 15 in, 2) 6.5 m, 3) 6.25 in^2, 4) 56 cm^2, 5) 32 ft^2,
6) 26 in^2, 7) 15 ft^2, 8) 24 m^2, 9) 62.8 ft, 10) 36π yd^3.

Lesson 6 Review 8 C

1) 44 pages, 2) 40 students, 32 students, 3) $\frac{4}{5}$, 10 pages, 4) 1.75
5) $ 20, 6) 2.5 hr, 7) 14.25 mi/hr, 8) 12 girls,
9) 25 machines, 10) 60%.

Summary 2 Answers Chapter 11

Lesson 7 Review 1 D

1) 1500, 2) 101, 3) 6 , 4) 8 , 5) 48,
6) 60, 7) 0 , 8) 3 , 9) 71, 10) 104.

Lesson 7 Review 2 D

1) 41.1, 2) 11.91, 3) 2, 4) 40.6, 5) 6.8,
6) 25, 7) 43.43, 8) 0, 9) 1 , 10) 20.

Lesson 7 Review 3 D

1) $\frac{1}{5}$, 2) 2.8, 3) 1.75, 4) 6, 5) 0.6,
6) 0.4, 7) 0.8, 8) 0.1, 9) 1.2 , 10) 6 .

Lesson 7 Review 4 D

1) $2\frac{7}{8}$, 2) 2.625 , 3) 2.75 , 4) 0.5 , 5) 3,
6) 0.76 , 7) 2.5 , 8) 36 , 9) 0.125 , 10) 3 .

Lesson 8 Review 5 D

1) - 0.6, 2) 2.6, 3) -0.3 , 4) -1.2 , 5) -0.6 ,
6) -6 , 7) -0.8 , 8) -0.5 , 9) -0.8 , 10) -9 .

Lesson 8 Review 6 D

1) 21 , 2) $-4x - 4y$, 3) 270, 4) 8009, 5) - 408,
6) 9 , 7) 7.5, 8) 2, 9) 14.1, 10) 4.

Lesson 8 Review 7 D

1) 18 in, 2) 7 m, 3) 2.25 in^2, 4) 35 cm^2 5) 16 ft^2
6) 28 m^2 , 7) 18 ft^2, 8) 54 in^2, 9) 125.6 ft, 10) 288π yd^3.

Lesson 8 Review 8 D

1) 9 pages, 2) 10 students ,3) 0.8, 8 pages , 4) 2.75 , 5) $ 16,
6) 2.1 hr. , 7) 18.25 mi/hr. , 8) 6 girls , 9) 25 machines , 10) 80%.

Lesson 9

Worksheet 9

1) 465433, 2) 462060 , 3) 427130, 4) 604, 5) 807 ,
6) 9009, 7) 5000, 8) 49043, 9) 2100, 10) 17.

Lesson 10

Worksheet 10

1) 476544, 2) 328962, 3) 243815, 4) 503, 5) 908,
6) 4004, 7) 5000, 8) 36031, 9) 1600, 10) 16.

Students!! Attention!!

Adults, help please students 1 – 4 grades to read and understand advice.

When I was in 5th grade I did not like math. I am not always understood math. I was 11 years girl. Next 6th grade I got a very good teacher. She explains very clearly and I understood her very well. I was worked very hard and I was not afraid to ask questions. Final Grade in Algebra in 6th grade I got an A. I said in class that I want to be a teacher of math. But some students said that I am not enough smart to be a math teacher. I said I will be work hard with math. I did it, I got a Master's degree in math and education, I love my students, I was teaching many years, in High School, part-time college math instructor, full-time university math lecturer.

Some people do a simple job, eat, sleep, go to work. People with a bad brain have not so a good job, not so good salary. If you look at TV how someone's dance, you will be not changing your brain. But if you take a dancing class this helps your body and brain ability. The same if you want to use the calculator, but the calculator not always with you. Much better if you learn how you need to do any operations with numbers without a calculator. If you read this math book you will know what you need to do. This teaches you how to improve your memory, attention, logic thinking, how to read your math books every day, your brain works better. You can work 50 minutes with math, 10 minutes break maybe dance. Do not afraid to do mistakes, fixed and move to the next step. Don't listen to anyone, choose what do you like more.

Students need to know if you work with the brain every day, your brain works better. You need a good brain in any job, in any class. When you finish High School, you need to get any profession, you need to go to college and get a good job and a better salary. You need to learn any subject, its work for the brain. Read the book, do the exercises yourself, try your best, and then check your answer at the end of the chapter. Be honest with yourself. You need to believe in yourself. The brain is like muscles. People with good brain lives longer. The professors at my university were over 80 years old, wise, and optimistic. It was a great pleasure to talk to them.

Made in United States
Cleveland, OH
05 November 2026

47656822R00276